Studies in Large Plastic Flow and Fracture

STUDIES IN LARGE
PLASTIC FLOW AND FRACTURE

WITH SPECIAL EMPHASIS ON
THE EFFECTS OF HYDROSTATIC PRESSURE

P. W. Bridgman

Late Higgins University Professor, Harvard University

HARVARD UNIVERSITY PRESS
Cambridge, Massachusetts
1964

© Copyright 1952 by the President and Fellows of Harvard College

All rights reserved

Distributed in Great Britain by Oxford University Press, London

Second Printing

Library of Congress Catalog Card Number 64-19952

PRINTED IN THE UNITED STATES OF AMERICA

FOREWORD

This volume was originally published in 1952 by the McGraw-Hill Book Company in its Metallurgy and Metallurgical Engineering Series under the editorship of Robert F. Mehl. Although nearly fifteen years have elapsed since its first appearance, the research it describes on plastic flow and fracture at high pressures still retains its freshness and significance, and basically it has not been superseded. At the time of Professor Bridgman's untimely death in 1961, the editorial committee appointed to carry forward the work begun by him on his collected papers advised the Harvard University Press to take advantage of an opportunity to acquire the original plates from McGraw-Hill and to undertake the reissue of the work as a project complementary to the larger collection. This seems especially appropriate since the present volume describes a good deal of experimental work not otherwise accessible in the open literature, and hence not adequately covered in the collected papers.

HARVEY BROOKS

February 1964

PREFACE

In the course of experiments extending over many years on the effect of high hydrostatic pressure on the properties of matter, in which the pressures were often deliberately pushed to the breaking point of the containers, I have had occasion to observe many fractures under unusual conditions. These fractures were often of an unanticipated nature and might be positively contrary to expectations based on engineering experience in a lower range of stresses. At first my interest in these fractures was a secondary one, and I was mainly concerned with attempting to understand the phenomena only insofar as was necessary for the design of my pressure apparatus. With the accumulation of material, however, and in particular with the discovery of the enormous effect of hydrostatic pressure in increasing ductility and in extending the domain of strain-hardening in steel, my interest grew in the subject for its own sake. Many experiments were therefore made with the explicit purpose of understanding better the nature of both the phenomena of fracture under conditions of high stress and the phenomena of the large plastic flow which often precedes such fractures. A number of these investigations were made during the war with practical applications in mind. The wartime investigations raised other questions which were later pursued further for their own interest.

In this book I present a coordinated exposition of all this experience under rather unusual conditions. I hope that, apart from any intrinsic interest, it may assist in leading to a better understanding of the difficult problems of flow and fracture, many of them still unsolved, in the narrower range of conditions more usually encountered in practical situations.

P. W. Bridgman

Cambridge, Mass.
November, 1950

CONTENTS

INTRODUCTION	1
PART I. TESTS UNDER HYDROSTATIC PRESSURE	9
1. The Effect of Nonuniformities of Stress at the Neck of a Tension Specimen	9
Conventional Cylindrical Tension Specimen	9
Two-dimensional Specimen	32
2. The Tension of Steel under Pressure	38
Experimental Method	38
Detailed Experimental Procedure	43
Brief Historical Survey	46
The Results	47
Phenomena of Plastic Flow	62
Fracture	71
Physical Characteristics of Tensile Fracture under Pressure	74
The Stresses at Fracture	80
The NDRC Tests on Armor Plate	85
3. Two-dimensional Tension under Pressure	87
Introduction	87
Experimental Method	87
Experimental Details	92
Radius of Curvature at the Neck	96
The Complete Stress System and the Conditions of Flow	97
Rupture	103
4. Tension Tests under Pressure on Materials Other Than Steel	106
Materials Normally Ductile	106
Aluminum	106
Copper	106
Brass	107
Bronze	107
Materials Normally Brittle	107
Introduction	107
Method	108
The Experiments	111
Glass	111
Carboloy	113
Beryllium	113
Phosphor Bronze	114
Al_2O_3 (Synthetic Sapphire)	115
Pipestone (Catlinite)	115
$NaCl$ (Rock Salt)	116

  Solenhofen Limestone 117
  Cast Iron 117
5. Simple Compression under Hydrostatic Pressure 118
 Ductile Materials 118
 Brittle Materials 119
  Glass (Pyrex) 119
  Al_2O_3 (Synthetic Sapphire) 120
  Sintered Carbides 121
  Possible Piston Material 123
 General Discussion of Brittle Fracture 124
6. Brinell Hardness under Hydrostatic Pressure 131
7. Punching under Pressure 134
 Introduction 134
 The Method 134
 The Measurements 136
 Discussion 140
8. The Collapse of Thick, Hollow Cylinders of Steel under External Pressure 142
 Introduction 142
 Experimental Results 143
 Mathematical Analysis 149
 Certain Early Results on the Collapse of Steel Cylinders under Pressure . . 160
9. The Effect of External Pressure on Cavities in Brittle Materials . . . 164
 Introduction 164
 Elastic Solutions for Spherical and Cylindrical Cavities 164
  Hollow Sphere under External Pressure 164
  Hollow Cylinder Closed with Caps over the Ends under External Hydrostatic
   Pressure 165
 Experimental Results 166
  Glass 166
  Materials Other Than Glass 169
  Compacting of Powders 172
10. Wire Drawing and Extrusion under Pressure 174

PART II. OTHER TESTS INVOLVING LARGE DEFORMATIONS 181

11. Simple Compression 181
 Introduction 181
 Experimental Method and Details 182
 Presentation of Results 184
 Discussion of Results 189
  The Strain-hardening Curve in Simple Compression 189
  Temporary Effects on Release and Reloading 190
  Fracture in Compression 190
  Hardness as a Function of Compressive Stress 190
  Comparison of Stress-strain Curves for Simple Compression with Simple
   Tension 191
12. Volume Changes in the Plastic Stages of Simple Compression 193
 Introduction 193
 Experimental Method 194

Experimental Results 200
 Soapstone 200
 Marble 201
 Diabase 202
 Quartz 204
 Mild Steel 205
 Norway Iron 208
 High-carbon Steel 209
 Cast Iron 210
 Stainless Steel 210
 Copper 210
 Brass 211
 Duralumin 212
Discussion and Summary 213

13. Two-dimensional Compression 215
 Introduction 215
 The Apparatus and Method 216
 Theoretical Background 220
 The Measurements 224
  Approximately Steady-state Measurements 224
  The Velocity of Flow 233
  The Generalized Strain-hardening Curve 239

14. Mixed Compression 243

15. Torsion Combined with Simple Compression 247
 Introduction 247
 The Apparatus 250
 Methods of Calculation 253
 Effect of Varying Compressive Load at Constant Speed 256
 Effect of Speed at Constant Compressive Load 258
 Effect of Heat-treatment 260
 Fracture 264
 Fracture without Load of Specimens Twisted under Load 268
 Experimental Results in Terms of Stress and Strain 270
 Discussion 271
 Correlation between Tension and Shear 271
 Agreement with Equations of Plasticity 274
 Isotropy after Plastic Flow 277

16. Shearing Combined with Approximately Hydrostatic Pressure . . . 279
 Introduction 279
 The Apparatus and Method 279
 The Qualitative Nature of the Results 281
 Quantitative Results 288
 Discussion 289

PART III. PLASTIC FLOW AND FRACTURE AFTER PRESTRAINING 293

17. Simple Tension after Prestraining in Tension 294
 Introduction 294
 Apparatus 295

The Tests	295
Fracture	303
General Considerations	305
18. Simple Tension after Prestraining in Simple Compression	307
Introduction	307
The Results	307
19. Simple Tension after Prestraining in Two-dimensional Compression	311
20. Simple Compression after Prestraining in Tension under Pressure	313
21. Simple Compression after Prestraining in Simple Compression	319
22. Simple Compression after Prestraining in Two-dimensional Compression	323
23. Torsion after Prestraining in Tension under Pressure	326
PART IV. GENERAL SURVEY	339
24. Gathering Up the Threads	339
Mathematical Background	339
Plastic Flow	348
Fracture	351
INDEX	357

Studies in Large Plastic Flow and Fracture

INTRODUCTION

The purpose of this book is to collect into a more or less coherent whole the experimental work on large plastic flow and fracture on which I have been engaged, mostly during the last dozen years. There would appear to be some need for this because a large part of the work was done during the war and was reported in government publications, some of which were classified. These reports have been made available since the war only in the form of nearly illegible photostat copies obtainable from the Superintendent of Public Documents in Washington. Recently various publications have appeared in the engineering literature in which this work was evidently not known, and there has already been some duplication of experiment. Since the war I have published a number of articles in more accessible places, principally in the *Journal of Applied Physics*, but I feel that even these would benefit by a more coordinated exposition in a place more accessible to engineers. In addition, there is a certain amount of earlier work that would benefit by fitting into the complete picture.

Most of this work had its origin and motivation in the discovery shortly before the outbreak of the war that ordinary steels increase enormously in ductility while exposed to hydrostatic pressures in the range between 300,000 and 450,000 psi. This phenomenon evidently connected directly with work at the Naval Research Laboratory at about the same time, in which it was shown, by photographing projectiles in the process of penetrating armor plate and determining the decelerations, that the stresses around the nose of projectiles in the process of penetration were in the same 300,000 to 450,000 range. It was obvious that here was a factor which had been little considered in analyses of the problem of armor penetration, namely, the alteration in the physical properties of the steel brought about by the action of those very stresses produced by the act of penetration. Here was a new property of steel, namely, the pressure coefficient of ductility, not disclosed by ordinary methods of testing, which was evidently of the greatest importance for ballistic problems. It was indicated that a systematic examination should be made of the magnitude of this new coefficient for the various steels used in the manufacture of projectiles and armor plate. A program was therefore initiated by the National Defense Research Committee (NDRC) of extensive investigation of the effect of pressure on the ductility of a variety of ballistic steels.

INTRODUCTION

In the early stages of this investigation it was hoped that a determination of this new parameter of steel, that is, the pressure coefficient of ductility, might yield the solution of a puzzling practical problem, namely, the reason for the frequent failure of correlation between the ordinary engineering data of a steel, such as tensile strength or impact toughness, and the ballistic behavior. With the entrance of a new parameter, it was hoped that a correlation might possibly now be found, thus permitting elimination of the expensive ballistic tests. Progress in finding the answer to this practical problem was unexpectedly retarded because of legalistic difficulties. Since any armor plates which fail in the ballistic tests are technically the property of the company supplying them, the Navy was bound by its contracts to return such defective plates to the manufacturers, with the result that none was available for my tests. By the time the red tape involved in this situation had been unwound the entire situation had altered, because it was found from other lines of evidence that ballistic failures are the result of "dirty" steel, and nothing so esoteric as the use of a new physical parameter was necessary to explain the capriciousness of the correlation. In the meantime many data had been accumulated on the effect of pressure on the tensile properties of a variety of different sorts of armor-plate steel of proved satisfactory ballistic performance. Furthermore, it was becoming evident at about this same time that any adequate understanding of armor penetration was too difficult to expect to obtain by analytical methods during the stress of war, and that the needs of the moments could be more effectively served by the accumulation of a large amount of actual ballistic test data. It was therefore realized by the NDRC that the military significance of these results had shifted from a short-range to a long-range affair, and in the latter capacity was perhaps beyond the proper scope of the NDRC. It is true that the beginning of a long-range program was initiated by the coordinated study in various laboratories of all the physical properties of a series of steels, selected and prepared under the direction of Dr. Seitz. The determination of the behavior under hydrostatic pressure in my laboratory was only part of this larger program. With the evident approach of the end of the war, however, such a program could not be justified within the NDRC. It was therefore presently abandoned, and all contracts of only long-range significance were terminated.

In the meantime similar work had been initiated under other auspices on a much less hand-to-mouth basis. The research department of the Watertown Arsenal was at that time under the direction of Col. H. H. Zornig, an officer of unusual breadth of training and liberality of view. His point of view was disclosed by his remark that one of the few advantages of a war was that "You college professors are willing to work for us and find out things for us that you otherwise wouldn't think of touching."

INTRODUCTION

In this spirit, and with the active interest and guidance of Drs. Clarence Zener and John H. Hollomon of the research department of the Arsenal, contracts were drawn and eight reports made to the Arsenal on various studies on the plastic flow of steel. These reports were never subject to any sort of classification. The primary purpose of this work was to exploit the effects of hydrostatic pressure, but studies were also made on a variety of other effects which seemed significant in throwing light on the general subject of large plastic deformation. Among these may be mentioned studies of the collapsing of heavy tubes under external hydrostatic pressure and of the plastic flow, under various sorts of stress, of small specimens cut in various orientations from specimens which had been subjected to previous plastic deformation of various sorts.

Toward the end of the war Col. Zornig was transferred elsewhere, and the honeymoon was over. A large number of questions had, however, been raised in my own mind to which I was anxious to find the answers. During the few years immediately after the war various investigations continued in my laboratory under the momentum acquired during the war, and the results were published in various professional journals. These are also collected in this book; for, although it cannot be claimed that they are inaccessible, nevertheless it seemed that they would benefit by some attempt at coordination. The same remark applies to some of my earlier work on large plastic deformation and fracture which will be included here, including a few minor matters not previously published.

With material so varied the question of the best method of presentation becomes serious. Since many of the tests were continued to the fracture of the specimen, one conceivable method of treatment would be first to consider the various phenomena of plastic flow and then to consider separately the phenomena of fracture. This has been discarded in favor of considering together the phenomena of both plastic flow and fracture associated with any one particular kind of testing. The greater number of tests were performed in the presence of hydrostatic pressure, and the chief interest in them was to determine the effect of pressure on the various flow and fracture characteristics. These tests with the cooperation of hydrostatic pressure are described first. Simplest of these are the tension tests under pressure. A complete description of these results involves the effect of pressure in raising the strain-hardening curve over its entire length and the raising of the fracture point, both as a function of pressure. The immediate experimental data given by these tests are the elongation-load curves measured with different mean hydrostatic pressures. In reducing the data to more significant form, as, for example, by converting the elongation-load curve to a stress-strain curve, an issue forces itself forward which can usually be ignored under more conventional conditions of tensile testing. This issue arises from the very great ductility which permits almost indefinite reductions of area, in one instance a reduction of 99.7 per cent having been measured. Associated with the great reduction of area is a nonuniformity in the distribution of stress across the section of the neck. Correction must be made for this nonuniformity to find the true correlations between stress and plastic flow and fracture. The first chapter of the book is devoted to a discussion of this preliminary problem.

In addition to the conventional tests on cylindrical tension specimens, two-dimensional tension has also been investigated to some extent under pressure. The two-dimensional condition was approximately realized by pulling longitudinally thin-walled tubes. The effect of pressure here is to increase the ductility greatly also. Corrections for the stress distribution at the neck are necessary in the two-dimensional case as well as in the three-dimensional case, and an analysis is also given in the first chapter.

Besides the tension tests several other sorts of test have been conducted under hydrostatic pressure. The results of these, although much less complete than the tensile tests, should give a qualitative picture of what to expect in general. These tests include simple compression, Brinell hardness, punching, and the collapse of thick-walled tubes under external hydrostatic pressure. In the same category are some fragmentary experiments, conducted at only a single pressure, on the effect of pressure on wires drawn entirely in a medium under pressure.

Large plastic strains may be realized in other ways than with the cooperation of hydrostatic pressure, and a few of the simplest cases of this sort were examined for the light which might be thrown on the general problem. Simple compression is perhaps the easiest of all stresses to apply, and it lends itself to the production of large strains without fracture. Routine testing with simple compression would doubtless be more common if it were not for the complication introduced by the friction on the ends of the compressed specimen, which, at strains at all high, results in barreling of the specimen and serious departures from uniformity in the stress distribution. A method of avoiding this had been used by Taylor and Quinney; this method consists in successive compressions by moderate amounts, with refiguring of the specimen between successive applications of stress so as to remove the barreling. Taylor and Quinney in this way determined the stress-strain curve in simple compression for copper up to compressions to one-fiftieth the original length. In this book will be found an application of the same method to steel up to compressions to one-twentieth the original length. The results for steel and copper turn out to be qualitatively different. Beside simple compression, two-dimensional compression has also been investigated here, that is, deformation by compression along one direction, equal extension in one of the orthogonal directions, and no change of dimension in the third orthogonal direction. Very few such tests seem to have been made hitherto. I believe that they are informative because of the very simple conditions under which plastic flow occurs, there being no complication, as there is in three-dimensional plastic flow, because of the interference of slip planes along mutually perpendicular directions.

Another sort of test which lends itself to the production of large strains is the torsion of a cylinder about its longitudinal axis combined with simple compressive stress along the same axis. The effect of the simple compression in this situation is much like the effect of hydrostatic pressure. Any incipient fractures which tend to form are healed before they can extend themselves, the effective ductility is greatly increased, and greatly increased angles of twist are tolerated without fracture. This sort of test does not seem to have been utilized previously to any extent, although there have been experiments on torsion combined with longitudinal extension instead of compression.

Another type of study included various sorts of tests performed on specimens which had been prestrained in one way or another. These tests were all conducted at atmospheric pressure. They were mostly on diminutive specimens cut in various orientations from the previously strained specimen. Part of the interest of the results was in the information thus afforded about the isotropy of the prestrained material. These tests included the following: simple tension after prestraining in tension under pressure, after prestraining in simple compression, and after prestraining in two-dimensional compression; simple compression after prestraining in tension under pressure, after prestraining in simple compression, and after prestraining in two-dimensional compression; and torsion after prestraining in simple tension under pressure.

Comparatively little attention was paid to time effects in all this work. However, in connection with two-dimensional compression an examination was made of some aspects of this problem under a restricted range of conditions, and these results are presented in Chap. 13 on two-dimensional compression. These results are all connected with what may be called "primary" flow or creep, that is, time effects in the interval immediately after the alteration of load, as distinguished from the slower longer range effects usually understood by "creep."

The work to be reported in the following would have been impossible without the invaluable help of my two assistants, Mr. L. H. Abbot and Mr. Charles C. Chase. Mr. Abbot is responsible for many of the actual observations and Mr. Chase for the construction of the apparatus and in many cases for setting up the experiment. In addition, during the several years of the NDRC contract I was indebted to Mr. Carmelo Lanza for conducting those experiments.

There follows now a bibliography of my own papers which are summarized in this book. Given first are the reports to the Watertown Arsenal and to the NDRC. These will be referred to in the following by their number as: Watertown 5 or NDRC 4. After these are given my other germane papers in the professional literature, arranged chronologically and given with full title. These will be referred to in the body of the book by a B followed by the number of the paper as given here. References to the work of others will be given as footnotes in the body of the text.

Watertown Arsenal reports

1. Tension Tests under Pressure, Watertown Arsenal No. 111/7, March, 1943.
2. Torsion Experiments, Watertown Arsenal No. 111/7-1, March, 1943.
3. The Effect of Prestraining in Tension on the Behavior of Steel under Tension, Torsion, and Compression, Watertown Arsenal No. 111/7-2, July 2, 1943.
4. The Collapse of Hollow Cylinders under External Pressure, Report No. WAL 111/7-3, Oct. 8, 1943.
5. The Shape of the Neck and the Fracture of Tension Specimens, No. WAL. 111/7-4, Jan. 28, 1944.
6. Tension Tests on Tubes under Hydrostatic Pressure, No. WAL. 111/7-5, June 5, 1944.
7. Plastic Properties of Steel: On the Effect of Prestraining in Tension on the Behavior of Steel in Tension, No. WAL. 111/7-6, June 20, 1944.
8. Effects of Large Strains in Simple Compression, and Coordinating Survey of Eight Reports upon Plastic Properties of Steel, No. WAL 111/7-7, Dec. 21, 1944.

NDRC reports

1. Plastic Deformation of Steel under High Pressure, Report No. A-95, Progress Report, Sept. 19, 1942.
2. Second Progress Report, on Plastic Deformation of Steel under High Pressure, Armor and Ordnance Report No. A-162, Mar. 22, 1943.
3. The Effect of Cold Working on the Ballistic Properties of Steel Plate, Armor and Ordnance Report No. A-177, Apr. 26, 1943.
4. Third Progress Report on Plastic Deformation of Steel under High Pressure, Armor and Ordnance Report No. A-218 (OSRD No. 1868), Division 2, Sept. 23, 1943.
5. Distortion of an Armor Plate Steel under Simple Compressive Stress to High Strains, Armor and Ordnance Report No. A-235 (OSRD No. 3019), Division 2, Dec. 4, 1943.
6. Final Report on the Plastic Properties of Steel under Large Strains and High Stresses, Armor and Ordnance Report No. A-294 (OSRD No. 4256), Division 2, Oct. 17, 1944.

Other papers by P. W. Bridgman on plastic flow and fracture

B1. The Collapse of Thick Cylinders under High Hydrostatic Pressure, *Phys. Rev.*, **34**, 1–24, 1912.

INTRODUCTION

B2. Breaking Tests under Hydrostatic Pressure and Conditions of Rupture, *Phil. Mag.*, July, 1912, 63–80.
B3. The Failure of Cavities in Crystals and Rocks under Pressure, *Am. J. Sci.*, **45**, 243–268, 1918.
B4. Stress-strain Relations in Crystalline Cylinders, *Am. J. Sci.*, **45**, 269–280, 1918.
B5. Some Mechanical Properties of Matter under High Pressure, *Proc. 2d Intern. Cong. Applied Mech.*, Zürich, 1926, 10 pp.
B6. Effects of High Shearing Stress Combined with High Hydrostatic Pressure, *Phys. Rev.*, **48**, 825–847, 1935.
B7. Shearing Phenomena at High Pressure of Possible Importance for Geology, *J. Geol.*, **44**, 653–669, 1936.
B8. Flow Phenomena in Heavily Stressed Metals, *J. Applied Phys.*, **8**, 328–336, 1937.
B9. Shearing Phenomena at High Pressures, Particularly in Inorganic Compounds, *Proc. Am. Acad. Arts Sci.*, **71**, 337–460, 1937.
B10. Reflections on Rupture, *J. Applied Phys.*, **9**, 517–528, 1938.
B11. Shearing Experiments on Some Selected Minerals and Mineral Combinations (with E. S. Larsen), *Am. J. Sci.*, **36**, 81–94, 1938.
B12. Considerations on Rupture under Triaxial Stress, *Mech. Eng.*, February, 1939, 107–111.
B13. Explorations toward the Limit of Utilizable Pressures, *J. Applied Phys.*, **12**, 461–469, 1941.
B14. On Torsion Combined with Compression, *J. Applied Phys.*, **14**, 273–283, 1943.
B15. The Stress Distribution at the Neck of a Tension Specimen, *Trans. ASM*, **32**, 553–574, 1944.
B16. Flow and Fracture, *Metals Technol.*, December, 1944, 32–39.
B17. Discussion of the above, *Metals Technol.*, April, 1945, Supplement to Technical Publication No. 1782.
B18. Discussion of a paper by Boyd and Robertson (mentions shearing with carboloy apparatus), *Trans. ASME*, January, 1945, p. 56.
B19. Effects of High Hydrostatic Pressure on the Plastic Properties of Metals, *Rev. Modern Phys.*, **17**, 3–14, 1945.
B20. The Tensile Properties of Several Special Steels and Certain Other Materials under Pressure, *J. Applied Phys.*, **17**, 201–212, 1946.
B21. Studies of Plastic Flow of Steel, Especially in Two-dimensional Compression, *J. Applied Phys.*, **17**, 225–243, 1946.
B22. The Effect of Hydrostatic Pressure on Plastic Flow under Shearing Stress, *J. Applied Phys.*, **17**, 692–698, 1946.
B23. The Rheological Properties of Matter under High Pressure, *J. Colloid Sci.*, **2**, 7–16, 1947.
B24. The Effect of Hydrostatic Pressure on the Fracture of Brittle Substances, *J. Applied Phys.*, **18**, 246–258, 1947.
B25. Large Plastic Flow and the Collapse of Hollow Cylinders, *J. Applied Phys.*, **19**, 302–305, 1948.
B26. Fracture and Hydrostatic Pressure, in "Fracturing of Metals," pp. 246–261, American Society for Metals, Cleveland, 1947.
B27. Volume Changes in the Plastic Stages of Simple Compression, *J. Applied Phys.*, **20**, 1241–1251, 1949.
B28. Effect of Hydrostatic Pressure on Plasticity and Strength, *Research*, London, **2**, 550–555, 1949.

Part I
TESTS UNDER HYDROSTATIC PRESSURE

CHAPTER 1

THE EFFECT OF NONUNIFORMITIES OF STRESS AT THE NECK OF A TENSION SPECIMEN[1]

Conventional Cylindrical Tension Specimen. Different levels of approximation may be recognized in the approach to the problem presented by the nonuniformities at the neck of a tension specimen. There is the first level of approximation, which may be called the zero level, in which even the existence of the neck is not recognized. This is the level of the technical "tensile strength," defined as the breaking load divided by the *initial* undeformed cross section. Such a definition obviously has its usefulness within a certain range of the conditions encountered in practice, but that it does not offer a very fundamental correlation with what is going on within the tensile specimen is obvious from the fact that the "tensile strength" depends to a certain extent on the original length of the specimen. In practice a closer correlation with the properties of the material as such is obtained by modifying the definition of tensile strength by specifying that the specimen must be longer than a certain limit. The second level of sophistication, which we may call the level of first approximation, takes account of the necking by using average values across the neck. At this level of approximation we have the "true stress" at the neck defined as the total tensile load on the specimen divided by the current area of cross section of the neck. In order to determine the "true stress" at all stages of a tension test, running measurements must be made of the diameter of the neck as well as of the load. One may in this way determine curves of true stress versus some significant strain parameter, such perhaps as the total elongation on a specified original length. Such a curve of true stress versus strain has in general an appearance quite different from that of a curve of load versus the same strain, the latter curve usually rising to a maximum and dropping off sharply to the fracture point, whereas the former continues rising to fracture. The strain versus true stress curve is often called the "strain-

---

[1] This chapter is based on material contained in B15, B16, and B17; Watertown fifth, sixth, and seventh reports; and NDRC fourth and sixth reports.

hardening" curve. However, the average stress across the neck is not adequate to describe all the significant physical phenomena, including the phenomena of fracture; for in general fracture begins in some local region and is presumably determined by the conditions at the point where it begins, and is not especially concerned with the average of conditions at other points. In particular, in the tensile test the evidence seems to point to the initiation of fracture on the axis of the specimen, and a characterization of the conditions adequate to determine when fracture occurs should therefore involve a specification of stress and strain on the axis as distinguished from the average stress and strain across the neck. The next level of approximation, that of the second approximation, may be defined as the level which seeks to determine the distribution of stress and strain across the neck with sufficient accuracy to distinguish the values on the axis from the average values and in this way to throw a clearer light on the conditions of fracture. The need for the second approximation will depend on the degree to which the average values depart from the values on the axis. This will depend in turn on the degree of necking. Under ordinary conditions of testing in which fracture occurs at reductions of area of the order of 50 per cent, the need for the second approximation is not particularly pressing. But, with the large reductions of area which may accompany the greatly enhanced ductility imparted by hydrostatic pressure, the need for the second approximation does become pressing, and an entirely different picture is presented of the conditions of fracture by taking into account the second approximation. My work on determining the effect of pressure on ductility had not proceeded far before the need for the second approximation became obvious, and some time was devoted to a specific study of this topic.

The "second" approximation is concerned only with the conditions at the neck. A complete solution of the problem here would demand a specification of the stress and the strain at all points in the plastically flowing tension specimen, and in particular would involve a complete determination of the contour of the specimen at points remote from the neck as well as in the neighborhood of the neck. As far as I know very little progress has been made on the solution of the general problem. I made some efforts in this direction, and to this end measured the complete contours of a large number of tension specimens; this work will be referred to further at the end of this chapter. No satisfactory line of attack on the general problem presented itself, however, and here we shall be concerned only with the approximation which seeks to determine approximately the stress and strain at the neck as a function of distance from the axis in the neck. For the purposes of this approximation the contour in the neighborhood of the neck may be characterized by a single parameter, the radius of curvature of the circle osculating the profile at the neck. A complete theory should give this radius in terms of the fundamental plastic parameters of the material, but we shall, in our theoretical deduction of the formulas, treat this radius as an experimentally determined parameter, to be independently measured in addition to the total load and diameter of the neck. Such measurements are not difficult, and the formulas are easily handled. It will then prove as a matter of experiment that, within limits which do not greatly restrict the use of the results for many purposes, the radius of curvature of the contour may be taken as a universal function of the reduction of area, thus simplifying the experimental procedure in many cases.

We now turn to the detailed mathematical analysis. We assume at first that the tension test is conducted under conventional conditions at atmospheric pressure, which may, to a sufficient approximation, be taken to be zero.

We assume rotational symmetry about the longitudinal axis z of the tension specimen, and furthermore assume that the specimen is symmetrical in the positive and negative z directions with respect to the plane perpendicular to the axis through the neck. We use conventional cylindrical coordinates r, θ, z, and use for the stress components the notation of Love's "Elasticity," or $\widehat{rr}$, $\widehat{\theta\theta}$, $\widehat{zz}$, $\widehat{\theta z}$, $\widehat{zr}$, $\widehat{r\theta}$. We introduce the additional symbols:

a = outside radius of cross section of neck
R = radius of curvature, at neck, of section through neck containing axis
A_0 = initial cross-sectional area of tension specimen
A = final cross-sectional area at neck

Any solution, whether in the elastic or the plastic range, is subject to the stress equations of equilibrium. Since the solution must have rotational symmetry, all derivatives with regard to θ vanish, and the $\widehat{r\theta}$ and $\widehat{\theta z}$ components of stress vanish for the same reason. The usual three stress equations of equilibrium reduce to two, which are

$$\frac{\partial \widehat{rr}}{\partial r} + \frac{\partial \widehat{rz}}{\partial z} + \frac{\widehat{rr} - \widehat{\theta\theta}}{r} = 0$$
$$\frac{\partial \widehat{rz}}{\partial r} + \frac{\partial \widehat{zz}}{\partial z} + \frac{\widehat{rz}}{r} = 0 \qquad (1\text{-}1)$$

The coordinates in the stress equations of equilibrium can usually be taken with sufficient approximation to be the coordinates of the unstrained body. Strictly, however, it is obvious that they are the coordinates of the strained body, and we shall have to so take them here, since we are dealing with large strains.

In addition to the stress equations of equilibrium, there are the boundary conditions. Referring to Fig. 1, these conditions reduce at the external curved surface, which is free from stress, to

$$\widehat{rr} \cos \alpha - \widehat{rz} \sin \alpha = 0$$
$$\widehat{rz} \cos \alpha - \widehat{zz} \sin \alpha = 0$$

These may be rewritten as

$$\left. \begin{array}{l} \widehat{rr} = \widehat{zz} \tan \alpha \\ \widehat{rz} = \widehat{zz} \tan \alpha \end{array} \right\} \text{ at the external surface}$$
(1-2)

There is the additional condition imposed by the total load

$$2\pi \int r\widehat{zz} \, dr = \text{load} = \text{const} \quad (1\text{-}3)$$

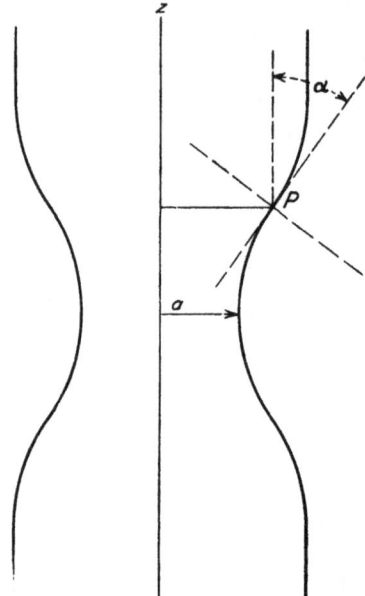

Fig. 1. Diagram for the boundary conditions at the neck of a tension specimen.

where the integration is to be extended across any plane section perpendicular to the axis.

At the neck the conditions simplify

$$\left. \begin{array}{l} \dfrac{\partial \widehat{zz}}{\partial z} = 0 \quad \text{for all } r \\ \widehat{rz} = 0 \quad \text{for all } r \\ \widehat{rr}_a = 0 \\ \widehat{rr}_0 = \widehat{\theta\theta}_0 \end{array} \right\} \text{ at the neck} \qquad (1\text{-}4)$$

The subscript a indicates a value at the outside surface of the neck and the subscript 0 the value on the axis. Of the above four conditions the first two are demanded by symmetry with respect to $+z$ and $-z$, the third by the condition at the free surface, and the fourth by the first of Eqs. (1-1) in order to avoid infinities on the axis.

It appears, then, that we are not concerned with $\widehat{rz}$ at the neck. We may now eliminate $\widehat{rz}$ between the two equations (1-1), giving a single equation between the three orthogonal stress components at the neck, namely,

$$\dfrac{\partial}{\partial r}(r\widehat{rr}) = \widehat{\theta\theta} + \int_0^r r \dfrac{\partial^2 \widehat{zz}}{\partial z^2} \, dr \qquad (1\text{-}5)$$

This single condition is obviously not enough to determine a solution. We must consider in addition the conditions imposed by the plastic behavior of the material. For this purpose we shall use the von Mises function $(\widehat{rr} - \widehat{\theta\theta})^2 + (\widehat{\theta\theta} - \widehat{zz})^2 + (\widehat{zz} - \widehat{rr})^2$. The value of this von Mises plasticity function varies with the conditions. If the body is "ideally" plastic, the function is a constant. This assumption would usually give a first approximation. But actually there is strain-hardening and the function depends on the strains. The precise dependence is most complicated and probably never has been completely determined for any actual material. We shall make a second approximation here, recognizing that the von Mises function depends on the strains but assuming that the strains vary so little across the neck that the von Mises function may be taken independent of r all the way across. The experimental justification for this assumption will be presented in due course.

In general, there are also conditions on the strains to be satisfied in the plastic range. Assuming isotropy of strain-hardening, these are

$$\begin{aligned} e_r &= \beta[\widehat{rr} - \tfrac{1}{2}(\widehat{\theta\theta} + \widehat{zz})] \\ e_\theta &= \beta[\widehat{\theta\theta} - \tfrac{1}{2}(\widehat{zz} + \widehat{rr})] \\ e_z &= \beta[\widehat{zz} - \tfrac{1}{2}(\widehat{rr} + \widehat{\theta\theta})] \end{aligned} \quad (1\text{-}6)$$

Here β is in general a function of the strains, to be determined by experiment.

These strains are subject to an additional condition if we suppose that, in conformity with our assumption about strain-hardening, e_z is constant across the section. The conditions of volume conservation give for the radial displacement ρ under these circumstances the equation

$$\rho = -\frac{e_z}{2} r + \frac{c}{r}$$

Here, in order to avoid infinities on the axis, $c = 0$, and $\rho = -(e_z/2)r$. This gives

$$e_r = e_\theta = -\tfrac{1}{2} e_z \quad (1\text{-}7)$$

The question now is: can we find a stress system which meets the various requirements thus far imposed? If we can find such a system, then we know by the usual physical and mathematical arguments that it is unique and hence *the* solution. Inspection shows that we can satisfy certain of these conditions in a rather simple way. If we set $\widehat{rr} = \widehat{\theta\theta}$, we can satisfy the condition on e_r and e_θ. In addition, the von Mises plasticity condition is satisfied if we set $\widehat{zz} = \widehat{zz}_a + \widehat{rr}$. This stress system is simply described. It is a longitudinal tension uniform all the way across the section of the neck, with a superposed hydrostatic tension ($\widehat{rr} = \widehat{\theta\theta} = \widehat{zz} - \widehat{zz}_a$) which vanishes at the outer boundary because of the condition $\hat{rr}_a = 0$ and presumably has its maximum on the axis. To complete the determination of the solution we have Eqs. (1-5), giving a connection between $\partial^2 \hat{zz}/\partial z^2$ and $\hat{rr}$, and Eq. (1-3). Because of the conditions already imposed, these are two equations for the two outstanding unknown functions $\partial^2 \hat{zz}/\partial z^2$ and $\hat{rr}$. Among other things, these equations will impose a condition on the contour of the neck, since Eq. (1-3) holds at other points than at the narrowest part of the neck.

No simple method presents itself for solving these equations rigorously. We shall therefore content ourselves with a solution by an approximate method valid in the immediate neighborhood of the neck where the contour of the neck may be approximated by its osculating circle and where one of the surfaces of principal stress may be approximated by a sphere. This approximation renounces the possibility, implied in Eq. (1-3), of obtaining some information about the shape of the contour of the neck, and treats the contour, as approximated by the osculating circle, as an independent parameter to be determined by experiment.

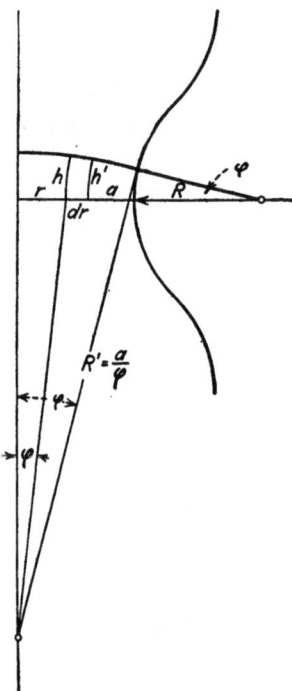

FIG. 2. Diagram for the approximate analysis of the stress conditions in the neighborhood of the neck of a tension specimen.

Figure 2 represents the state of affairs in a neighborhood close enough to the neck to permit the external contour to be represented by a circle. The radius of the neck a and the radius of curvature R of the contour are given. At the external surface we can see from Eqs. (1-2) that one of the lines of principal stress is normal to the surface, and we know that on the axis the lines are normal to the axis. We may, therefore, approximate to the complete lines of principal stress in the immediate neighborhood of the neck by circles, with centers on the axis. Consider the particular stress circle that cuts the contour at such a point as to subtend the small angle ϕ at the center of curvature of the contour. The radius of curvature R' of the stress circle is a/ϕ. Consider now a point at the neck distant r from the axis. Let the length r subtend an angle ϕ' at the center of the circle R'. Then $\phi' = (r/a)\phi$. Through the point r, perpendicular to the radius, there passes one member of the family of lines of principal stress of which the external contour is another member. This line is perpendicular to the radius r and the circle of radius R'. We may assume that this line is also a circle, and we might compute its radius. It happens, however, that to the order of small quantities in which we are interested this refinement is not necessary, and it will be sufficient to treat this line of principal stress as a straight line perpendicular to the radius at r. Consider now the element of volume bounded by two axial planes including between them the small angle θ, by the cylinders with radii r and $r + dr$, by the plane perpendicular to the axis at the neck, and by the spherical surface of radius R'. By construction the forces across the faces of this element are entirely normal and are given by the principal stress components. The condition that the net component of force in the r direction on the six faces of this element vanishes gives the equation

$$\left(\widehat{zz} + h\frac{\partial \widehat{zz}}{\partial z}\right) \sin \phi' \left(r + \frac{dr}{2}\right) \theta \, dr$$
$$= \widehat{rr} h r \theta - \left(\widehat{rr} + dr \frac{\partial \widehat{rr}}{\partial r}\right)(r + dr)h'\theta + \widehat{\theta\theta} \sin \theta \, h \, dr$$

We have further
$$h = R\phi + R'[\cos \phi' - \cos \phi]$$
$$h' = R\phi + R'[\cos (\phi' + d\phi') - \cos \phi]$$

Expansion of this equation, keeping only terms of the lowest order, gives

$$\widehat{zz} \frac{r^2}{a} = \widehat{rr} \left(\frac{3}{2}\frac{r^2}{a} - \frac{a}{2} - R\right) - r\frac{d\widehat{rr}}{dr}\left(R + \frac{1}{2}\frac{a^2 - r^2}{a}\right)$$
$$+ \widehat{\theta\theta}\left(R + \frac{1}{2}\frac{a^2 - r^2}{a}\right) \quad (1\text{-}8)$$

Into this equation involving the three stress components we may now substitute the particular relations imposed above, namely, $\widehat{\theta\theta} = \widehat{rr}$, and $\widehat{zz} = \widehat{rr} + \widehat{zz}_a$. $\widehat{zz}_a$ is the longitudinal stress at the external surface where the other two components of stress $\widehat{rr}$ and $\widehat{\theta\theta}$ vanish, and is therefore the flow stress for an ordinary tension test at the particular elongation under homogeneous conditions. We replace the notation $\widehat{zz}_a$ by F, for flow stress. Equation (1-8) now becomes an equation for the single component $\widehat{rr}$:

$$\frac{d\widehat{rr}}{dr}\left(R + \frac{1}{2}\frac{a^2 - r^2}{a}\right) + \frac{r}{a}F = 0 \quad (1\text{-}9)$$

The variables are separated, and the equation may be at once integrated, giving, together with the boundary condition, the closed solution

$$\widehat{rr} = F \log \frac{a^2 + 2aR - r^2}{2aR}$$
$$\widehat{zz} = F\left(1 + \log \frac{a^2 + 2aR - r^2}{2aR}\right) \quad (1\text{-}10)$$

The connection with the load is given by

$$\text{Load} = \int_0^a 2\pi r \widehat{zz}\, dr = \pi F(a^2 + 2aR) \log\left(1 + \frac{1}{2}\frac{a}{R}\right)$$

Also

$$\widehat{zz}_{\text{av}} = \frac{\text{load}}{\pi a^2} = F\left(1 + 2\frac{R}{a}\right)\log\left(1 + \frac{1}{2}\frac{a}{R}\right) \quad (1\text{-}11)$$

By substituting the explicit expression for $\widehat{rr}$ into Eq. (1-8) it is possible by successive differentiations to obtain an explicit expression for $\partial^2 \widehat{zz}/\partial z^2$, but there seems no particular interest in writing this out.

The solution at which we have thus arrived satisfies the equations of equilibrium, the boundary conditions, and the von Mises plasticity condition. Mathematically, these conditions uniquely determine a solution. We arrived at this solution by setting the strains constant across the section and the hoop stress zero at the periphery, but we would not have been able to find a solution with these properties unless these were in fact the properties of the solution. That is, constancy of strain and vanishing of peripheral hoop stress are to be described as discoveries rather than as assumptions. This point is emphasized because there has been some misunderstanding of the matter (see in particular B17).

We now examine the general nature of the solution. The stresses at the neck at which we have arrived may be expressed as the sum of two stress systems. The first of these, or the stress, system (1), is a constant tension $\widehat{zz}$ along the axis, equal to $\widehat{zz}_a$ or F; that is,

$$\widehat{rr}_1 = \widehat{\theta\theta}_1 = 0 \qquad \widehat{zz}_1 = F$$

The second stress system (2) is a hydrostatic tension

$$\widehat{rr}_2 = \widehat{\theta\theta}_2 = \widehat{zz}_2$$

which varies across the section from the value zero at the outer surface to a maximum on the axis. The total $\widehat{zz}$ is therefore a maximum on the axis and a minimum at the free surface. The solution is thus qualitatively different from the elastic case. The rigorous solution for a hyperboloid of revolution has been worked out by Neuber[1] for the elastic case. Here $\widehat{zz}$ is a maximum at the outer surface, $\widehat{rr}$ is of course zero at the outer

[1] H. Neuber, *Z. angew. Math. Mech.*, **13**, 439–442, 1933.

surface after passing through positive values in the interior, and $\hat{\theta}\theta$ has its maximum positive value at the outer surface. If the longitudinal load on a hyperboloid of revolution is increased into the plastic region it may be expected that plastic yield will begin first at the outside. As plastic yield increases it would be natural to expect that the solution found above would be approached as a limiting case, modified perhaps by some residual memory of the initial yield at the outside. But for an ordinary tensile specimen, which is initially a cylinder, the first stages of plastic yield are known to be uniform up to the beginning of necking, since the specimen preserves its cylindrical figure up to this point, and there is no reason to think that the elastic solution for the hyperboloid has any effect on the final plastic solution.

We now consider the evidence that the strain is actually uniform across the neck. If it is found as a physical fact that the strain is not uniform this can mean only that the conditions postulated in the mathematical solution are not exactly applicable. Since the equations of equilibrium and the boundary conditions are secure, this could mean only that the von Mises plasticity condition is not exactly applicable. The question of uniformity of strain may be approached from two points of view, theoretical and experimental. At the time that I wrote my paper in the 1944 volume of *Transactions of the ASM*, I had only the theoretical argument. I showed in that paper by a detailed examination of the nature of the equations themselves and of the possible solutions that one was pretty much forced to conclude that the strains were uniform and certainly could not be qualitatively like the strains in the elastic case. This analysis I regard as essentially sound, but I am not sure that it can be expected to be very convincing for anyone who does not take the very considerable labor of following the analysis in full detail for himself. It was obvious that an experimental approach, if it could be devised, would be better. This I was not at first able to do, but in the seventh report to the Watertown Arsenal a method was found.

It is evidently required that there be some way of identifying points in the interior of the tension specimen and following them as they are displaced to new positions as the specimen necks down. No method of complete identification of all the interior points presents itself, but a method of partial identification for the purpose is to core the tension specimen as indicated by the section shown in Fig. 3. This cored specimen is then pulled in tension to the desired amount of necking, then split in two longitudinally along the axis, and the contour of the core and the outside of the specimen determined. A condition for uniformity of strain is obviously that the ratio of external diameter to diameter of the core be the same before and after pulling. If several different specimens with different core diameters are used, the uniformity of strain across the section of the neck may be tested.

If the core is merely a mechanical fit for the outside of the specimen the two parts will not deform together, but necking in core and sleeve will usually start in different places and the two parts will pull away from each other. Some attachment is therefore necessary between core and sleeve. An attachment by sweating with soft solder is not sufficiently strong. If the core and sleeve are soft-soldered together they probably begin to neck at about the same point on the axis, but in the final stages before fracture the solder ruptures and the sleeve pulls away from the core. This incidentally demonstrates the reality of the radial component of tension inside the neck deduced by the mathematical analysis. Silver solder, however, proved strong enough to withstand this tension, and the specimen deforms as if it were a single piece. When the specimen is split after necking, the core is identified by a fine enveloping line of silver, which may be emphasized by appropriate etching. The core may easily be made to fit the sleeve within 0.001 in., so that the silver line is at most 0.0005 in. wide and at the neck may be considerably less. Since it is necessary to raise the specimens to a red heat in order to apply the silver solder, heat-treated specimens cannot be examined in this way. However, since the effect must be entirely geometrical, it would appear sufficient to establish the result for a single soft steel.

Fig. 3. Cored tension specimen for determining the distribution of strain across the neck.

Six cored specimens were systematically examined. These were all made from the same length of commercial cold-rolled steel. The cores were from the same rod, each core from the length contiguous to its own sleeve to ensure maximum uniformity. Two specimens each were made of three different core diameters such that the area of the core was approximately 0.50, 0.25, and 0.12 of the total section. One set of three different core diameters was pulled at atmospheric pressure to a reduction of area of 42 per cent; the second set of three was pulled under a hydrostatic pressure of approximately 350,000 psi to a reduction of area of approximately 92 per cent. Several specimens had previously been pulled at atmospheric pressure to complete fracture. The distortion produced by the fracture at the neck of these specimens reduced the accuracy of the measurements, but within the greater irregularity of the readings the results were qualitatively the same as for the nonfractured specimens.

Figures 4 and 5 are photographs of the longitudinal section of the two specimens pulled to 92 per cent reduction of area in which the core originally occupied approximately 12 and 25 per cent of the section. It is evident to the eye that the ratio of core diameter to external diameter is approximately the same at the neck as in the comparatively undeformed regions remote from the neck, where the ratio is the same as the initial ratio before pulling. This was confirmed by micrometer microscope measurement of all six specimens at points spaced along the axis. The

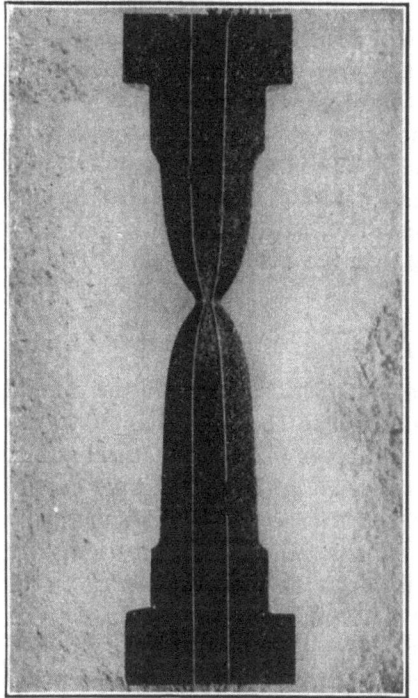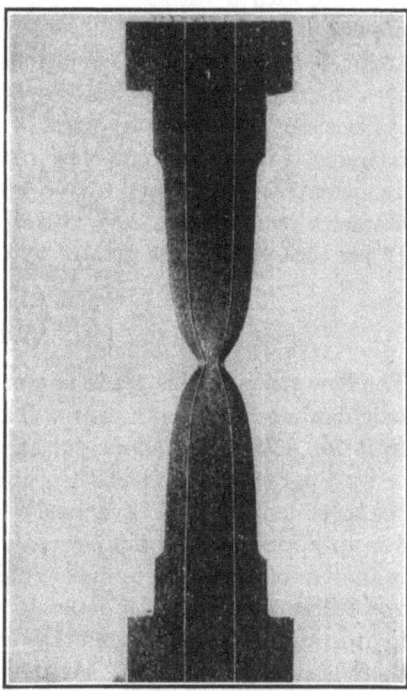

FIG. 4. Longitudinal section of the cored tension specimen after pulling to a reduction of area of 92 per cent. The core originally occupied 12 per cent of the section.

FIG. 5. Longitudinal section of the cored tension specimen after pulling to a reduction of area of 92 per cent. The core originally occupied 25 per cent of the section.

micrometer measurements show in some cases a small variation of ratio of external diameter to core diameter. The direction of variation is such that the ratio has a maximum at the neck. As is to be expected the variation is greatest in the specimens pulled to the greatest reduction of area and with the smallest core. At the neck of the specimen pulled to 92 per cent reduction in which the area of the core was initially 12 per cent of the total area, the ratio of external to core diameter was 3.18 against the initial value 2.88. For the corresponding specimen in which the core area was approximately 25 per cent of the total section, the ratio of external diameter to core diameter at the neck was 2.09 against 1.95

initial. For the corresponding specimen in which the core area was initially approximately 50 per cent of the cross section the ratios were 1.46 and 1.43. For the three specimens pulled to only 42 per cent reduction of area any increase of ratio of external diameter to core diameter at the neck was beyond experimental error.

It would thus appear that the assumption of uniformity of strain is justified with a rather good degree of approximation for reductions of area as high as 92 per cent, which is much beyond that which can be attained under ordinary conditions, and that for ordinary reductions of area the departure from uniformity is hardly measurable.

Consider now the magnitude of the error which we may expect to be introduced in the extreme case by departure of strain from uniformity. The greatest effect found above, a change in the ratio of external to core diameter from 3.18 to 2.88, corresponds to a natural strain of the inner 12 per cent of the area greater by 0.2 than the average strain

$$\left[\log_e \left(\frac{3.18}{2.88} \right)^2 = 0.2 \right]$$

The flow stress of this grade of steel at a reduction of area in the general neighborhood of 92 per cent will be shown in the next chapter to rise from 190,000 to 200,000 psi for an increment of 0.2 in the natural strain, or by 5 per cent. This is, then, the variation in the inner 12 per cent of the specimen; the effect averaged over the entire section is correspondingly less, or approximately 0.5 per cent. Superposed on this correction there is another of the opposite sign. This is due to the effect of pressure on the von Mises function. In general, we shall find that this increases with hydrostatic pressure, or, in other words, pressure increases flow stress or there is strain-hardening. At the axis of the neck there is a superposed hydrostatic tension, which therefore will produce a softening effect in the direction to compensate for the hardening effect just considered. Data to be presented later will show that this effect in the negative direction is of the general order of half the positive effect at the strain of 92 per cent. All in all, therefore, it would appear that for ordinary reductions of area any effects arising from nonuniformity of strain distribution would be quite beyond experimental error; and, even for the extreme strains dealt with here, the error must be so small as to be almost beyond experimental error and certainly cannot be large enough to affect any of our qualitative conclusions.

There is independent experimental evidence that the strains at the neck are distributed with approximate uniformity. N. N. Davidenkov and Miss N. I. Spridonova,[1] in a paper in which they object to my

[1] N. N. Davidenkov and N. I. Spridonova, *Proc. ASTM*, **46**, 1–12, 1946.

"assumption" of uniform strain across the neck, apparently not being familiar with my experimental justification as described above, themselves offer experimental evidence for the uniformity of the strain. This consists in statistical analysis of a great many measurements of the shape of the deformed grains at various positions in the neck of tension specimens. They can find no consistent evidence of outstanding distortion in any preferred direction, and they infer that the strain must be uniform. This independent confirmation is welcome, but it seems to me that the method is much less sensitive than my method of coring.

The procedure in the last step of my solution above, from Fig. 2 on, was suggested by the method used by Siebel[1] in an earlier solution of the problem. The nature of the approximations made by Siebel was such that it is difficult to estimate the degree of approximation to be expected of his final solution. He solved the problem in two dimensions, in rectangular coordinates x and z, the y coordinate running to infinity and not entering the solution. The Y_y component of stress was set equal to zero. Under these conditions his solution was approximate because he effectively set the h and h' of the analysis above equal to each other, thus discarding a finite term, and he also effectively set $\widehat{zz}$ = const, for a first integration. The resultant solution was applied to the cylindrical case merely by substituting r for x in the formula and integrating around the circle, and without any consideration of the effect of the stress component $\widehat{\theta\theta}$ which must of necessity appear when passing to cylindrical coordinates. Siebel's final formula was

$$\widehat{zz}_{\text{av}} = F\left(1 + \frac{1}{4}\frac{a}{R}\right)$$

Siebel's formula for the hydrostatic tension on the axis is a less good approximation than his formula for $\widehat{zz}_{\text{av}}$.

Davidenkov and Spridonova also arrive at the same formula for $\widehat{zz}_{\text{av}}$ as Siebel. Their analysis also involves assumptions the effect of which is not easy to estimate.

We next consider the qualitative effect of taking into account the nonuniformity of the stress distribution as given in formulas (1-10) for $\widehat{rr}$ and $\widehat{zz}$. There will be an effect both on our picture of the conditions of flow or the stress-strain curve, and on our picture of the conditions of fracture. If no account is taken of stress nonuniformity, the stress-strain curve or the "strain-hardening" curve is constructed from the tensile test data by plotting the longitudinal strain at the neck against the "true stress" at the neck. The latter is by definition load per unit

---

[1] E. Siebel, Berichte der Fachausschüsse des Vereins deutscher Eisenhüttenleute, Werkstoffausschuss, *Ber.* 71, 1925.

area of the neck, or load/πa^2, which is the same as $\widehat{zz}_{av}$. Now the stress, which is thus correlated with the flow, varies from point to point, and any correlation between flow and stress thus found is strictly of significance only when the variation of stress from point to point is also specified. As necking proceeds, there is a gradual shift in the significance of the "true stress" with respect to flow. Before necking starts, true stress is also the average stress and is also the only component of stress, assuming no hydrostatic pressure on the outside. As necking proceeds the average stress across the neck acquires other components, so that flow is taking place under gradually changing components of stress. It would seem to be of more significance to correlate flow at all stages with the same simple stress system which prevails in the initial stages of flow. This can be accomplished by using the "flow stress" above, which is defined as the stress $\widehat{zz}$ at the outer surface of the neck, where it is the only component of stress, and where flow is therefore taking place under the same simple conditions as before necking begins.

The problem now arises whether one may properly speak of *a* flow stress when flow occurs with more than one component of stress. I think the implication in speaking of *a* flow stress is that the flow is determined by a single component of stress and not by the three components which together determine the complete stress system. Now we shall presently want to speak of a pressure coefficient of a flow stress when the tension test is conducted in a medium under pressure. The experimental picture presented by this phraseology is that the tension specimen is subjected to a hydrostatic pressure, and that then superposed on this pressure an additional simple tensile stress is applied, as by hanging on a weight, and the flow stress under pressure will then be that tensile stress, as given by the weight, which produces flow in extension. Now any stress system in which two of the three components are equal may obviously be reduced to a single stress component plus a superposed hydrostatic pressure; and, if flow is taking place under the complete stress system, then the flow stress may properly be defined as the single stress component remaining after the hydrostatic pressure has been subtracted off. Formulated analytically, if the stress system has the components X_x, Y_y ($= X_x$), Z_z, then the stress system which is left after subtracting off the hydrostatic pressure X_x, X_x, X_x is 0, 0, $Z_z - X_x$, and the flow stress, if flow is taking place, is defined as $Z_z - X_x$. Notice that the hydrostatic pressure which is subtracted off is *not* the mean hydrostatic pressure $\frac{1}{3}(X_x + Y_y + Z_z)$. This would not leave a stress system with two vanishing components. According to this understanding it would not be proper to speak of *a* flow stress when flow is taking place under a stress system with three unequal components. This usage is, however, by no means universally accepted, and it is not uncommon, for instance, to designate as the flow stress the numerically largest of the three unequal principal stress components when flow is taking place. This usage, which may be innocent and natural enough in a narrow range of conditions, leads to unwanted statements when dealing with as wide a range of stress conditions as we shall encounter in this book. For example, it is easy to devise examples in which this usage would force one to say that a material was flowing in extension under a compressive flow stress.

We shall therefore in this book understand by the "flow stress" the single component of stress which survives after two of the principal components of stress have been reduced to zero by subtracting off a hydrostatic component (either hydrostatic pressure or hydrostatic tension), and we shall not speak of a flow stress in more complicated cases. In such more complicated cases it would be perfectly legitimate, however, to speak of a flow-stress system, if flow is taking place.

For the sake of formal completeness we may now introduce explicitly into the solution for the stresses the hydrostatic pressure P under which the test is conducted. It is obvious that the solution already obtained applies with the simple addition of the hydrostatic pressure. The reason is that the von Mises plasticity condition and the stress equations of equilibrium are unaffected by a uniform hydrostatic pressure. The complete stress system in the tension specimen is now the resultant of three superposed stress systems: the hydrostatic pressure, a component $\widehat{zz}$ constant across the section and equal to the flow stress F, and a hydrostatic tension varying from zero at the outer surface of the neck to a maximum on the axis. Written out formally the stresses are

$$\widehat{rr} = -P + 0 + F \log \frac{a^2 + 2aR - r^2}{2aR}$$
$$\widehat{\theta\theta} = -P + 0 + F \log \frac{a^2 + 2aR - r^2}{2aR} \qquad (1\text{-}12)$$
$$\widehat{zz} = -P + F + F \log \frac{a^2 + 2aR - r^2}{2aR}$$

The connection between F and the tensile load (weight on the specimen in the "pressurized" laboratory) is as before

$$F\left(1 + 2\frac{R}{a}\right) \log \left(1 + \frac{1}{2}\frac{a}{R}\right) = \frac{\text{load}}{\pi a^2} = \text{average tension} \qquad (1\text{-}13)$$

Returning now to the distribution of stress at the neck, the flow stress at the outside surface as given by Eqs. (1-12) is merely F. At other points of the neck the flow stress is to be found by subtracting off the hydrostatic component $\widehat{rr}$ ($= \widehat{\theta\theta}$), two of the stress components thus being equal as required above. This leaves for the flow stress, according to Eqs. (1-12), the constant value F all the way across the neck, independent of r. In constructing a stress-strain or a strain-hardening curve, it is this flow stress which is to be plotted against the strain. The flow stress is less

TABLE I

$\dfrac{a}{R}$	Correction factor	Siebel's $\dfrac{1}{1 + \frac{1}{4}(a/R)}$	$\dfrac{\widehat{rr}_0}{F}$
0	1.000	1.000	0
⅓	0.927	0.923	0.154
½	0.897	0.889	0.223
1	0.823	0.800	0.405
2	0.722	0.667	0.693
3	0.656	0.571	0.916
4	0.606	0.500	1.099

than the "average tension," which is ordinarily used. The factor by which the average tension is to be multiplied to obtain the flow stress F may be called the "correction factor" and by Eq. (1-13) is equal to $\dfrac{1}{[1 + 2(R/a)] \log [(1 + \frac{1}{2}(a/R)]}$. It is given in Table I as a function of a/R and is shown graphically in Fig. 6.

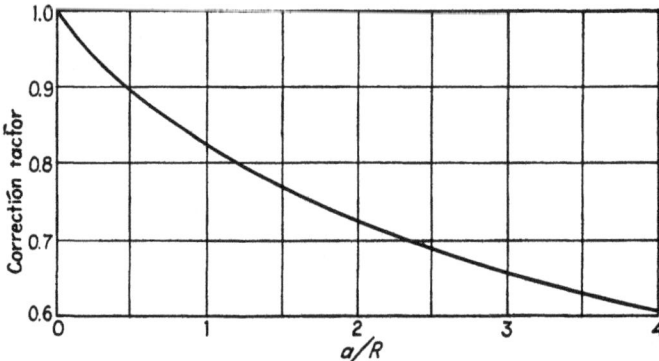

FIG. 6. The "correction factor" as a function of the radius of curvature of the contour at the neck.

Since the correction factor is less than unity, the strain-hardening curve will rise less rapidly with strain when flow stress is plotted against strain than when the average tension is plotted against strain. Under the extreme elongations reached in this work, we shall see later that the lowering of the stress-strain curve thus demanded amounts to about 30 per cent.

Next consider the effect of stress distribution on fracture. Fracture without doubt is initiated on the axis. There is experimental evidence for this in the work of other experimenters, and it is exactly what would be expected from the distribution of stress arrived at above. We have seen that, except for the imposed hydrostatic pressure, the stress system consists of a uniform tension F along the axis, on which is superposed a system of hydrostatic tension rising to a maximum on the axis. Now in general hydrostatic pressure increases ductility, often by large amounts, as will be shown later in this book. Hydrostatic tension would therefore be expected to decrease ductility; and, since the maximum hydrostatic tension is on the axis, the maximum decrease of ductility would be expected to be there, so that there is where fracture should begin. In specifying the stress conditions under which fracture occurs it is obvious that it is the stress on the axis which should be specified and not the average stress across the neck. The amount by which the hydrostatic tension on the axis may differ from the flow stress is shown also in Table I, in the fourth column. This factor is markedly larger than the correction factor, so that it is more important to consider the stress distribution when we are dealing with conditions of fracture than it is when we are dealing with the strain-hardening curve. In cases of large strain, consideration of the stress distribution may very materially alter our description of the conditions under which fracture occurs, as we shall see in detail later.

Use of the formulas which we have derived for the stress distribution at the neck demands a knowledge of the parameter a/R in addition to the total load and the radius of the neck a, which are the parameters conventionally determined. If the specimen is not pulled to fracture, it is not difficult to determine the R of the specimen after release of load. I have used several methods. One was calculation from five or more measurements of the diameter of the specimen at intervals equally spaced along the axis. Another method was to draw a much magnified image of the neck with a camera-lucida arrangement, and find by cut-and-try manipulations with a compass the circle which is the best fit for the neck. Probably the best method is to slide a long slightly tapering cone across the neck perpendicular to the axis, watching under a microscope the pattern of light shining between cone and specimen. It is easy to determine rather sharply the point of transition between too great and too small a radius of the cone. A micrometer measurement of the radius of the cone then gives R at once. Strictly, R should be determined from measurements while the load is acting. The error in proceeding as above must be very small, since the only effect is the elastic effect on releasing stress. No attempt was made to apply any sort of correction for this elastic effect. It would obviously have been of prohibitive difficulty to attempt to measure the R of specimens pulled under pressure while load and pressure were acting.

If the tension specimen has been pulled to fracture it is not so easy to get a good value for a/R. The two fractured halves may usually be roughly fitted together and a measurement of R of sorts obtained by one of the procedures just described. In many cases, however, it seems that there is very appreciable distortion of the contour produced by the act

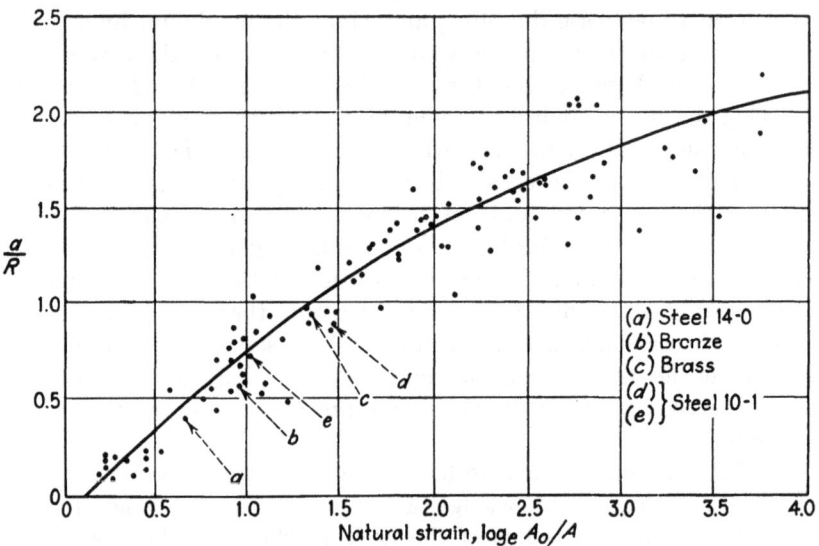

Fig. 7. a/R as a function of natural strain for nearly all the tension specimens pulled under pressure. These include various ordinary and stainless steels, and a few other materials as indicated.

of fracture itself, so that the R obtained by measurement of the broken pieces is not the R which prevailed at the moment of fracture. More consistent results can be obtained for the R at fracture by plotting R against a for a series of specimens pulled to varying amounts short of fracture and extrapolating to the R corresponding to the a measured at fracture. The latter does not seem so susceptible to falsification by the process of fracture itself.

Many measurements were made on a/R for different sorts of steel of different heat-treatments, pulled at various hydrostatic pressures, and also for other metals. The detailed data will be presented in the next chapter. Evidence gradually accumulated that, when all the measurements of a/R are plotted against the strain at the neck, for which a convenient measure is $\log_e (A_0/A)$, as will appear later, they lie on a band centering about a mean curve from which they do not depart by large amounts. In Fig. 7 are shown a number of the experimental points thus determined. If the formulas for the stress distribution are consulted, it will be seen that for most purposes, within experimental error, it is sufficiently accurate to take the value of a/R for the particular strain from Fig. 7, instead of using the directly measured value. This is not always within experimental error, and consistent departures from the curve occur, but nevertheless for most purposes the approximation is good enough, and it is evidently a matter of much convenience to use it.

Included in the materials plotted in Fig. 7 are steels described in detail in Chap. 2 and designated there as 11-0, 12-0, 13-0, 9-0, 9-1, 9-2, 9-3, 9-4,

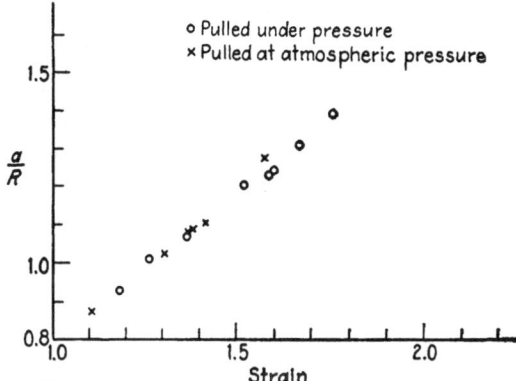

FIG. 8. a/R against strain for tempered martensite (steel 9-8). The diagram shows that a/R is independent of the pressure of pulling. The exact values of the pressures corresponding to the various points in the diagram are given in Table V.

9-5, 9-6, 10-1, 14-0, 15-0, 16-0, 17-0, 18-0, and various armor plates. In addition to the steels, brass and bronze are also included, as indicated.

Explicit measurements, tabulated in Chap. 2 under steels 9-7 and 9-8, showed that a/R does not depend perceptibly on the pressure of pulling. Figure 8 shows the degree of independence of pressure. Furthermore, if a specimen is pulled under high pressure, pressure released, and pulling resumed at atmospheric pressure, the a/R values will lie on the same curve. It was also established that a/R falls on the same curve for specimens fractured in tension after a prestraining in simple compression. The single curve thus appears to reproduce the experimental results over a wide range of conditions. The only exception encountered was one of the stainless steels, which falls out of line with most other steels in other respects also.

If it is regarded as a good enough approximation for the purpose in hand to dispense with the direct measurement of a/R for the particular specimen concerned and to take the mean a/R from the curve of Fig. 7, then the corrections for the stress distribution at the neck may be specified in terms of the natural strain at the neck. Table II shows such results.

TABLE II

$\log \dfrac{A_0}{A}$	Correction factor	$\log\left(1 + \dfrac{1}{2}\dfrac{a}{R}\right)$	Ratio of hydrostatic tension on axis to average longitudinal stress
0.1	1.000	0.000	0.000
0.5	0.920	0.165	0.152
1.0	0.853	0.325	0.277
1.5	0.803	0.448	0.360
2.0	0.776	0.533	0.414
2.5	0.752	0.601	0.451
3.0	0.736	0.649	0.477
3.5	0.724	0.690	0.499
4.0	0.715	0.718	0.513

The figures of the second column of Table II are shown graphically in Fig. 9.

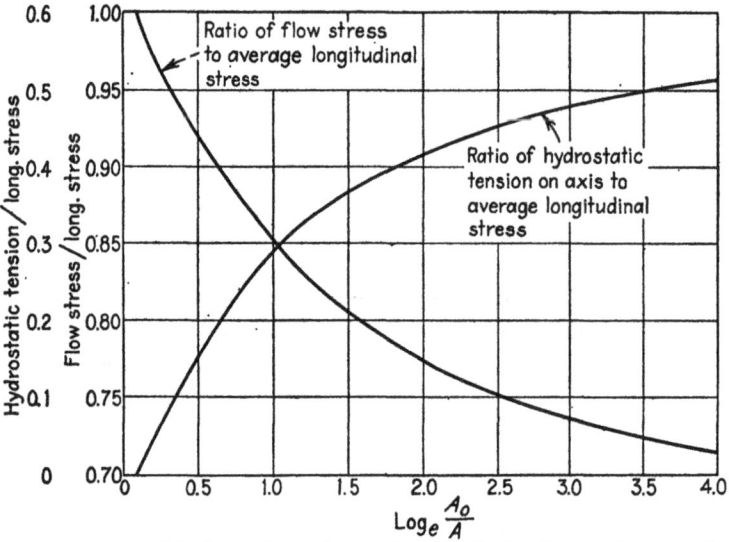

FIG. 9. Flow stress and hydrostatic tension on the axis of a tension specimen as a function of strain at the neck.

Table II also gives $\log[1 + \frac{1}{2}(a/R)]$, which is the factor by which the *flow stress* is to be multiplied to obtain the hydrostatic tension on the axis. By combining the second and third columns of the table the ratio of the hydrostatic tension on the axis to the average longitudinal stress is obtained. This is shown in the fourth column of the table and also in Fig. 9.

As an example of the use of the table, consider a specimen pulled out to a strain of 2 (elongation at the neck of 7.4-fold), with an average stress at the neck (total load divided by the area of the neck) of 200,000 psi. Take conventional cylindrical coordinates, z along the axis of the specimen and r and θ at right angles. Then the flow stress, or $\widehat{zz}$ at the outer surface of the neck, is 155,200 psi (= 200,000 × 0.776). $\widehat{zz}$ increases from the value 155,200 at the outer surface to the value

$$155{,}200 + 155{,}200 \times 0.533 = 237{,}900 \text{ psi on the axis}$$

The stress components $\widehat{rr}$ and $\widehat{\theta\theta}$ both increase from zero at the outer surface to 82,700 psi (= 155,200 × 0.533) on the axis. For large reductions of area it appears therefore that these corrections are important.

An indication of the probable error in using mean values for a/R may be obtained from Fig. 7. At a strain of 2, for example, the figure shows that a deviation of a/R by as much as 0.2 in either direction from the curve would be very unusual. The corresponding uncertainty in a/R, when applied to the example of the last paragraph, gives 150,800 and 159,200 psi as the limits of the maximum range of uncertainty for the flow stress. The similar range of uncertainty for the longitudinal component of stress on the axis is between 241,700 and 233,700 psi, respectively. It would appear that for most purposes it is amply good enough to use the curve of Fig. 7 and dispense with the direct measurement of a/R.

The curve of Fig. 7 and Table II based on it is slightly different from the corresponding figure and table published in my original paper on the corrections for the stress distribution at the neck. The reason for this is that at the time of publication of that paper I did not have so many data on which to base the construction of the curve of Fig. 7.

Table II has been carried only to the value 4 for the natural strain. The reason for this is that, if the specimen is pulled to higher strains, as is easily possible if the pulling is conducted under hydrostatic pressure, the neck loses its geometrical regularity, the section is no longer circular, and the contour no longer has rotational symmetry. The reason for this is doubtless that individual grains monopolize the cross section. The effect is much more marked in copper and aluminum than in steel. In aluminum the cross section has been found to become approximately square at longitudinal extensions of not more than 15- or 20-fold.

The problem of the stress distribution at the neck is obviously only a small part of the more general problem of completely describing the state of affairs at all points of the tension specimen. The solution of the general problem will give the stress at all points, and the deformation at all points, including in particular the complete contour of the whole specimen. The general problem is of great complexity, and no promising method of attack has presented itself. Into the solution must go in general a knowledge of the strain-hardening effects for the most general types of deformation; this will include a knowledge of the nonisotropies accompanying various types of deformation. One might hope, however, that such an exhaustive knowledge might not be necessary and that empirical regularities might disclose themselves which would make possible partial conclusions. In particular, when I started this work, the question presented itself whether perhaps it might be possible from a study of the complete contour of the specimen to derive information about the strain-hardening curve in simple tension over the entire stress range. Such an expectation is not implausible, because obviously parts of the specimen at a distance from the neck were flowing at an earlier stage of the process, and these parts, which are frozen at later stages of the process, thus leave a permanent record in the final shape of the tension specimen. The problem is how to interpret the shape. In order to find some answer to this question an elaborate study was made of the shape of the complete tension specimen as a function of the degree of advancement of the pulling process. This study was made in the early part of the work for the NDRC and consisted in drawing on a much enlarged scale with the help of a camera obscura the entire contour of tension specimens of a number of different grades of armor plate in different degrees of advancement of the drawing process. The specimens with large reductions of area were drawn under pressure. To further facilitate the study, a very shallow screw thread was turned on the surface of the virgin specimen. On pulling, the spires of the helix were pulled apart and by their extension gave a direct indication of the longitudinal extension at the exterior surface in addition to the knowledge about the average radial strain across the section which was obtainable from a measurement of the diameter.

One very simple generalization emerged from this study. Not far from the neck the contour of a tension specimen exhibits a point of inflection. The observation was that the parts of the specimen behind the point of inflection are approximately frozen as extension progresses, the region of sensible movement being the region between the point of inflection and the neck. The result is that as drawing progresses the contour of the specimen folds itself down along a fixed curve, as shown by the dotted lines in Fig. 10. Furthermore, the curve along which the contour folds itself is the same for all the grades of steel examined and in all degrees of heat-treatment. A particular consequence of this general fact would be that a/R is the same function of the extension for all steels, as we have found to be approximately the case. It would be a further consequence that the contour of the tension specimen would be expected to have the same shape for any other metals whose a/R falls on the same curve as for steel. In particular, this should apply to copper and bronze, as suggested by Fig. 7. This was not checked. It was definitely established by a rather elaborate study for lead, however, that the shape of the contour is not the same for it as for steel. Lead draws out more nearly uniformly, and the point of inflection is situated relatively much farther from the neck than for steel. In fact, flow can be shown to be taking place for lead through a range of area of cross section of at least 2-fold; that is, plastic flow continues over a stress range of at least 2-fold for lead.

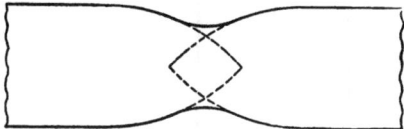

FIG. 10. Showing how the contour in the neighborhood of the neck develops by folding along a fixed curve as pulling proceeds.

The strain-hardening curve thus rises much less rapidly for lead than for steel.

The example of lead indicates how complex a complete solution of the problem of the shape of the contour of a tension specimen would be. The shape is determined by the relative velocity of flow at different places, and this factor of relative rates of flow as a function of stress and strain is not recognized in any simple formulation of the plasticity conditions such as that of von Mises, for example. The von Mises condition may be expected to give approximate results only when the velocity of flow falls off very rapidly when the stresses exceed the von Mises limit. Later experiments on two-dimensional compression will show that this condition does hold approximately for steel.

It is probable that the experimental material which I obtained for the NDRC has not been fully exploited. In particular, I believe that more could be got out of a detailed study of the way in which the helical markings on the external surface of the tension specimens were distorted. Unfortunately, I could not find time to exploit all the possibilities.

A problem closely analogous to that treated thus far in this chapter presents itself with respect to two-dimensional tension. Such a state of affairs occurs when a sheet in the form of a rectangle of width much greater than its length is pulled by tension forces applied along the long edge. The constraints are here such that the width remains approximately constant. The deformation is an extension in the direction of the tensile force and an equal contraction in the direction perpendicular to the face of the sheet. Such a sheet eventually reaches an unstable condition, a two-dimensional neck begins to form, and eventual fracture occurs at the neck, just as in the more usual three-dimensional case where the tension specimen has rotational symmetry about the axis of pull. In order to get a correct picture of the relations it is necessary to apply corrections for the deviation of the stress distribution from uniformity at the neck, just as in the three-dimensional case. In the present work the rigorous two-dimensional case was approximated by pulling thin-walled tubes. There is, under these conditions, some contraction of the circumference of the tube, so that the strain is not strictly two-dimensional. The deviation is not large, however, and in the experimental work the corrections were applied which are applicable to the strictly two-dimensional case.

Two-dimensional Specimen. The analysis for the two-dimensional case goes as follows. Figure 11 represents a section of the sheet through the neck. The direction of tension in the sheet is along the Z axis; the X axis is taken perpendicular to the face of the sheet, and the Y axis is in the sheet, perpendicular to X and Z and perpendicular to the plane of the paper. The sheet is taken as infinitely wide in the Y direction, and edge effects are neglected; that is, neither stress nor strain is a function of y, and all derivatives with regard to y in the fundamental equations vanish.

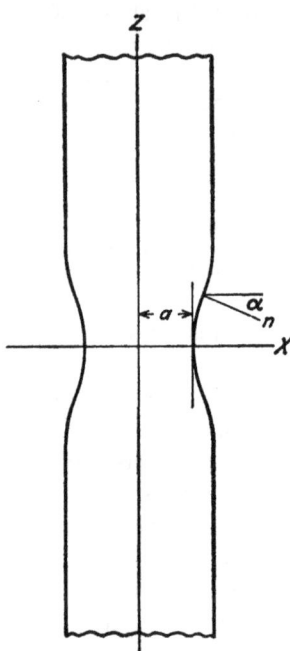

FIG. 11. The relations in the neighborhood of the neck for two-dimensional tension of a sheet.

The three usual stress equations of equilibrium reduce to the following two in virtue of the vanishing of the y derivatives and also because $Y_z = X_y = 0$ by symmetry:

$$\frac{\partial X_x}{\partial x} + \frac{\partial X_z}{\partial z} = 0$$
$$\frac{\partial X_z}{\partial x} + \frac{\partial Z_z}{\partial z} = 0 \quad (1\text{-}14)$$

Elimination of X_z gives

$$\frac{\partial^2 X_x}{\partial x^2} - \frac{\partial^2 Z_z}{\partial z^2} = 0 \quad (1\text{-}15)$$

X_x is symmetric about the origin, so that $\left.\dfrac{\partial X_x}{\partial x}\right|_{x=0} = 0$. Hence integration gives

$$\frac{\partial X_x}{\partial x} = \int_0^x \frac{\partial^2 Z_z}{\partial z^2}\, dx \quad (1\text{-}16)$$

The boundary conditions become

$$X_z \cos \alpha - X_x \sin \alpha = 0$$
$$X_x \cos \alpha - Z_z \sin \alpha = 0 \qquad (1\text{-}17)$$

where $\cos xn = \cos \alpha$, $\cos yn = 0$, and $\cos zn = -\sin \alpha$ (see Fig. 11). These two equations may be rewritten

$$X_x = Z_z \tan^2 \alpha$$
$$X_z = Z_z \tan \alpha \qquad (1\text{-}18)$$

The solution is not determined by the stress equations of equilibrium only, but in general the conditions of plasticity must also be satisfied. In general these conditions should be written for all points of the body and should be written in their general form for all six components of stress and strain. We shall not be able to obtain a complete solution of the problem, however, but shall have to content ourselves with a partial solution in which the shape of the specimen is assumed and the distribution of stress and strain is determined only in the plane of the neck itself. For this solution we shall essentially apply the stress equations of equilibrium throughout a band on both sides of the plane of the neck but shall apply the conditions of plasticity only in the plane of the neck itself. In this plane the principal stresses are along the three coordinate axes. In particular, X_z vanishes by symmetry, and the equations take a simplified form. We assume the von Mises plasticity condition is satisfied

$$(X_x - Y_y)^2 + (Y_y - Z_z)^2 + (Z_z - X_x)^2 = \text{const}$$

and we assume the flow conditions in the simple form applicable when there is no change in direction of the principal axes during flow:

$$e_x = \gamma[X_x - \tfrac{1}{2}(Y_y + Z_z)]$$
$$e_y = \gamma[Y_y - \tfrac{1}{2}(Z_z + X_x)] \qquad (1\text{-}19)$$
$$e_z = \gamma[Z_z - \tfrac{1}{2}(X_x + Y_y)]$$

Here γ is a factor of proportionality. For our purposes it is not specified further except that it is the same in all three equations.

Subject to the limitations thus far imposed we may distinguish two limiting sorts of flow.

1. The width of the plate in the Y direction may be taken as infinite compared with that in the Z direction, and the tension Z_z may be thought of as applied through suitable clamps at the two ends of the plate so that the dimensions in the Y direction are unaltered. Under these conditions $e_y = 0$, and at the neck $Y_y = \tfrac{1}{2}(Z_z + X_x)$ by Eq. (1-19). Actually, of

course, at the edges $Y_y = 0$, and the expressions just found for Y_y can be applicable only in the central portions of the plate sufficiently remote from the edges. In the borders of the plate there must be shearing stresses building up the value of Y_y in the central regions. The transition zones near the edges may be neglected in comparison with the width of the plate, and the conditions given in the equation may be taken to represent the average conditions throughout the plate. At the outer faces of the plate $X_x = 0$, so that at the neck $(Y_y = \frac{1}{2}Z_z)_{x=\pm a}$, where $2a$ is the thickness of the plate.

2. The second limiting case is when the length of the plate is infinite in comparison with the infinite width, so that any clamps through which the tension may be applied at the ends exert no effective constraint in the central regions. In the initial stages of flow, before necking starts, we have $Y_y = 0$, giving

$$e_x = e_y = -\tfrac{1}{2} e_z$$

After necking starts there will be local constraint, independent of the initial proportions, and the actual deformation will be intermediate between cases 1 and 2. Nevertheless it will be profitable to set up the limiting cases for separate treatment.

In both cases the solutions are qualitatively the same and qualitatively similar to the solution for the conventional solid test bar. It is only the detailed form of the equations which differs. It is evident on inspection that the two sets of plasticity conditions will be satisfied if (a) the strain is uniform across the neck, and (b) the stress consists of a system constant across the neck plus a superposed hydrostatic tension which may vary with the coordinate X across the neck. With regard to condition (a), experimental proof has been given that the assumption of uniform strain is a sufficiently close approximation for the conventional three-dimensional case of a solid cylindrical test piece. Since the direct experimental proof for the present two-dimensional case would involve serious experimental difficulty, at least if attempted along the same lines as for the three-dimensional case, we shall simply take over the assumption without further attempt at proof. At least there seems to be no reason to think that the assumption is not adequate here also. With regard to condition b, the constant stress system in case 1 is obviously $X_x = 0$, $Y_y = \frac{1}{2}Z_z = \text{const.}$ For case 2 the constant stress system is $X_x = Y_y = 0$, $Z_z = \text{const.}$

The further steps in the solution run exactly parallel to those for the cylindrical case. The same diagram (Fig. 2) applies, which now represents a section through the X-Z plane of a system of rectangular coordinates in which the system is infinite in the Y direction, instead of a sec-

STRESS DISTRIBUTION AT NECK

tion through the z axis in the r-z plane of a system of cylindrical coordinates in which the coordinate θ plays no part.

The condition that the total force in the X direction acting on the element of thickness dx and heights h and h' along the Z axis vanishes is

$$\left(Z_z + h\frac{\partial Z_z}{\partial z}\right)\sin\phi'\,dx = X_x h - \left(X_x + dx\frac{\partial X_x}{\partial_z}\right)h' \quad (1\text{-}20)$$

where

$$h = R\phi + R'(\cos\phi' - \cos\phi)$$

$$= R\phi + \frac{a}{\phi}\left[\left(1 - \frac{1}{2}\frac{x^2}{a^2}\phi^2\right) - \left(1 - \frac{1}{2}\phi^2\right)\right]$$

$$= \phi\left[R + \frac{a}{2}\left(1 - \frac{x^2}{a^2}\right)\right] \quad (1\text{-}21)$$

$$h' = \phi\left\{R + \frac{a}{2}\left[1 - \left(\frac{x + dx}{a}\right)^2\right]\right\}$$

$$= h - \phi\frac{x}{a}dx \quad (1\text{-}22)$$

Substituting back, using the condition that $\partial Z_z/\partial z = 0$ at the neck by symmetry, and dividing by dx, gives

$$Z_z\frac{x}{a} = \frac{x}{a}X_x - \left[R + \frac{a}{2}\left(1 - \frac{x^2}{a^2}\right)\right]\frac{\partial X_x}{\partial x} \quad (1\text{-}23)$$

But now our assumption about the stresses $Z_z = Z_{za} + X_x$ gives

$$\frac{\partial X_x}{\partial x} = -\frac{Z_{za}}{a}\frac{x}{R + (a/2)[1 - (x^2/a^2)]} \quad (1\text{-}24)$$

Integration, subject to the condition $X_x|_{x=0,} = 0$, gives at once

$$X_x = Z_{za}\log\left[1 + \frac{1}{2}\frac{a}{R}\left(1 - \frac{x^2}{a^2}\right)\right] \quad (1\text{-}25)$$

The complete stress system may at once be written down.
Case 1:

$$\begin{aligned}X_x &= Z_{za}\log\left[1 + \frac{1}{2}\frac{a}{R}\left(1 - \frac{x^2}{a^2}\right)\right]\\ Y_y &= Z_{za}\left\{\frac{1}{2} + \log\left[1 + \frac{1}{2}\frac{a}{R}\left(1 - \frac{x^2}{a^2}\right)\right]\right\}\\ Z_z &= Z_{za}\left\{1 + \log\left[1 + \frac{1}{2}\frac{a}{R}\left(1 - \frac{x^2}{a^2}\right)\right]\right\}\\ e_x &= -e_z \qquad e_y = 0\end{aligned} \quad (1\text{-}26)$$

Case 2:

$$X_x = Z_{za} \log\left[1 + \frac{1}{2}\frac{a}{R}\left(1 - \frac{x^2}{a^2}\right)\right]$$
$$Y_y = Z_{za} \log\left[1 + \frac{1}{2}\frac{a}{R}\left(1 - \frac{x^2}{a^2}\right)\right]$$
$$Z_z = Z_{za}\left\{1 + \log\left[1 + \frac{1}{2}\frac{a}{R}\left(1 - \frac{x^2}{a^2}\right)\right]\right\}$$
$$e_x = e_y = -\tfrac{1}{2}e_z$$

(1-27)

In case 2 the formulas are formally exactly the same as for the cylindrical case, merely replacing r by x. In case 1 the only formal difference is the inclusion of the term $\tfrac{1}{2}$ in Y_y.

The variable hydrostatic tension superposed on the constant stress reaches its maximum value, $Z_{za} \log\left(1 + \frac{1}{2}\frac{a}{R}\right)$, on the axis. Again the formal expression is exactly the same as for the cylindrical case. Differences appear, however, in the mathematical expressions when a connection is sought with the total load or the average stress. This is obviously to be expected because of the different form of the elements of cross-sectional area. We have in the present problem

$$\frac{1}{2}\frac{\text{load}}{\text{width}} = aZ_{z\,\text{av}} = \int_0^a Z_z \, dx \qquad (1\text{-}28)$$

The explicit value for Z_z may be substituted and the integral evaluated in closed form, giving for both cases 1 and 2

$$Z_{za} = \frac{Z_{z\,\text{av}}}{\left(1 + 2\frac{R}{a}\right)^{\frac{1}{2}} \log\left[1 + \frac{a}{R} + \left(\frac{2a}{R}\right)^{\frac{1}{2}}\left(1 + \frac{1}{2}\frac{a}{R}\right)^{\frac{1}{2}}\right] - 1}$$
$$\equiv Z_{z\,\text{av}} \times (\text{correction factor}) \qquad (1\text{-}29)$$

The correction factor, defined by the identity, is the factor by which $Z_{z\,\text{av}}$ is to be multiplied to obtain Z_{za}. The formal expression for the correction factor is more complicated than for the cylindrical case.

The correction factor depends only on the parameter a/R, so that to determine it the radius of curvature of the external contour at the neck must be determined. It is given in Table III, in which the factor for the cylindrical case is reproduced for convenience of reference. The correction for the sheet is somewhat less than for the cylinder, so that for the same value of a/R it is even more important to apply the correction for stress distribution to sheets than to the conventional round tension specimens.

TABLE III

$\dfrac{a}{R}$	Correction factor	
	Sheet	Cylinder
0.1	0.965	0.976
0.2	0.941	0.956
0.3	0.915	0.931
0.5	0.875	0.896
0.7	0.828	0.864
1.0	0.782	0.823
1.5	0.717	0.768
2.0	0.669	0.723

The next step, in analogy with the solution for the cylindrical case, would be to examine the possibility of dispensing with a measurement of a/R and expressing the correction factor in terms of the strains. Any such substitution would have to rest on empirically established connections between a/R and the strain and would involve the collection of extensive experimental material. Since only a few tests were made in two-dimensional tension, there would have been no profit in taking this step, but instead the correction was determined by measuring a/R for each specimen.

CHAPTER 2

THE TENSION OF STEEL UNDER PRESSURE[1]

Experimental Method. The tension tests were conducted in what amounts to a miniature testing laboratory completely flooded with a liquid on which the pressure could be raised by any arbitrary amount up to a maximum of about 450,000 psi (30,000 kg/cm^2). It is as if the entire testing laboratory were transported to the bottom of an ocean of variable and controllable depth. The "tension" exerted on the specimen in the following is to be understood as the stress acting in addition to the hydrostatic pressure; it can be visualized as the stress produced by a weight hung on the specimen in the ocean. Specifically, the tension would be this weight divided by the cross-sectional area of the tension specimen.

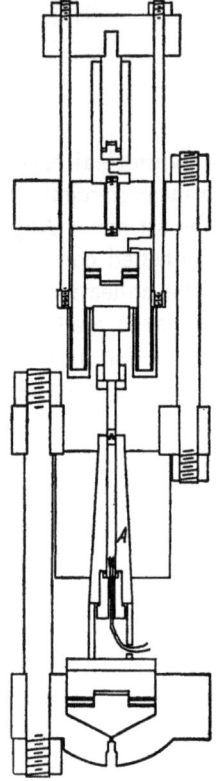

FIG. 12. General arrangement of pressure apparatus for generating 450,000 psi in which the tension experiments were conducted.

The "pressurized" laboratory was the interior of the cylinder with which many experiments had already been conducted up to 450,000 psi maximum. It is shown in Fig. 12. The pressure chamber in which the specimen is mounted is 0.5 in. in diameter and of the order of 4 in. long. As indicated in the figure, the exterior of the pressure vessel is given a conical shape, and it is surrounded by heavy supporting rings with the same angle as the cone. The cylinder is mounted, as shown, between two opposed hydraulic presses. Internal pressure is produced by the upper press, with piston 3.5 in. in diameter, which pushes a suitably packed piston of carboloy of 0.5 in. diameter into the bore of the vessel. The lower press, with piston 6 in. in diameter, simultaneously with the production of internal pressure, pushes the entire pressure vessel into the supporting rings, thus generating, by the action of the cone, an external pressure on the vessel which increases proportionally with the increase of

---

[1] This chapter is based on material contained in Watertown first, third, fifth, and seventh reports; NDRC first, second, fourth, and sixth reports; and B19, B20.

internal pressure. The two presses may be coupled together so as to exert pressures automatically in the correct ratio, or, more simply, the two presses may be driven by two independent hand pumps, the pressure being maintained at the correct ratio by manual control.

Inside the "pressurized laboratory" it is required to apply the tensile load to the tensile specimen and to make the corresponding measurements, which include, in addition to a measurement of the pressure prevailing within the laboratory, a measurement of the load and of the extension of the specimen. The physical manipulation of applying a load to the specimen might conceivably be done through auxiliary pistons reaching into the pressure chamber through suitable stuffing boxes, but the technical difficulties of multiple pistons would be very great at such high pressures. A method was found of applying tensile load to the specimen with the same piston as that with which the hydrostatic pressure within the chamber is produced. To this end the tensile specimen itself was mounted in a double-yoke arrangement, indicated schematically in Fig. 13, the geometry of which is such that when the over-all length of the yoke is decreased by pushing on its ends the length of the tension specimen is increased. The yoke is made of hardened steel, ground carefully to dimensions all over in order to avoid buckling effects. The yoke is shortened by the advance of the piston with which pressure is generated. In the initial stages of advance of the pressure-generating piston the yoke is not in contact with it, and advance of the piston serves only to increase pressure by compressing the fluid with which the pressure chamber is filled. When the piston has advanced by an amount sufficient to come into contact with the yoke, further advance of the piston both increases pressure and transmits an increasing tensile load to the specimen. The pressure at which contact is first made with the yoke may be controlled by suitably choosing the amount of liquid with which the pressure vessel is initially filled.

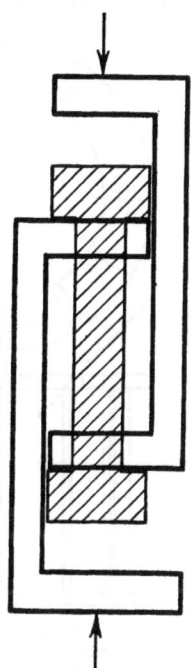

FIG. 13. Idealized scheme of the yoke arrangement by which tension is applied to the specimen within the pressure vessel.

It is an obvious disadvantage of this method that the pressure rises during the progress of the tensile test, so that a complete tensile test cannot be conducted under a single predetermined pressure. By suitably choosing the various dimensions, however, the rise of pressure during the progress of the test may be made comparatively small. The variation of pressure during the final stages of the test, which are the critical stages, is in fact so small that no correction is necessary, and the test was usually characterized by the pressure of the final reading. In some cases any possible effect of pressure variation during the test was further minimized by using a special apparatus. In much of the work for the NDRC the upper limit of pressure contemplated was 225,000 psi. For these measurements a specially long pressure vessel was used, which, by starting with more liquid than usual, made the change of pressure during the progress of the test less than usual.

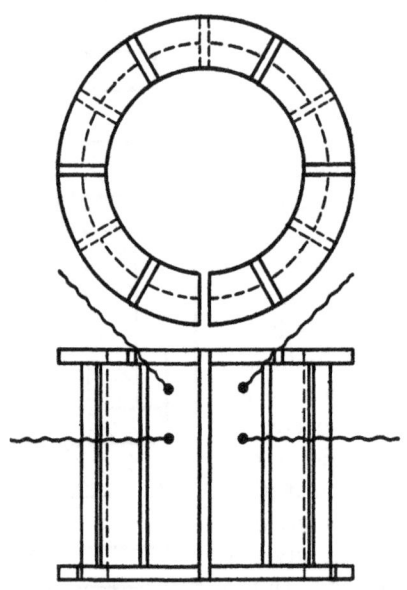

Fig. 14. The "grid" by which the tension on the specimen is measured by means of the change of electrical resistance induced by the stress in a member supporting the specimen.

The changes of length of the tension specimen were given with sufficient accuracy by measurements of the position of the pressure-generating piston. This was measured with an Ames 1/10,000-in. dial gauge directly attached to the piston. Any corrections arising from elastic distortion of other parts of the apparatus could be determined by blank experiments and proved to be negligible for most purposes. The accuracy was not high enough to give elongations in the elastic part of the range.

The tensile load on the specimen might in principle also be determined from measurements on the pressure-generating piston, namely, by measuring the excess force required to advance the piston over the force required to overcome the pressure alone. This excess force was, however, always a comparatively small fraction of the total force and was therefore subject to prohibitively large percentage errors arising from the irregular friction of the piston. It was therefore necessary to measure the tensile load inside the pressure vessel instead of from the outside. This was done by interposing between the end of the yoke and the bottom of the pressure chamber, which eventually received the thrust on the tension specimen, a so-called "grid," which was simply a member whose electrical resistance changed with the thrust. The idea of measuring a thrust by measuring a change of electrical resistance is not a new one, and various devices can be found in the literature, such as flat coils of manganin embedded between mica. All the devices with which I was familiar suffered, however, from one or another disadvantage, such as error from friction and hysteresis in the case of the mica, or of too great deformability in the case of certain other devices; so that it proved necessary to develop a new device. After much preliminary work the "grid" was developed. This is shown in Fig. 14. It is essentially a short cylindrical thin-walled annulus of steel, which by its configuration is strong and stiff against axial compressive forces, and which at the same time is cut in such a way by a system of slots that the path of the electrical current flowing through it is long, with consequent comparatively high resistance and comparatively high sensitivity to small fractional changes of resistance. The percentage change of resistance of such an arrangement is proportional to the intensity of the compressive stresses; high sensitivity demands a small area of cross section, and this demands carrying the working stress in the grid as near as possible to the rupture point. This pretty definitely indicated steel as the best metal. Furthermore, the demand that the relation between change of resistance and compressive stress be linear and without hysteresis imposed restrictions on the steel suitable for the grid. A high-strength tool steel, heat-treated to a high hardness, known under the trade name of "Teton" and manufactured by the Ludlum Steel Co., was found after trial to satisfy all demands. Such a steel will exhibit a decrease of resistance of the order of 1.5 per cent under a simple compressive stress of the order of 200,000 psi, which is not high enough to exceed the elastic limit in compression.

Such a grid can be calibrated with known weights at atmospheric pressure. In use, the resistance of the grid changes because of two effects: increase of hydrostatic pressure and increase of simple compression which is transmitted to it by the tension in the tension specimen. The effect of hydrostatic pressure is determined by independent calibration. The total change of resistance of the grid is corrected for the effect of pressure by subtracting off the pressure effect as calculated from the calibration, leaving a residual change of resistance due to the simple compression. The compression then may be calculated, but a correction must obviously be applied in this calculation for the effect of pressure on the compression coefficient. Ideally the pressure coefficient in simple compression might be derived from measurements of the extra force required to drive the pressure-generating piston, but any single determination of this suffers, as already explained, from too great error due to excessive friction. However, error due to irregularities of this sort should be capable of elimination if averages of many readings are taken, and in fact in all the work done during the war a value was used for the pressure coefficient of the compression coefficient which was determined from the mean of some 600 measurements. The correction so determined was about 20 per cent at the top pressure of 450,000 psi, the tension given without applying the correction being smaller than the correct tension. Later, however, a simple independent method was found for determining this pressure coefficient of the compression coefficient. The method is indicated in Fig. 15. The tensile specimen is made in two halves as indicated. The opposing surfaces of the two halves are carefully worked flat and are surrounded by a thin copper collar to keep the pressure-transmitting liquid from penetrating into the space between them. The compound specimen is pulled as an ordinary tension specimen. It is obvious that, except for the slight resistance offered by the copper sheath, the two halves will be pulled apart when the tensile load is equal to the hydrostatic pressure. By determining the apparent fracture point of the compound specimen with the uncorrected grid, and comparing with the known hydrostatic pressure, the proper correction for the resistance of the grid may be determined. The correction thus determined agreed exactly, within the sensitiveness of the measurements, with the correction otherwise determined from the average of 600 measurements of the force on the piston.

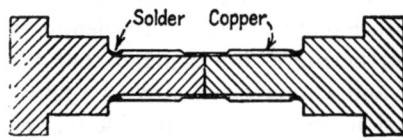

FIG. 15. Device for determining the pressure coefficient of the compression coefficient of the grid of Fig. 14.

The remaining measurement, that of hydrostatic pressure, was made by the method conventional in all my high-pressure work, namely, by determining the change of resistance of a coil of manganin. The details of this pressure measurement need not be elaborated. They will be found, for example, in my book, "The Physics of High Pressure."

Since the measurement of the resistance of the grid demands a potentiometer method, four independently insulated leads are necessary altogether, with a common ground for grid and manganin coil. The plug with which four insulated leads are carried into the pressure chamber is shown in Fig. 16. To avoid complication, only a single lead is shown in this figure. Much development work was involved in this plug, but it is now satisfactory to the extent that 100 or more applications of the highest pressure can be made without loss of insulation or development of mechanical leak.

The general dimensions of the tension specimens are indicated in Fig. 17. These were subject to considerable variation in the diameter depending on the strength of the specimen, but not much variation was possible in the length.

Detailed Experimental Procedure. The experimental details in conducting a tension test were as follows. The specimen, after its dimensions were measured, was mounted in the pressure vessel, which was then filled with the amount of liquid which previous experience had shown would be

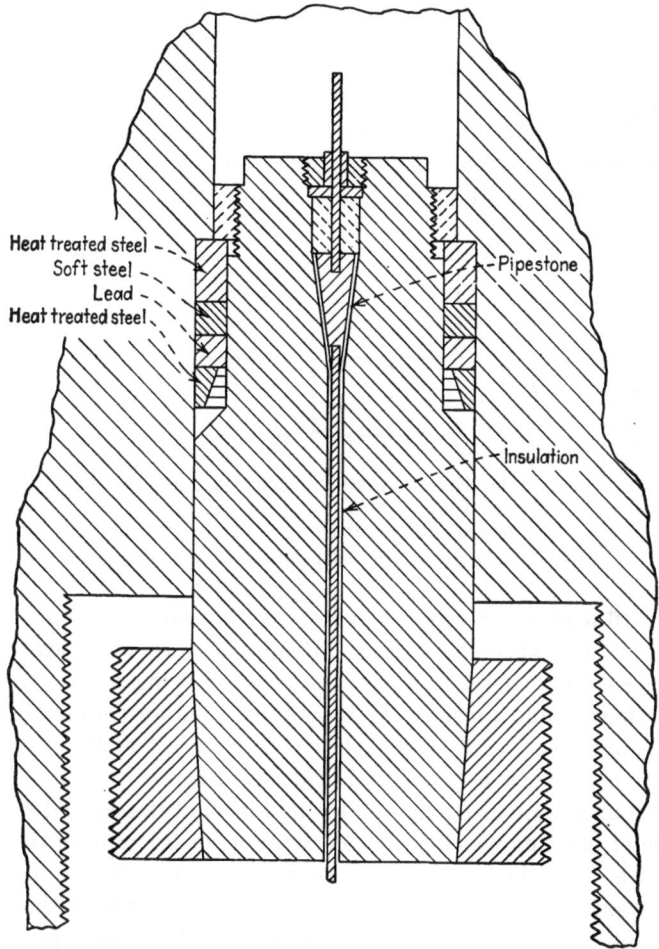

FIG. 16. Illustrates the method of taking insulated leads into the pressure vessel.

necessary to make the initial contact between piston and specimen at the desired initial pressure. In order to control the pressure of initial contact, the effective length of the yoke might be extended by inserting into the pressure vessel dummy plugs of suitable lengths. Liquid was then introduced and air bubbles removed by gentle exhausting. The final amount of liquid was controlled by first completely filling the vessel and then removing excess liquid by inserting into the upper end of the vessel a plug of length adjusted to 0.001 in. with a nut. The pressure of first contact, other things being constant, depends somewhat on the temperature of the apparatus, that is, on the temperature of the room, since the compressibility of the transmitting liquid depends on its temperature. This liquid was a mixture of iso- and normal pentanes. The use of such a light liquid is necessary in order to avoid freezing under pressure. It was in general possible to control the pressure of first contact within 30,000 psi, or usually much better. This was good enough.

The apparatus now having been filled, the piston was inserted and pressure increased by the operation of the hand pumps connected to the

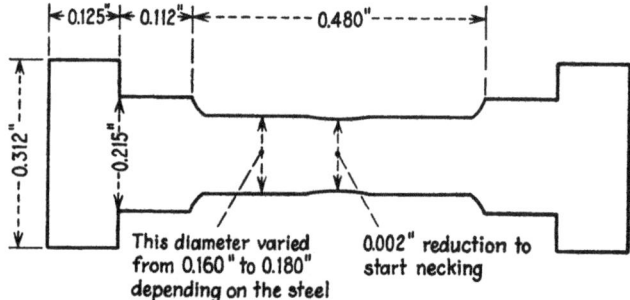

FIG. 17. Shows the dimensions of the tension specimens.

presses until a pressure was reached a few thousand atmospheres below the pressure of expected contact. Readings were now begun of grid resistance as a function of pressure, the readings being so spaced that four such readings would presumably be obtained before contact was made. These readings were plotted on a large scale as fast as they were made. The relation between pressure and grid resistance is linear, and the line was drawn in as soon as enough points had been obtained to determine it. The scale of plotting was 0.5 mm on the plot for 0.1 mm on the potentiometer wire with which resistance was measured, and the accuracy of the readings was such that they lay on a straight line to better than 0.5 mm on the plot. In the neighborhood of expected, contact, pressure was increased in small steps until contact had been made as indicated by the grid resistance lying off the extrapolated straight line. The tensile test proper now began. The advance of the piston was manipulated with the hand pump in conjunction with the Ames dial gauge attached to the piston; this provided a more sensitive control than manipulation by pressure increments. It was easily possible in this way to control the advance of the piston and so the length of the specimen to any desired ten-thousandth of an inch. The initial stages of the tension test were in the elastic range; the displacements were so controlled as to give perhaps half a dozen readings in this range. Although it was possible to control the readings to the requisite amount in the elastic range, nevertheless the accuracy was not sufficient to give the elastic extension of the specimen itself in this range, since by far the largest part of the total displacement of the piston is due to distortions of various parts of the apparatus, and these did not prove to be reproducible enough to permit determination by blank runs. For this reason, it was not possible to find from these measurements the effect of pressure on the Young's modulus of the specimen, something that would be of much interest. Other methods would be necessary for this, preferably dynamic methods.

Readings were continued until the proportional limit was reached, as indicated by the running plots. Readings were then continued beyond the elastic limit to the maximum load, which marks the beginning of necking, and then on to rupture or to any desired extension short of rupture, depending on the particular purpose in view. In the range above the elastic limit the readings were in general spaced for much wider intervals of elongation than below the limit. After the completion of the tensile test the piston was withdrawn, at first by small steps, and readings were taken during the initial part of the withdrawal. The function of these readings was to obtain a couple of grid readings after contact was lost between piston and specimen. These grid readings did not in general fall on the same straight line as the first readings with increasing pressure before contact was established. The reason is that the grid is temperature-sensitive and the temperature of the room, and so the temperature of the grid, usually changed during the course of the test. The total temperature effect during the run is given by the difference between the final points and the initial straight line. A correction was then applied for temperature drift during the run by distributing the total drift between the different readings in proportion to the time elapsed from the initial readings, the time of every reading having been recorded with this purpose in view. The total duration of the run was of the order of 2 hr, and there were some 20 readings altogether. The total correction for temperature drift of the grid was seldom as much as 5 per cent of the maximum load on the specimen, so that the assumption of a linear dependence of drift on time should not have introduced appreciable error. The limiting accuracy was set by the grid resistance, the total load corresponding to about 2 cm on the potentiometer wire. The grid readings were consistent to better than the sensitiveness of a single reading, 0.1 mm, a reading of load being the result of three potentiometer readings, made with alternating directions of current flow, to eliminate error from drift in the parasitics.

Although the extensions of the specimen could not be determined in the elastic range with adequate accuracy, above the elastic limit, where the extensions are larger, the accuracy was entirely adequate to correct the observed readings by the readings obtained on the blank runs and so to obtain the extensions of the specimen corrected for distortion in the apparatus. The total extension might on occasion be as much as 0.2 in.; this is about 500 times as much as the elastic extension at the elastic limit.

Brief Historical Survey. The first tensile tests under pressure were made in March, 1941, on an annealed high-carbon tool steel. The great increase of ductility and strength at high pressures was observed. The military significance of the results was called to the attention of Dr. Richard C. Tolman, then with the NDRC in Washington, and at his instigation further explorations were made. By the first of May, tensile tests had been made on several armor-plate steels under pressure and the existence of the same sort of effect established. In particular, one armor-plate steel, which fractured at atmospheric pressure with a 27 per cent reduction of area at a "true stress" of 186,000 psi, fractured under a hydrostatic pressure of 400,000 psi with 70 per cent reduction of area at a "true stress" of 485,000 psi. Another softer plate, which at atmospheric pressure fractured with 58 per cent reduction of area, fractured under 420,000 psi with a reduction of area immeasurably close to 100 per cent. The Brinell hardness of the same armor plates was also measured under pressure and found to increase materially, although the effects were not nearly so large as on the tensile properties. A contract was then drawn with the NDRC, new apparatus constructed, and in December, 1941, the systematic investigation begun of the effect of pressure on the tensile properties of a large variety of armor-plate steels. Most of the readings under the NDRC contract were made by Mr. Carmelo Lanza. Eventually some 50 different specimens of armor plate were examined at several different pressures and often in directions parallel and perpendicular to the direction of rolling or perpendicular to the face of the plate. As already stated, the primary purpose of these measurements was practical, as a means of investigating the properties of armor plate and making improvements. The scientific value of the results was relatively much less than could have been obtained by the same amount of properly directed effort, because the specimens of armor plate were subject to all the imperfections arising from the exigencies of their manufacture. This work for the NDRC came to the attention of the Watertown Arsenal, which was interested in conducting a longer range program on steels of controlled and definite properties in order to obtain information about the properties of steels as such. In July, 1942, measurements were started under contract with the Watertown Arsenal of the tensile properties under pressure of several series of steels of known composition and of various heat-treatments. Most of the readings under the Arsenal contract were made by Mr. L. H. Abbot. All the reductions of observations and calculations for both contracts were made by me personally. Steels of some six different compositions were examined, in some 10 different heat-treatments. Work for the Arsenal continued into August, 1944. In the meantime, as much information as was profitable with regard to armor plate had been collected for the NDRC, and a beginning was made on a more systematic long-range program on several steels of known heat-treatments prepared under the direction of Professor Frederick Seitz, this being part of a larger program involving the cooperation of several other laboratories. This more systematic work for the NDRC was begun in April, 1943, and was terminated in August, 1944, when it became evident that the probable duration of the war would not justify the expense. After the termination of the war I continued on my own the investigation of several questions not cleared up by the work for the government during the war, in particular the effect of pressure on the tensile properties of several stainless steels, which proved to be somewhat out of line from the more ordinary steels, and the pressure effect on tensile properties of steels of hardness greater than Rockwell C 48, these not having been satisfactorily handled in the preceding work. Later, a number of normally brittle substances were examined under pressure; some of these acquired high ductility under pressure.

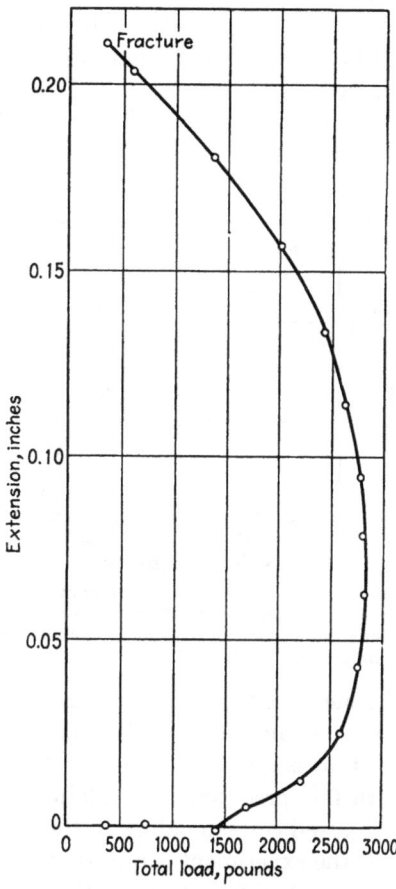

Fig. 18. Sample load-extension curve for a tension specimen pulled under pressure. This specimen was of steel 9-3, pulled under a pressure of 390,000 psi.

The Results. Figures 18 and 19 show typical results for a hard and a soft heat-treatment of the same steel. In these diagrams the total load in pounds is plotted against total extension in inches, the measured extension being corrected by a term linear in the total load such as to make the corrected extension below the elastic limit vanish. In these two diagrams extension of the specimen was continued to fracture. More often than not, however, the test was concluded before fracture. In such cases the final diameter of the specimen at the neck was measured on removal from the pressure apparatus. If the specimen had been fractured, the best estimate was made from the fractured pieces as to the probable neck diameter at fracture.

Diagrams like Fig. 18 and 19 yield several pieces of information. In the first place, the load at which the elastic limit is exceeded, or at which there begins to be an appreciable corrected extension, can obviously be determined with some precision. This point is a function of pressure, so that the diagrams give information on the effect of pressure on the elastic limit. The maximum load is also given by the diagram. This is the load at which necking starts. The maximum load, divided by the original cross section, gives the so-called "tensile strength," so that the effect of pressure on tensile strength can be obtained from the diagram. If the test is terminated short of fracture, the final load, combined with the measured cross section, gives the so-called "true stress" at the final load.

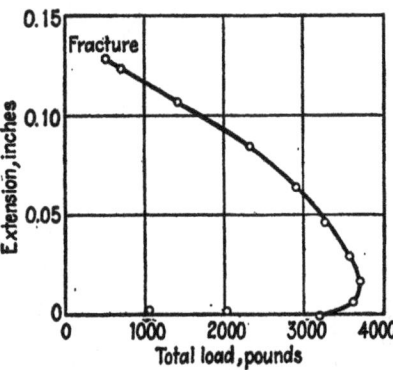

FIG. 19. Sample load-extension curve for a tension specimen pulled under pressure. This specimen was of steel 9-2, pulled under a pressure of 390,000 psi.

If, in addition, the shape of the contour as given by a/R is measured, or is assumed in terms of the total reduction of area, as it may be with fair precision, as explained in Chap. 1, the "flow stress" may be computed corresponding to the final reading.

If the experiment is repeated at the same pressure, terminating the test at several different elongations, then the "true stress" and the "flow stress" may be found as a function of reduction of area for this particular pressure, and in this way the strain-hardening curve may be determined for this pressure. If the same procedure is repeated at several different pressures, the effect of pressure on the entire strain-hardening process is established. If the test is continued to fracture, the effect of pressure on the true breaking stress is determined, making correction for stress distribution in terms of a/R as already discussed.

If the experiments had been conducted under the conditions of an ordinary tensile test, it would have been possible to make running measurements of the diameter of the neck during the course of the test and so to determine the complete strain-hardening curve at a single pressure from a single test at that pressure. As it is, it is possible to do this approximately by assuming that the neck diameter is known in terms of the overall measured elongation. This dependence can be approximately established by measurements on a range of materials. It turned out, however, that the reduction of area is not approximately enough a universal function of the extension to give much confidence in results obtained in this way, and it was therefore preferred to adopt the longer procedure of establishing the strain-hardening curves from a number of runs, using only the final points.

Plots like those in Figs. 18 and 19, when analyzed as just explained, yield finally the strain-hardening curves as a function of pressure and curves of breaking stress as a function of pressure or reduction of area. This information is of immediate interest in defining the properties of the material, and it is with this that we shall be concerned in the future. We shall be primarily concerned with the results obtained on steels of well-defined properties, reserving for the end a description of some of the qualitative features shown by some of the armor plates.

The parameter used to define the strain in all the following is the "natural strain" at the neck, defined as $\log_e (A_0/A)$, where A_0 is the initial and A the final area of the neck. If the constancy of volume of the material at the neck is assumed, $\log_e (A_0/A)$ is equal to $\log_e (l/l_0)$, where l is the final length of an infinitesimal element of the axis at the neck and l_0 is the initial length of the same element. The advantage of using natural strains is that many quantities are linear in terms of them. For convenience of reference a table is given of the natural strain in terms of other parameters (Table IV). The increasing insensitiveness at high strains of the conventional measure of distortion, percentage reduction of area, is obvious.

A summary of all the experimental determinations of the effect of pressure on the tension of steels is now presented in the next pages in Table V. This includes only those steels with well-defined properties and does not include most of the armor plates investigated for the Navy. Each line in Table V summarizes a complete tension run involving readings at 20 or more loads. The first column designates the specimen. The designation is given by three figures: the first shows the variety of steel, the second the heat-treatment of that variety, and the third the particular individual specimen of the indicated steel and treatment. The steels and their heat-treatments are given in Table VI beginning on page 60. The second column in Table V shows the pressure in pounds per square inch under which the test was conducted. This is the final pressure, that is, the pressure at the last reading of the tensile test, at the maximum elongation. The initial pressure at which the tensile test began was somewhat lower, as already explained. The difference between initial and final pressure usually amounted to only two or three thousand atmospheres. At first a precise record was kept of the initial as well as of the final pressure, but after experience had accumulated that the results were not appreciably affected by the range of pressure during the run, only final pressures were recorded. It is this pressure obviously which is significant, being the pressure either of maximum strain or of fracture.

TABLE IV

$\frac{A_0}{A}$	Percentage reduction of area	Natural strain $\log_e \frac{A_0}{A}$
1.0	0.0	0.000
1.5	33.3	0.405
2.0	50.0	0.693
3.0	66.7	1.099
5.0	80.0	1.609
10.0	90.0	2.303
20.0	95.0	2.996
35.0	97.1	3.555
50.0	98.0	3.912
100.0	99.0	4.605

The third column in Table V gives the natural strain at the final point of the test. It was determined from measurements of the diameter of the neck of the specimen after it had been removed from the apparatus. It is not so accurate for those specimens which fractured as for the non-fractured specimens, there usually being some distortion incidental to the act of fracture. However, the error should not be large. The fourth column indicates whether the specimen was carried to fracture or not. In the case of fractured specimens, the values listed in the other columns are of course those prevailing in the moment of fracture, manipulations being made circumspectly in small steps in approaching the fracture point in order to get these values. The fifth column gives the average tensile stress at the maximum total tensile load, as given by the maximum load in curves like Figs. 18 and 19. It is the so-called "engineering stress" and is the maximum load divided by the original cross section; it is otherwise called the "tensile strength." The "true stress" at the maximum load is about 23 per cent greater than the engineering stress, since necking

THE TENSION OF STEEL UNDER PRESSURE

TABLE V. COLLECTED RESULTS FOR TENSION TESTS ON STEELS UNDER PRESSURE
Part I. Tests for the NDRC

Specimen	Pressure, psi	Strain, $\log_e \frac{A_0}{A}$	Fracture or not	Average stress at max load, psi	Average final stress, psi	a/R	Correction factor	Final flow stress F, psi	Ratio of areas at fracture	Remarks
1-0-1	176,000	2.71	n.f.	109,000	....	1.87	0.73	....		
1-0-2	344,000	0.43	n.f.	116,000	138,000	0.12	0.97	133,000		
1-0-3	40,000	1.35	f.	106,000	219,000	(1.03)	0.82	164,000	0.32	
1-0-4	298,000	4.33	f.	111,000	....	(2.17)	0.71	....	0	
1-0-5	Atmos.	0.91	f.	105,000	169,000	(0.70)	0.87	146,000	0.43	
1-0-6	404,000	2.04	n.f.	120,000	294,000	1.28	0.79	231,000		
1-0-7	207,000	3.93	f.	106,000	....	....	....	....	0	
1-0-8	398,000	4.53	f.	115,000	....	....	....	....	0	
1-0-9	227,000	0.82	n.f.	112,000	....	0.43	0.91	....		
1-0-10	140,000	3.01	f.	102,000	385,000	(1.83)	0.74	284,000	0.06	
2-0-1	280,000	3.99	f.	136,000	....	....	....	....	0	
2-0-2	188,000	3.55	f.	137,000	470,000	(1.99)	0.72	340,000	0.09	
2-0-3	390,000	1.63	n.f.	139,000	268,000	1.29	0.79	213,000		
2-0-4	Atmos.	0.77	f.	123,000	191,000	(0.59)	0.88	169,000	0.52	
2-0-5	397,000	4.46	f.	138,000	....	....	....	....	0	
2-0-6	117,000	2.45	f.	131,000	314,000	(1.63)	0.75	236,000	0.08	
2-0-7	54,000	1.34	f.	128,000	250,000	(1.02)	0.82	204,000	0.23	
2-1-1	Atmos.	0.86	f.	106,000	197,000	(0.85)	0.84	166,000	0.46	
2-1-2	53,000	1.35	f.	111,000	256,000	(1.22)	0.80	204,000	0.08	
2-1-3	168,000	3.02	f.	116,000	296,000	(1.71)	0.75	221,000	0.002	
2-1-4	87,000	1.84	f.	113,000	320,000	(1.44)	0.77	247,000	0.02	
2-1-5	334,000	2.86	n.f.	139,000	405,000	1.69	0.75	301,000		
2-1-6	336,000	2.05	n.f.	139,000	316,000	1.36	0.78	247,000		
2-1-7	295,000	0.57	n.f.	138,000	162,000	0.27	0.94	152,000		
2-1-8	269,000	4.15	f.	136,000	....	....	....	....	0.00	
2-2-1	Atmos.	0.73	f.	118,000	174,000	0.75	0.86	150,000	0.48	
2-2-2	201,000	3.23	f.	126,000	422,000	(1.75)	0.74	314,000	0.03	
2-2-3	58,000	1.24	f.	124,000	213,000	(1.15)	0.80	170,000	0.32	
2-2-4	99,000	1.02	n.f.	124,000	211,000	0.69	0.87	183,000		
2-2-5	272,000	1.33	n.f.	128,000	233,000(?)	0.89	0.84	194,000(?)		
2-2-6	311,000	3.94	f.	142,000	....	....	....	....	0	
2-2-7	395,000	1.96	n.f.	139,000	318,000	1.37	0.78	248,000		
2-2-8	372,000	4.03	f.	135,000	....	....	....	....	0	
2-3-1	Atmos.	0.50	f.	113,000	149,000	(0.52)	0.89	132,000	0.68	
2-3-2	227,000	0.81	n.f.	125,000	182,000	0.64	0.87	159,000		
2-3-3	47,000	0.86	f.	120,000	182,000	(0.85)	0.84	153,000	0.11	
2-3-4	126,000	2.10	f.	121,000	305,000	(1.50)	0.77	234,000	0.09	
2-3-5	358,000	2.89	n.f.	131,000	355,000	1.91	0.73	258,000		
2-3-6	315,000	0.76	n.f.	130,000	189,000	0.59	0.88	167,000		
2-3-7	343,000	4.21	f.	129,000	....	....	....	....	0	
2-3-8	259,000	3.52	f.	128,000	....	....	....	....	0	

| | | | | | | | | Flow stress | | |
								At max	Final	
2-4-1-a	329,000	0.07	n.f.	Not reached	93,000	0	1	....	93,000	Successive pullings of same specimen, all but one having cracked. Readings for 2-4-1-d lost
2-4-1-b	386,000	0.27	n.f.	Not reached	116,000	0	1	....	116,000	
2-4-1-c	407,000	0.51	n.f.	121,000	144,000	0.36	0.92	121,000	133,000	
2-4-1-e	344,000	1.26	n.f.	150,000	211,000	0.97	0.83	136,000	174,000	
2-4-1-f	338,000	1.89	n.f.	204,000	262,000	1.35	0.78	169,000	204,000	
2-4-1-g	310,000	4.28	f.	242,000	....	....	....	190,000	....	

52 DEFORMATION UNDER HYDROSTATIC PRESSURE

TABLE V. COLLECTED RESULTS FOR TENSION TESTS ON STEELS UNDER PRESSURE
Part I. (Continued)

Specimen	Pressure, psi	Strain, $\log_e \frac{A_0}{A}$	Fracture or not	Average stress at max load, psi	Average final stress, psi	$\sigma\frac{a}{R}$	Correction factor	Final flow stress F, psi	Ratio of areas at fracture	Remarks
2-5-1	Atmos.	0.84	f.	291,000	372,000	(0.65)	0.87	326,000	0.30	
2-5-2	41,000	1.11	f.	297,000	406,000	(0.85)	0.84	343,000	0.20	
2-5-3	99,000	1.58	f.	284,000	520,000(?)	(1.18)	0.80	417,000(?)	0.20	
2-5-4	69,000	1.19	f.	305,000	419,000	(0.91)	0.83	350,000	0.16	
2-5-5	257,000	2.35	f.	314,000	663,000	(1.59)	0.76	502,000	0.02	
2-5-6	327,000	0.94	n.f.	315,000	399,000	0.92	0.83	334,000		
2-5-7	280,000	0.85	n.f.;	310,000	384,000	0.85	0.84	323,000		
2-5-8	323,000	1.85	n.f.	390,000	532,000	1.56	0.76	405,000	....	Second application to 2-5-7
2-6-1	Atmos.	0.68	f.	268,000	410,000	(0.51)	0.90	367,000	0.27	
2-6-2	42,000	1.01	f.	300,000	427,000	(0.78)	0.85	365,000	0.21	
2-6-3	106,000	1.58	f.	290,000	480,000	(1.18)	0.80	384,000	0.14	
2-6-4	148,000	2.05	f.	295,000	657,000	(1.44)	0.77	503,000	0.13	
2-6-5	273,000	2.75	f.	302,000	684,000	(1.75)	0.74	510,000	0.02	
2-6-6	313,000	1.10	n.f.	305,000	415,000	1.00	0.82	340,000		
2-6-7	380,000	2.37	n.f.	312,000	586,000	1.72	0.74	453,000		
2-7-1	Atmos.	0.95	f.	192,000	261,000	(0.73)	0.86	224,000	0.49	
2-7-2	42,500	1.26	f.	196,000	307,000	(0.97)	0.83	251,000	0.42	
2-7-3	127,000	1.94	f.	196,000	432,000	(1.38)	0.78	338,000	0.26	
2-7-4	221,000	2.56	f.	204,000	510,000	(1.67)	0.75	384,000	0.12	
2-7-5	310,000	0.92	n.f.	202,000	284,000	(0.71)	0.86	245,000		
2-7-6	339,000	1.91	n.f.	207,000	393,000	(1.37)	0.78	308,000		
2-7-7	406,000	0.90	n.f.	221,000	295,000	(0.70)	0.87	255,000		
3-0-1	Atmos.	0.46	f.	157,000	194,000	(0.32)	(0.93)	180,000	0.64*	* Fracture irregular; significance of last column doubtful
3-0-2	185,000	2.84	f.	167,000					0.10	
3-0-3	280,000	3.96	f.	169,000					0.00	
3-0-4	410,000	3.73	n.f.	173,000	551,000	2.19	0.71	390,000		
3-0-5	96,000	1.39	f.	166,000	303,000	(1.05)	0.81	247,000	0.22	
3-0-6	334,000	0.44	n.f.	171,000	198,000	0.38	0.92	183,000		
3-0-7	391,000	4.29	f.	172,000	787,000(?)	(2.16)	0.71	557,000(?)	0.00	
3-0-8	26,000	0.61	f.	159,000	201,000	(0.45)	0.90	182,000	0.37	
4-0-1	Atmos.	0.23	f.	172,000	169,000	(0.10)	0.97	164,000	1.00	
4-0-2	401,000	2.18	n.f.	195,000	416,000	1.72	0.75	311,000		
4-0-3	200,000	2.20	f.	191,000	422,000	(1.41)	0.78	326,000	0.03	
4-0-4	112,000	1.03	f.	188,000	284,000	(0.80)	0.85	241,000	0.18	
4-0-5	395,000	3.76	f.	194,000	522,000	(2.05)	0.72	375,000	0.00	
4-0-6	28,000	0.40	f.	181,000	207,000	(0.27)	0.94	193,000	0.40	
4-0-7	394,000	2.68	n.f.	193,000	465,000	1.60	0.76	353,000		
4-0-8	394,000	0.26	n.f.	200,000	205,000	1.29	0.79	162,000		
4-1-1	Atmos.	0.33	f.	183,000	206,000	(0.32)	0.93	191,000	0.76	
4-1-2	230,000	1.75	n.f.	194,000	411,000	1.43	0.77	318,000		
4-1-3	108,000	1.14	f.	193,000	320,000	(1.09)	0.81	260,000	0.25	
4-1-4	365,000	1.92	n.f.	204,000	389,000	1.49	0.77	298,000		
4-1-5	355,000	3.96	f.	201,000	....	....	....	....	0.00	
4-1-6	244,000	3.01	f.	192,000	444,000	(1.83)	0.74	326,000	0.07	
4-1-7	285,000	1.17	n.f.	197,000	308,000	0.91	0.83	257,000		

TABLE V. COLLECTED RESULTS FOR TENSION TESTS ON STEELS UNDER PRESSURE
Part I. *(Continued)*

Specimen	Pressure, psi	Strain, $\log_e \frac{A_0}{A}$	Fracture or not	Average stress at max load, psi	Average final stress, psi	$\frac{a}{R}$	Correction factor	Final flow stress F, psi	Ratio of areas at fracture	Remarks
4-2-1	Atmos.	0.33	f.	138,000	156,000	(0.33)	0.92	143,000	Not cup-cone	
4-2-2	185,000	1.75	f.	150,000	334,000	1.40	0.78	260,000	0.09	
4-2-3	99,000	0.88	f.	152,000	220,000	(0.88)	0.84	184,000	0.31	
4-2-4	26,000	0.44	f.	148,000	176,000	(0.45)	0.90	159,000	0.58	
4-2-5	270,000	2.77	f.	158,000	339,000	(1.75)	0.74	253,000	0.05	
4-2-6	340,000	1.62	n.f.	157,000	312,000	1.17	0.80	250,000		
4-2-7	340,000	3.02	n.f.	163,000	(?)	1.79	0.74	(?)		
4-2-8	257,000	1.27	n.f.	155,000	284,000	0.83	0.84	240,000		
4-3-1	Atmos.	0.12	f.	140,000	140,000	0.00	1.00	140,000	1.00	
4-3-2	230,000	1.72	n.f.	161,000	352,000	1.61	0.76	266,000		
4-3-3	144,000	0.79	f.	159,000	238,000	(0.79)	0.85	203,000	Too irregular	
4-3-4	54,000	0.25	f.	154,000	159,000	(0.22)	0.95	151,000	0.57	
4-3-5	330,000	1.06	n.f.	164,000	258,000	0.70±	0.87	223,000		
4-3-6	312,000	3.08	f.	163,000	572,000	(1.85)	0.73	420,000	0	
4-3-7	380,000	1.66	n.f.	168,000	331,000	1.38	0.78	259,000		
4-3-8	383,000	2.56	n.f.	170,000	350,000	1.77	0.74	259,000		
4-4-1	Atmos.	0.89	f.	112,000	175,000	(0 69)	0.87	150,000	0.56	
4-4-2	97,000	1.65	f.	110,000	261,000	(1.22)	0.80	207,000	0.34	
4-4-3-a	407,000	0.80	n.f.	129,000	198,000	(0.62)	0.88	175,000		
4-4-3-b	323,000	2.36	n.f.	180,000	339,000	(1.60)	0.76	258,000		
4-4-4-a	226,000	0.55	n.f.	126,000	162,000	(0.40)	0.91	148,000		
4-4-4-b	220,000	2.04	n.f.	157,000	298,000	(1.44)	0.77	231,000		
4-4-5-a	299,000	0.46	n.f.	124,000	153,000	(0.32)	0.93	143,000		
4-4-5-b	299,000	1.56	n.f.	153,000	258,000	(1.17)	0.80	207,000		
4-4-5-c	319,000	4.36	f.	259,000	594,000	(2.18)	0.71	421,000	0.11	
4-4-6	56,000	1.34	f.	114,000	221,000	(1.03)	0.82	182,000	0.40	
4-5-1	Atmos.	0.13	f.	Not reached	315,000	0	1.00	315,000	0.63	
4-5-2	40,000	0.23	f.	341,000	326,000	(0.10)	0.97	318,000	0.49	
4-5-3	58,000	0.27	f.	340,000	345,000	(0.15)	0.96	332,000	0.40	
4-5-4	250,000	1.34	n.f.	347,000	493,000	1.29	0.79	389,000		
4-5-5	341,000	2.24	n.f.	383,000	634,000	1.84	0.74	470,000		
4-5-6	318,000	2.27	f.	359,000	710,000	(1.55)	0.76	540,000	0.06	
4-5-7	258,000	1.77	f.	355,000	585,000	(1.29)	0.79	460,000	0.11	
4-5-8	199,000	1.19	f.	367,000	496,000	(0.91)	0.84	415,000		
4-6-1	Atmos.	0.49	f.	227,000	252,000	(0.34)	0.92	234,000	0.52	
4-6-2	62,000	0.80	f.	233,000	305,000	(0.62)	0.88	268,000	0.42	
4-6-3-a	314,000	0.22	n.f.	250,000	236,000	(0.00)	1.00	237,000		
4-6-3-b	400,000	0.53	n.f.	245,000	295,000	(0.39)	0.92	268,000		
4-6-3-c	360,000	2.05	n.f.	308,000	498,000	1.52	0.76	381,000		
5-0-1	Atmos.	1.12	f.	125,000	214,000	(0.74)	0.86	184,000	0.38	
5-0-2	83,000	1.75	f.	126,000	265,000	(1.19)	0.80	213,000	0.25	
5-0-3	298,000	3.65	f.	136,000	198,000	(1.96)	0.73	143,000	0	
5-0-4-a	270,000	0.17	n.f.	129,000	125,000	0	1.00	125,000		Longitudinal direction
5-0-4-b	342,000	0.70	n.f.	141,000	187,000	(0.41)	0.91	170,000		
5-0-4-c	329,000	2.18	n.f.	177,000	330,000	1.48	0.77	252,000		
5-0-4-d	345,000	2.71	n.f.	330,000	366,000	1.76	0.74	272,000		
5-0-5	227,000	2.98	f.	138,000	411,000	(1.78)	0.74	305,000	0.03	

TABLE V. COLLECTED RESULTS FOR TENSION TESTS ON STEELS UNDER PRESSURE
Part I. (*Continued*)

Specimen	Pressure, psi	Strain, $\log_e \frac{A_0}{A}$	Fracture or not	Average stress at max load, psi	Average final stress, psi	a/R	Correction factor	Final flow stress F, psi	Ratio of areas at fracture	Remarks
5-0-6	Atmos.	1.04	f.	127,000	206,000	(0.70)	0.86	177,000	0.37	⎫
5-0-7-a	340,000	1.15	n.f.	144,000	243,000	0.79	0.85	207,000		⎪ Radial
5-0-7-b	270,000	2.19	n.f.	209,000	323,000	1.42	0.78	251,000		⎬ direction
5-0-8	339,000	4.33	f.	143,000					0	⎪
5-0-9	152,000	2.38	f.	129,000	345,000	(1.53)	0.76	262,000	0.14	⎭
5-0-10	Atmos.	1.03	f.	124,000	202,000	(0.69)	0.87	175,000	0.39	⎫
5-0-11	142,000	2.46	f.	133,000	408,000	(1.56)	0.76	311,000	0.16	⎪ Circumfer-
5-0-12-a	271,000	0.67	n.f.	139,000	186,000	0.36	0.92	170,000		⎬ ential
5-0-12-b	298,000	2.69	n.f.	182,000	407,000	1.60	0.76	309,000		⎪ direction
5-0-13	325,000	4.35	f.	139,000	582,000	(2.10)	0.72	415,000	0	⎭
6-0-1	Atmos.	1.14	f.	104,000	183,000	(0.88)	0.84	153,000	0.41	⎫ Parallel to
6-0-2	335,000	4.42	n.f.	121,000	558,000	2.09	0.72	(401,000)		⎪ face of plate
6-0-3-a	286,000	0.51	n.f.	116,000	144,000	0.19	0.95	138,000		⎪ and parallel
6-0-3-b	322,000	1.54	n.f.	145,000	221,000	1.16	0.80	177,000		⎬ to direction
6-0-3-c	309,000	3.05	n.f.	182,000	319,000	1.87	0.73	234,000		⎪ of rolling
6-0-4	172,000	3.03	f.	114,000	165,000	(1.84)	0.74	121,000	0.05	⎭
6-0-5	Atmos.	1.03	f.	104,000	177,000	(0.80)	0.85	150,000	0.44	⎫ Parallel to
6-0-6	149,000	2.89	f.	112,000	364,000	(1.79)	0.74	270,000	0.07	⎪ face of
6-0-7	395,000	2.71	n.f.	116,000	340,000	1.50	0.77	260,000		⎪ plate and
6-0-8-a	355,000	0.21	n.f.	124,000	124,000	0.00	1.00	124,000		⎬ perpendic-
6-0-8-b	311,000	1.27	n.f.	120,000	214,000	0.93	0.83	177,000		⎪ ular to
6-0-8-c	320,000	2.71	n.f.	207,000	360,000	1.64	0.75	270,000		⎪ direction of
										⎭ rolling
7-0-1	¹Atmos.	1.10	f.	155,000	236,000	(0.84)	0.84	200,000	0.38	⎫ Parallel to
7-0-2	115,000	2.22	f.	163,000	347,000	(1.52)	0.76	265,000	0.12	⎪ face of
7-0-3	155,000	2.71	f.	180,000	411,000	(1.73)	0.74	308,000	0.4(?)	⎬ plate and
7-0-4-a	157,000	0.81	n.f.	163,000	213,000	0.65	0.87	186,000		⎪ parallel to
7-0-4-b	189,000	2.97	n.f.	219,000	389,000	(1.81)	0.74	288,000		⎪ direction of
										⎭ rolling
7-0-5	Atmos.	0.88	f.	159,000	223,000	(0.68)	0.87	193,000	0.37	⎫ Parallel to
7-0-6	217,000	3.37	f.	176,000	452,000	(1.94)	0.73	330,000	0	⎪ face of
7-0-7-a	218,000	0.87	n.f.	142,000	244,000	0.66	0.87	213,000		⎬ plate and
7-0-7-b	257,000	3.24	n.f.	227,000	518,000	1.87	0.73	380,000		⎪ perpendic-
7-0-8	116,000	2.24	f.	164,000	405,000	(1.54)	0.76	310,000	0	⎪ ular to
										⎭ direction of rolling
8-0-1	Atmos.	1.09	f.	152,000	256,000	(0.83)	0.85	216,000	0.38	
8-0-2	88,000	2.00	f.	156,000	352,000	(1.42)	0.77	271,000	0.17	
8-0-3-a	360,000	0.19	n.f.	186,000	179,000	0.06	0.99	177,000		
8-0-3-b	290,000	1.12	n.f.	162,000	262,000	0.89	0.84	221,000		
8-0-3-c	338,000	2.37	n.f.	266,000	389,000	1.76	0.74	291,000		
8-0-4	158,000	2.51	f.	159,000	366,000	(1.65)	0.75	275,000	0	
8-0-5	Atmos.	0.79	f.	142,000	221,000	(0.60)	0.88	194,000	0.43	
8-0-6	390,000	2.27	n.f.	180,000	420,000	1.45	0.77	324,000		
8-0-7	237,000	0.42	n.f.	164,000	183,000	0.40	0.91	167,000		

THE TENSION OF STEEL UNDER PRESSURE

TABLE V. COLLECTED RESULTS FOR TENSION TESTS ON STEELS UNDER PRESSURE
(Continued)
Part II. Tests for the Watertown Arsenal

Specimen	Pressure, psi	Strain, $\log_e \frac{A_0}{A}$	Fracture or not	Average stress at max load, psi	Average final stress, psi	$\frac{a}{R}$	Correction factor	Final flow stress F, psi	Ratio of areas at fracture
9-0-1	Atmos.	0.78	f.	99,000	176,000	.....	0.88	155,000	
9-0-2	398,000	2.74	n.f.	109,000	384,000	2.06	0.71	275,000	
9-0-3	410,000	0.89	n.f.	113,000	210,000	0.54	0.89	187,000	
9-0-4	401,000	2.01	n.f.	113,000	330,000	.....	0.78	257,000	
9-0-5	389,000	1.40	n.f.	112,000	265,000	.....	0.83	220,000	
9-0-6	430,000	3.48	n.f.	111,000	(?)	0.97	.....	(?)	
9-0-7	300,000	2.10	n.f.	114,000	324,000	.....	0.77	250,000	
9-0-8	405,000	2.30	n.f.	115,000	355,000	.....	0.76	272,000	
9-0-9	Atmos.	0.38	n.f.		137,000	.....	0.94	128,000	
9-0-10	Atmos.	0.45	n.f.		149,000	.....	0.93	138,000	
9-0-11	Atmos.	0.66	n.f.		163,000	.....	0.89	145,000	
9-0-12	155,000	0.28	n.f.		139,000	.....	0.96	133,000	
9-0-13	153,000	0.46	n.f.		157,000	.....	0.92	144,000	
9-0-14	162,000	0.72	n.f.		187,000	.....	0.87	163,000	
9-0-15	375,000	0.24	n.f.		142,000	.....	0.96	137,000	
9-0-16	360,000	0.51	n.f.		186,000	.....	0.91	169,000	
9-0-17	365,000	1.17	n.f.		318,000	.....	0.83	262,000	
9-0-18	363,000	1.54	n.f.		.....	.....	.....	213,000	
9-0-19	302,000	1.64	n.f.		.....	.....	.....	203,000	
9-0-20	252,000	1.53	n.f.		.....	.....	.....	213,000	
9-0-21	266,000	2.20	n.f.		.....	.....	.....	270,000	
9-0-22	221,000	0.90	n.f.		.....	.....	.....	170,000	
9-0-23	260,000	1.65	n.f.		.....	.....	.....	201,000	
9-0-24	Atmos.	0.42	n.f.		.....	.....	.....	130,000	
9-1-1	Atmos.	0.98	f.	169,000	320,000	.....	0.85	272,000	
9-1-2	411,000	3.17	n.f.	195,000	622,000	.....	0.73	455,000	
9-1-3	373,000	0.89	n.f.	194,000	324,000	.....	0.87	281,000	
9-1-4	408,000	1.99	n.f.	194,000	468,000	.....	0.77	362,000	
9-1-5	393,000	1.38	n.f.	189,000	386,000	.....	0.81	314,000	
9-2-1	Atmos.	0.88	f.	183,000	305,000	.....	0.86	264,000	
9-2-2	395,000	3.34	n.f.	210,000	655,000	.....	0.72	475,000	
9-2-3	350,000	0.75	n.f.	210,000	324,000	.....	0.88	286,000	
9-2-4	345,000	1.63	n.f.	205,000	455,000	.....	0.79	361,000	
9-2-5	Atmos.	0.89	f.		296,000	.....	0.87	258,000	
9-2-6	112,000	1.63	f.		382,000	.....	0.80	304,000	
9-2-7	208,000	2.57	f.		575,000	.....	0.75	431,000	
9-2-8	383,000	3.73	f.		773,000	.....	0.72	557,000	
9-2-9	357,000	1.09	n.f.		355,000	0.92	0.84	296,000	
9-2-10	357,000	0.22	n.f.		206,000	0.13	0.97	200,000	
9-2-11	346,000	2.43	n.f.		518,000	1.67	0.75	388,000	
9-2-12	218,000	1.92	n.f.		445,000	1.45	0.77	343,000	
9-2-13	221,000	1.02	n.f.		335,000	0.84	0.79	282,000	
9-2-14	217,000	0.19	n.f.		218,000	0.10	0.98	213,000	
9-2-15	115,000	0.88	n.f.		295,000	0.76	0.86	253,000	
9-2-16	116,000	1.35	n.f.		348,000	1.18	0.80	280,000	
9-2-17	95,000	0.22	n.f.		210,000	0.19	0.95	200,000	
9-2-18	Atmos.	0.85	f.		337,000	.....	0.87	294,000	

TABLE V. COLLECTED RESULTS FOR TENSION TESTS ON STEELS UNDER PRESSURE
Part II. (Continued)

Specimen	Pressure, psi	Strain, $\log_e \frac{A_0}{A}$	Fracture or not	Average stress at max load, psi	Average final stress, psi	$\frac{a}{R}$	Correction factor	Final flow stress F, psi	Ratio of areas at fracture
9-2-19	330,000	2.35	n.f.	....	609,000	1.66	0.75	457,000	
9-2-20	130,000	1.54	f.	....	416,000	.....	0.81	335,000	
9-2-21	365,000	0.57	n.f.	....	321,000	0.53	0.89	286,000	
9-2-22	173,000	2.11	f.	....	570,000	.....	0.77	440,000	
9-2-23	394,000	3.95	f.						
9-2-24	60,000	1.28	f.	....	384,000	.....	0.82	317,000	
9-2-25	269,000	2.98	f.	....	796,000	.....	0.74	588,000	
9-2-26	273,000	2.76	f.	....	680,000	.....	0.75	507,000	
9-3-1	Atmos.	0.92	f.	114,000	214,000	.....	0.86	184,000	
9-3-2	405,000	2.75	n.f.	131,000	415,000	.....	0.75	310,000	
9-3-3	340,000	0.79	n.f.	131,000	269,000	.....	0.87	183,000	
9-3-4	388,000	1.39	n.f.	136,000	288,000	.....	0.82	235,000	
9-3-5	186,000	2.78	f.	....	340,000	.....	0.74	253,000	
9-3-6	269,000	3.87	f.						
9-3-7	145,000	2.37	f.	....	474,000	.....	0.76	358 000	
9-3-8	385,000	2.25	n.f.	....	360,000	1.77	0.74	267,000	
9-3-9	34,000	1.10	f.	....	250,000	.....	0.84	210,000	
9-3-10	387,000	4.61	f.						
9-3-11	Atmos.	0.81	f.	....	206,000	.....	0.87	180,000	
9-3-12	302,000	0.51	n.f.	....	176,000	0.21	0.95	167,000	
9-3-13	366,000	1.59	n.f.	....	303,000	1.14	0.80	243,000	
9-3-14	384,000	3.22	n.f.	....	457,000	1.80	0.74	328,000	
9-3-15	385,000	1.17	n.f.	....	250,000	0.81	0.85	213,000	
9-4-1	Atmos.	0.89	f.	105,000	189,000	.....	0.86	163,000	
9-4-2	405,000	2.65	n.f.	114,000	385,000	.....	0.75	288,000	
9-4-3	350,000	0.77	n.f.	115,000	201,000	.....	0.89	178,000	
9-4-4	393,000	1.49	n.f.	115,000	270,000	.....	0.80	217,000	
9-4-5	199,000	1.42	n.f.	....	250,000	0.85	0.85	212,000	
9-4-6	350,000	0.43	n.f.	....	144,000	0.21	0.94	136,000	
9-4-7	18,000	1.13	f.	....	149,000	.....	0.84	125,000	
9-4-8	263,000	4.14(?)	f.						
9-4-9	116,000	2.14	f.	....	313,000	.....	0.77	240,000	
9-4-10	410,000	3.38	n.f.	....	435,000	1.67	0.77	337,000	
9-4-11	227,000	1.69	n.f.	....	296,000	0.96	0.83	245,000	
9-4-12	Atmos.	0.90	f.	....	189,000	.....	0.86	163,000	
9-4-13	74,000	1.56	f.	....	225,000	.....	0.80	180,000	
9-4-14	182,000	2.83	f.	....	396,000	.....	0.74	295,000	
9-4-15	389,000	4.80	f.						
9-4-16	389,000	2.41	n.f.	....	310,000	1.52	0.76	235,000	
9-4-17	383,000	0.96	n.f.	....	207,000	0.58	0.88	183,000	
9-4-18	386,000	2.04	n.f.	....	280,000	1.29	0.79	221,000	
9-4-19	315,000	2.81	n.f.	....	269,000	1.54	0.75	203,000	
9-4-20	333,000	1.06	n.f.	....	226,000	0.52	0.89	202,000	
9-5-1	390,000	3.43	n.f.	....	(577,000)	1.95	0.72	(414,000)	
9-5-2	390,000	0.90	n.f.	....	305,000	0.87	0.84	255,000	
9-5-3	380,000	1.78	n.f.	....	417,000	1.41	0.78	325,000	

THE TENSION OF STEEL UNDER PRESSURE

TABLE V. COLLECTED RESULTS FOR TENSION TESTS ON STEELS UNDER PRESSURE
Part II. (*Continued*)

Specimen	Pressure, psi	Strain, $\log_e \frac{A_0}{A}$	Fracture or not	Average stress at max load, psi	Average final stress, psi	$\frac{a}{R}$	Correction factor	Final flow stress F, psi	Ratio of areas at fracture
9-6-1	Atmos.	0.98	f.	....	227,000	.....	0.85	194,000	
9-6-2	382,000	2.28	n.f.	....	408,000	1.60	0.76	310,000	
9-6-3	63,000	1.39	f.	....	305,000	.....	0.81	248,000	
9-6-4	348,000	0.74	n.f.	....	223,000	0.48	0.91	202,000	
9-6-5	159,000	2.10	f.	....	369,000	.....	0.77	284,000	
9-6-6	407,000	5.05	f.						
9-6-7	295,000	3.67	f.						
9-7-1	Atmos.	1.04	f.	....	230,000	0.95	0.84	193,000	0.33
9-7-2	100,000	1.10	n.f.	....	234,000	0.87	0.84	197,000	
9-7-3	189,000	1.25	n.f.	....	260,000	1.07	0.81	210,000	
9-7-4	290,000	1.14	n.f.	....	223,000	0.87	0.84	187,000	
9-7-5	365,000	1.25	n.f.	....	271,000	0.99	0.82	222,000	
9-7-6	Atmos.	0.13	n.f.	....	135,000	0.04	1.00	135,000	
9-7-7	365,000	2.19	n.f.	....	356,000	1.57	0.76	270,000	
9-7-8	370,000	2.77	n.f.	....	427,000	1.75	0.74	316,000	
9-7-9	352,000	2.37	n.f.	....	375,000	1.58	0.76	285,000	
9-7-10	355,000	1.73	n.f.	....	315,000	1.33	0.78	246,000	
9-7-11	365,000	2.15	n.f.	....	366,000	1.53	0.76	278,000	
9-7-12	369,000	3.29	n.f.	....	451,000	2.03	0.72	325,000	
9-8-1	Atmos.	1.24	f.	....	262,000	1.16	0.80	209,000	0.35
9-8-2	110,000	1.69	n.f.	....	307,000	1.31	0.78	241,000	
9-8-3	211,000	1.67	n.f.	....	311,000	1.30	0.79	245,000	
9-8-4	307,000	1.56	n.f.	....	305,000	1.27	0.79	241,000	
9-8-5	400,000	1.72	n.f.	....	338,000	1.30	0.79	266,000	
9-8-6	Atmos.	1.20	f.	....	248,000	1.10	0.81	201,000	0.35
9-8-7	365,000	2.22	n.f.	....	397,000	1.57	0.76	302,000	
9-8-8	367,000	3.15	n.f.	....	449,000	1.87	0.73	330,000	
9-8-9	112,000	2.09	f.	....	....	.....	.....	....	0.19
9-8-10	100,000	1.30	n.f.	....	256,000	1.02	0.82	210,000	
9-8-11	196,000	1.41	n.f.	....	285,000	1.10	0.81	231,000	
9-8-12	281,000	1.38	n.f.	....	304,000	1.08	0.81	246,000	
9-8-13	390,000	1.11	n.f.	....	270,000	0.873	0.84	227,000	
9-8-14	380,000	2.95	n.f.	....	444,000	1.85	0.73	326,000	
10-1-1	Atmos.	0.07	f.						
10-1-2	82,000	0.22	f.	....	351,000	.....	0.97	340,000	
10-1-3	270,000	0.98	f.	....	610,000	0.74	0.86	522,000	
10-1-4	334,000	1.43	f.	....	640,000	0.89	0.81	518,000	
10-2-1	370,000	0.07	f.	....	510,000	0.00	1.00	510,000	0.74
11-0-1	359,000	0.22	n.f.	....	167,000	0.20	0.95	159,000	
11-0-2	360,000	0.88	n.f.	....	218,000	0.70	0.87	190,000	
11-0-3	267,000	1.73	n.f.	....	310,000	1.37	0.78	242,000	
11-0-4	370,000	2.39	n.f.	....	366,000	1.69	0.75	274,000	
12-0-1	325,000	0.27	n.f.	....	198,000	0.19	0.95	188,000	
12-0-2	345,000	0.95	n.f.	....	274,000	0.81	0.85	233,000	
12-0-3	340,000	1.71	n.f.	....	363,000	1.31	0.78	283,000	
12-0-4	352,000	2.54	n.f.	....	450,000	1.62	0.75	337,000	

TABLE V. COLLECTED RESULTS FOR TENSION TESTS ON STEELS UNDER PRESSURE
Part II. (Continued)

Specimen	Pressure, psi	Strain, $\log_e \frac{A_0}{A}$	Fracture or not	Average stress at max load, psi	Average final stress, psi	$\frac{a}{R}$	Correction factor	Final flow stress F, psi	Ratio of areas at fracture
13-0-1	340,000	0.33	n.f.	....	233,000	0.17	0.96	223,000	
13-0-2	370,000	0.82	n.f.	....	303,000	0.70	0.86	261,000	
13-0-3	365,000	1.51	n.f.	....	379,000	1.20	0.80	303,000	
13-0-4	360,000	2.05	n.f.	....	452,000	1.51	0.77	348,000	
14-0-1	410,000	1.36	n.f.	68,000	171,000	.....	0.82	140,000	
14-0-2	425,000	1.71	n.f.	69,000	185,000	.....	0.79_2	147,000	
14-0-3	420,000	2.27	n.f.	70,000	225,000	.....	0.76_2	171,000	
14-0-4	420,000	2.90	n.f.	69,000	286,000	.....	0.73_9	214,000	
14-0-5	Atmos.	0.92	f.	56,000	112,000	.....	0.86_2	97,000	
14-0-6	Atmos.	0.93	f.	59,000	114,000	.....	0.86_1	99,000	

TABLE V. COLLECTED RESULTS FOR TENSION TESTS ON STEELS UNDER PRESSURE (Continued)
Part III. Tests Other than NDRC or Watertown Arsenal

Specimen	Pressure, psi	Strain, $\log_e \frac{A_0}{A}$	Fracture or not	Average stress at max load, psi	Average final stress, psi	$\frac{a}{R}$	Correction factor	Final flow stress F, psi	Ratio of areas at fracture
15-0-1	194,000	4.21	f.	99,000	480,000	1.18(?)	0.71	340,000	All shear
15-0-2	190,000	2.25	n.f.	98,000	237,000	1.49	0.76	182,000	
15-0-3	93,000	2.92	f.	93,000	324,000	1.73	0.74	240,000	0.106
15-0-4	265,000±	5.20	f.	101,000	1,000,000−	......	....	....	All shear
15-0-5	197,000	1.59	n.f.	99,000	199,000	1.14	0.80	160,000	
15-0-6	187,000	1.83	n.f.	99,000	219,000	1.26	0.78	172,000	
16-0-1	186,000	3.52	f.	106,000	362,000	1.44	0.72	264,000	
16-0-2	202,000	2.23	n.f.	108,000	280,000	1.54	0.76	215,000	
16-0-3	262,000	4.38	f.	109,000	460,000	1.11(?)	0.71	327,000	
16-0-4	95,000	2.57	f.	102,000	313,000	1.65	0.75	235,000	
16-0-5	198,000	2.84	n.f.	107,000	314,000	2.03	0.74	234,000	
16-0-6	195,000	1.47	n.f.	111,000	232,000	1.04	0.81	188,000	
17-0-1	109,000	2.09	f.	235,000	470,000	1.05	0.77	363,000	0.22
17-0-2	225,000	2.29	f.	256,000	519,000	1.29	0.76	395,000	All shear
17-0-3	350,000	2.76	f.	284,000	740,000	0.80(?)	0.74	552,000	
17-0-4	340,000	2.25	n.f.	285,000	509,000	1.69	0.76	390,000	
17-0-5	319,000	1.88	n.f.	271,000	465,000	1.59	0.78	364,000	
17-0-6	330,000	1.50	n.f.	280,000	420,000	0.94	0.81	340,000	
18-0-1	104,000	2.51	f.	194,000	465,000	1.49	0.75	350,000	0.34
18-0-2	226,000	2.69	f.	218,000	585,000	1.30	0.75	439,000	All shear
18-0-3	346,000	3.00	f.	255,000	1,040,000(??)	1.36	0.74	740,000(??)	All shear
18-0-4	354,000	2.85	n.f.	241,000	597,000	1.69	0.74	444,000	
18-0-5	349,000	2.26	n.f.	241,000	528,000	1.40	0.76	404,000	
18-0-6	340,000	1.92	n.f.	239,000	456,000	1.37	0.78	355,000	

THE TENSION OF STEEL UNDER PRESSURE

TABLE V. COLLECTED RESULTS FOR TENSION TESTS ON STEELS UNDER PRESSURE
Part III. (*Continued*)

Specimen	Pressure, psi	Strain, $\log_e \frac{A_0}{A}$	Fracture or not	Average stress at max load, psi	Average final stress, psi	a/R	Correction factor	Final flow stress F, psi	Ratio of areas at fracture
19-1-1	Atmos.	0.59	f.	310,000	345,000	......	0.90	313,000	0.36
19-1-2	146,000	1.30	f.	317,000	503,000	......	0.82	416,000	0.22
19-1-3	198,000(?)	1.59	f.	(?)	(?)	......	....	....	0.16
19-1-4	375,000	2.60	f.	350,000	1,320,000(?)	......	0.75	990,000(?)	0
19-1-5	338,000	2.37	f.	360,000	718,000	......	0.76	546,000	0.083
19-1-6	236,000	1.88	f.	330,000	597,000	......	0.78	465,000	0.104
19-2-1	Atmos.	0.49	f.	375,000	391,000	......	0.92	360,000	0.45
19-2-2	115,000	0.91	f.	359,000	465,000	......	0.86	404,000	0.28
19-2-3	240,000	1.36	f.	395,000	660,000	......	0.82	542,000	0.15
19-2-4	359,000	1.64	f.	401,000	619,000	......	0.80	495,000	0
19-3-1	Atmos.	0.25	f.	....	398,000	......	0.96	381,000	0.62
19-3-2	340,000	1.45	f.	440,000	650,000	......	0.81	528,000	0
19-3-3	Atmos.	0.14	f.	341,000	394,000	......	0.97	385,000	0.61
19-3-4	363,000	1.37	f.	458,000	718,000	......	0.82	590,000	Shearing fracture
19-3-5	114,000	0.41	f.	425,000	450,000	......	0.93	419,000	0.37
19-3-6	225,000	0.79	f.	459,000	614,000	......	0.88	541,000	Shear
19-4-1	Atmos.	0.07	f.	....	414,000	......	0.99	410,000	0.75
19-4-2	400,000	0.98	f.	593,000(?)	697,000	......	0.85	597,000	All shear
19-4-3	112,000	0.32	f.	490,000	520,000	......	0.94	493,000	0.60
19-4-4	239,000	0.68	f.	504,000	627,000	......	0.89	561,000	0.30
19-5-1	344,000	0.07	f.	582,000	582,000	......	0.99	576,000	0.72
19-5-2	Atmos.	0.006	f.	309,000	309,000	......	1.00	309,000	0.9±
20-0-1	Atmos.	0.77	f.	....	186,000	......	0.88	164,000	
20-0-2	410,000	2.74	n.f.	....	397,000	1.43	0.74	296,000	
20-1-1	Atmos.	0.33	f.	....	450,000	......	0.94	425,000	
20-1-2	339,000	1.44	f.	....	875,000	......	0.81	710,000	
20-1-3	345,000	0.57	n.f.	....	526,000	......	0.91	480,000	

usually starts at a natural strain of approximately 0.2 (= $\log_e^{-1} 1.23$). It is to be recalled that necking starts at the maximum load. The sixth column gives the average final stress, that is, the total tensile load at the final point (weight hanging on the specimen in the "pressurized laboratory") divided by the actual cross section at the final point as given by column 3. This is often called the "true stress." The seventh column contains the measured values of a/R in those cases where it was measured. It is the data of this column which are collected in Fig. 7 of Chap. 1. Column 8 contains the correction factor by which the average stress of column 6 is converted into the flow stress of column 9. This

TABLE VI. DESIGNATION OF STEELS

A steel will be designated by two figures, and a particular specimen of that steel by the addition of a third figure. The first figure in the designation of the steel indicates the composition and the second figure the heat-treatment. These are now described in detail, the grouping being by composition.

Steel 1 is steel No. 1 of the sixth NDRC report, of composition: C 0.34, Mn 0.75, P 0.017, S 0.033, Si 0.18. This was used only in the "as-received" condition and is designated as 1-0.

Steel 2 is steel No. 2 of the sixth NDRC report. The composition is: C 0.45, Mn 0.83, P 0.016, S 0.035, Si 0.19. It was used in eight heat-treatments, as follows:

2-0 "as received"
2-1 normalized at 1650°F for ½ hr
2-2 annealed, fine-grained. At 1500°F for ½ hr
2-3 annealed, coarse-grained. At 1900°F for ½ hr
2-4 brine-quenched and spheroidized. Held at 1500°F for ½ hr, quenched, and held at 1300°F for 10 hr
2-5 brine-quenched and tempered. Held at 1500°F for ½ hr, quenched, and held at 600°F for 1 hr
2-6 brine-quenched and tempered. Held at 1800°F for ¼ hr, quenched, and held at 600°F for 1 hr
2-7 brine-quenched and tempered. Held at 1500°F for ½ hr, quenched, and held at 900°F for 1 hr

---

Steel 3 is steel No. 3 of the sixth NDRC report. The composition is: C 0.68, Mn 0.71, P 0.013, S 0.028, Si 0.19. It was used only in the "as-received" condition and is designated as 3-0.

Steel 4 is steel No. 4 of the sixth NDRC report. The composition is: C 0.90, Mn 0.47, P 0.015, S 0.036, Si 0.11. It was used in seven heat-treatments, as follows:

4-0 "as received"
4-1 normalized at 1650°F for ½ hr
4-2 annealed, fine-grained. At 1450°F for ½ hr
4-3 annealed, coarse-grained. At 1900°F for ½ hr
4-4 brine-quenched and spheroidized. Held at 1450°F for ½ hr, quenched, and held at 1300°F for 10 hr
4-5 brine-quenched and tempered. Held at 1800°F for 1 hr, quenched, and held at 600°F for 1 hr
4-6 brine-quenched and tempered. Held at 1450°F for ½ hr, quenched, and held at 900°F for 1 hr

---

Steel 5 is a Navy gun steel, designated by them as No. 301, and described in the sixth NDRC report. The composition is specified as: C 0.55 to 0.60, Mn 0.25 to 0.45, S 0.040, P 0.040, Si 0.15 to 0.25, Ni 3.35 to 3.65, Cr 2.15 to 2.45. It was used in the "as-received" condition and is designated as 5-0.

Steel 6 is Navy gun steel No. 302, described in the sixth NDRC report. The composition is given as: C 0.30, Mn 0.18, S 0.020, P 0.011, Si 0.07, Ni 2.98, Cr 1.18.

TABLE VI. DESIGNATION OF STEELS. (*Continued*)

The specimens used were cut from a block which had been cut from a gun. It was used "as received" and is designated as 6-0.

Steel 7 is Navy steel No. 16919 described in the sixth NDRC report. The composition is given as: C 0.26, Mn 0.23, S 0.025, P 0.02, Si 0.06, Ni 3.04, Cr 1.40. The specimens used were cut from a block cut from an armor plate. They were used in the "as-received" condition and are designated as 7-0.

Steel 8 is Navy steel No. 19797 described in the sixth NDRC report. The composition is given as: C 0.30, Mn 0.24, S 0.031, P 0.024, Si 0.08, Ni 2.96, Cr 1.29. It was a piece cut from an armor plate, used in the "as received" condition and designated as 8-0.

Steel 9 is a SAE steel 1045 provided by the Watertown Arsenal. It was used in nine heat-treatments, as follows:
9-0 "as received"
9-1 quenched into water from 1575°F and drawn to 400 to 500°F
9-2 quenched into water from 1575°F and drawn to 800°F. Its Rockwell C hardness was 40.3
9-3 quenched into salt at 800°F from 1575°F. The Rockwell B hardness was 91.7
9-4 quenched into salt at 1100°F from 1575°F. The Rockwell B hardness was 85.5
9-5 quenched into water from 1450°F and drawn at 900°F
9-6 quenched into water from 1575°F and drawn at 1100°F. The Rockwell C hardness was 21.0
9-7 tempered pearlite, Rockwell C hardness 25.8
9-8 tempered martensite (650°F), Rockwell C hardness 28.3

Steel 10 is Ketos tool steel, obtained from the Halcomb Steel Co. A typical composition is: C 0.90, Mn 1.20, Si 0.25, Cr 0.50, W 0.50. It was used in two heat-treatments:
10-1 quenched into oil and drawn at 650°F, Rockwell C 52.5
10-2 quenched into oil and not drawn, Rockwell C 65.0

Steel 11 is SAE steel 1315 (manganese steel), provided by the Arsenal and used in the "as-received" condition, which was described as "annealed." It is designated as 11-0. The Rockwell C is 21.1.

Steel 12 is SAE steel 2320 (nickel), provided by the Arsenal and used in the "as-received" condition, described as "annealed." It is designated as 12-0. The Rockwell C was 29.1.

Steel 13 is SAE steel 4140 (chromium), provided by the Arsenal, and used in the "as-received" condition, described as "annealed." It is designated as 13-0. The Rockwell C was 34.5.

TABLE VI. DESIGNATION OF STEELS. (*Continued*)

Steel 14 is SAE steel 1020 provided by the Arsenal and used in the dead soft condition, designated as 14-0.

Steel 15 is a stainless steel obtained from Dr. A. L. Feild of the Rustless Iron and Steel Corp. The composition is: C 0.094, Mn 0.36, P 0.023, S 0.022, Si 0.35, Cr 12.26, Ni 0.46, Mo 0.50. It was used in the "as-received" condition and is designated as 15-0. Its Rockwell B hardness was 87.

Steel 16 is a stainless steel from Dr. Feild of the Rustless Iron and Steel Corp. The composition is: C 0.067, Mn 0.050, P 0.020, S 0.030, Si 0.51, Cr 17.49, Ni 0.41. It was used "as received" and is designated as 16-0.

Steel 17 is a stainless steel from Dr. Feild of the Rustless Iron and Steel Corp., of composition: C 0.058, Mn 0.70, P 0.030, S 0.013, Si 0.85, Cr 18.51, Ni 8.95, Cu 0.20. It was used "as received" and is designated as 17.0.

Steel 18 is a stainless steel from Dr. Feild of the Rustless Iron and Steel Corp., of composition: C 0.051, Mn 0.59, P 0.030, S 0.020, Si 0.47, Cr 18.31, Ni 10.27, Cu 0.20. It was used "as received" and is designated as 18-0.

Steel 19 was provided by the Arsenal, described by them only as a "high-carbon" steel. It was used in five heat-treatments. These were all quenched into oil at room temperature after being held at 1575°F for 30 min. They were then drawn back to various hardnesses:
    19-1   drawn to Rockwell C 48
    19-2   drawn to Rockwell C 51
    19-3   drawn to Rockwell C 56
    19-4   drawn to Rockwell C 60
    19-5   drawn to Rockwell C 63

Steel 20 is SM steel from the Carpenter Steel Co. A typical analysis is: C 0.60, Mn 0.70, Si 1.90, P 0.03 (max), S 0.03 (max). It was used in two treatments:
    20-0   "as received," "annealed"
    20-1   quenched into oil from 1650F and drawn to 500°F

matter is fully explained in Chap. 1. The correction factor was obtained from the measured a/R when it was determined and in other cases was taken from Fig. 7 of Chap. 1 in terms of the natural strain. Column 10 gives a rough characterization of the nature of the fracture in those cases where the fracture was of the cup-cone type. This will be discussed in greater detail in connection with the whole subject of fracture.

The experimental material of Table V is discussed in the first place with regard to the phenomena of plastic flow below the fracture point, and secondly with regard to the phenomena of fracture.

Phenomena of Plastic Flow. If the final flow stress F of column 9 is plotted against the natural strain of column 3 it will be found that for any one steel and heat-treatment the points lie roughly on a single curve, the scatter off a single curve being no more than the usual scatter in tests of this sort. This curve represents the plastic properties of the material. If the von Mises plasticity criterion were applicable the curve would be a straight line at a single constant stress parallel to the strain axis. The stress, however, rises with increasing strain, and for this reason we shall refer to these curves as "strain-hardening" curves. Typical strain-hardening curves of this kind, as constructed from the data of Table V, are shown in Fig. 20 for the three steels 11-0, 12-0, and 13-0, and in Fig. 21 for the 9-4 series. In Fig. 20 the points lie on straight lines with a rather unusual precision, and in Fig. 21 on a straight line with consider-

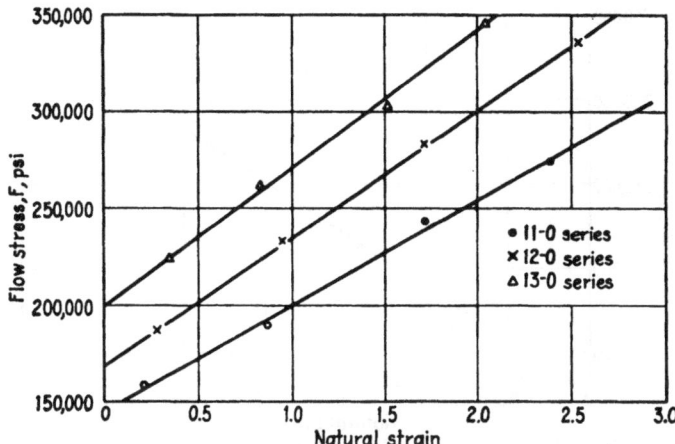

FIG. 20. Strain-hardening curves (flow stress versus natural strain) for steels 11-0, 12-0, and 13-0.

ably more scatter, this corresponding to the fact that, although all the specimens of Fig. 21 are of ostensibly the same steel, the different specimens were taken from different original bars.

The linear relation shown in Figs. 20 and 21 is found on detailed plotting to hold almost without exception for all the other material of Table V. This leads to the first generalization from these experiments, namely, that to a first approximation the strain-hardening curve, that is, the curve of flow stress F against natural strain, is linear. The linear relation also holds within experimental error between strain and the average tensile stress, or load/πa^2, the difference between F and average stress being unimportant in this connection. The linear relation is seen to hold up to strains very much higher than can be reached in ordinary tensile testing at atmospheric pressure, where the natural strains are usually restricted to values less than 1. The range of strains now available within which the linear relation approximately holds is in general somewhere between 3 and 4. Although still higher strains can be reached without fracture under pressure (once a value for A_0/A of 300, or a natural strain of 5.7 was observed without fracture), the measurements of such high strains become increasingly inaccurate for two reasons. In the first place the total tensile load corresponding to the very small cross section becomes too small to measure accurately, and in the second place the cross section loses its circular figure, the whole section becoming dominated by what was originally one grain.

In the last paragraph the strain-hardening curve was described as linear to a "first approximation." One reason for this approximation is at once apparent, namely, the pressures at which the strain-hardening

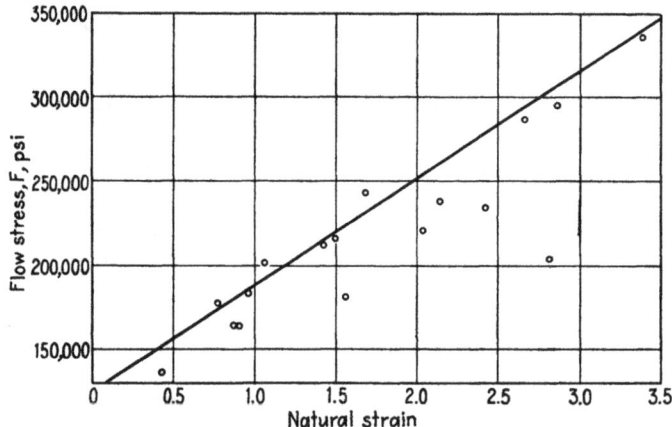

FIG. 21. Strain-hardening curve (flow stress versus natural strain) for steel 9-4.

points have been determined have been haphazard. Under these conditions one has a right not to expect a curve at all, but to expect the points to be spread over a region. It is only when all the measurements are made at a single pressure that one has a right to expect a strain-hardening *curve*. The fact that a curve is obtained with haphazard pressures indicates that the effect of pressure as such on the strain-hardening is unimportant, the role of pressure being merely to permit the large strains without fracture which determine the strain-hardening. This is indeed the case to a first approximation. In nearly all the work tabulated above, no consistent correlation was apparent between pressure and the stress-strain points, in view of the sometimes large scatter arising from other factors. By the time the last series of measurements was being made under the Arsenal contract, however, skill in making the measurements had so increased, and probably also the homogeneity of the material of the specimens had also increased because of care in preparation, that it was possible to establish a definite effect of pressure on the strain-hardening curve. The data showing this are for steels 9-7 and 9-8. The effects are 2-fold, and both are in the expected direction. In the first place, the strain-hardening for a given strain increases somewhat with increasing pressure, or, in other words, F at constant natural strain is somewhat higher at the higher hydrostatic pressures. In the second place, the strain-hardening curve is not linear but falls off a few per cent at the upper end of the strain range. Both these matters will be discussed in more detail presently.

Going back now to the first approximation, within which the strain-hardening curve may be taken as linear, the parameters for the strain-hardening lines of all steels are collected into Table VII. The parameters

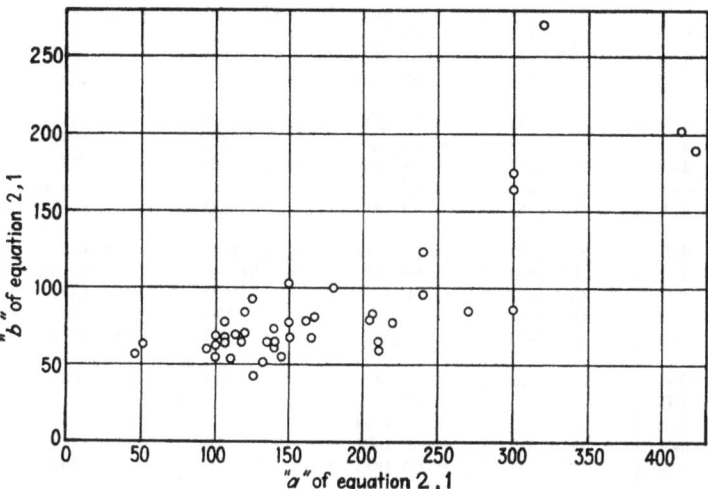

FIG. 22. Plot of the parameter b of the strain-hardening curve versus the parameter a for all the steels listed in Table VII.

of the strain-hardening lines listed in the table are the coefficients of the strain-hardening equation

$$F = a + bS \qquad (2\text{-}1)$$

where S is the natural strain. These parameters were found from plots similar to Figs. 20 and 21, made from the data of Table V. The significance of these parameters is apparent. a denotes the flow stress at zero strain and is roughly a measure of the elastic limit or the "strength" of the steel. b is the slope of the strain-hardening line and represents the rate at which the steel hardens with increasing strain. A high value for b indicates a steel which rapidly strain-hardens, that is, a steel for which the von Mises plasticity function fails to be constant by a large amount.

One might perhaps suspect some sort of correlation between the coefficients a and b. To find whether there is such a correlation, a is plotted against b in Fig. 22 for all the steels of Table VII. It appears that there

TABLE VII. COLLECTED DATA FOR STRAIN-HARDENING AND DUCTILITY LINES

Steel	Strain-hardening $F = a + b$ (strain)		Ductility $P = \alpha + \beta$ (strain)$_f$	
	a	b	α	β
1-0	+100,000	+ 62,000	− 57,000	+ 67,000
2-0	130,000	+ 51,000	− 33,000	+ 63,000
2-1	105,000	+ 78,000	− 57,000	+ 77,000
2-2	105,000	+ 67,000	− 64,000	+ 90,000
2-3	110,000	+ 53,000	− 37,000	+ 87,000
2-4	95,000	+ 60,000		
2-5	240,000	+ 97,000	−135,000	+160,000
2-6	270,000	+ 85,000	− 80,000	+115,000
2-7	166,000	+ 81,000	−121,000	+129,000
3-0	150,000	+ 69,000	− 28,000	+ 78,000
4-0	150,000	+ 77,000	− 15,000	+105,000
4-1	160,000	+ 79,000	− 19,000	+ 93,000
4-2	120,000	+ 83,000	− 23,000	+113,000
4-3	125,000	+ 92,000	Not linear	Not linear
4-4	100,000	+ 69,000	− 66,000	+ 91,000
4-5	300,000	+ 87,000	− 22,000	+166,000
4-6	208,000	+ 83,000	Not linear	Not linear
5-0	112,000	+ 69,000	− 96,000	+102,000
6-0	100,000	+ 55,000	− 90,000	+ 84,000
7-0	140,000	+ 63,000	− 81,000	+ 87,000
8-0	140,000	+ 65,000	− 75,000	+ 83,000
9-0	118,000	+ 65,000		
9-1	204,000	+ 79,000		
9-2	180,000	+100,000	−110,000	+130,000
9-3	135,000	+ 64,000	− 67,000	+ 77,000
9-4	107,000	+ 66,000	− 83,000	+ 93,000
9-5	210,000	+ 60,000		
9-6	140,000	+ 73,000	−102,000	+108,000
9-7	120,000	+ 70,000		
9-8	140,000	+ 65,000	−164,000	+132,000
10-1	300,000	+175,000	−210,000	+270,000
11-0	144,000	+ 55,000		
12-0	164,000	+ 68,000		
13-0	210,000	+ 65,000		
14-0	44,000	+ 58,000		
15-0	50,000	+ 65,000	−156,000	+ 88,000
16-0	126,000	+ 42,000	−132,000	+ 88,000
17-0	220,000	+ 77,000	−660,000	+380,000(??)
18-0	150,000	+103,000	−940,000	+350,000(??)
19-1	240,000	+123,000	− 91,000	+179,000
19-2	300,000	+165,000	− 40,000	+286,000
19-3	420,000	+190,000	− 36,000	+286,000
19-4	412,000	+202,000	− 34,000	+432,000
20-1	320,000	+270,000	−200,000	+400,000(?)

is a rough correlation, large values of the one parameter tending to be associated with large values of the other. Further, b tends to be proportional to a, the two vanishing together. This is, perhaps, as might be expected. The correlation is not, however, nearly so close as it appeared at the time of writing the first Arsenal report, when it was suspected on the basis of data for a single steel that the correlation might perhaps be exact.

We next consider more in detail the effect of pressure on the strain-hardening curve. It has already been stated that to a first appproxima-

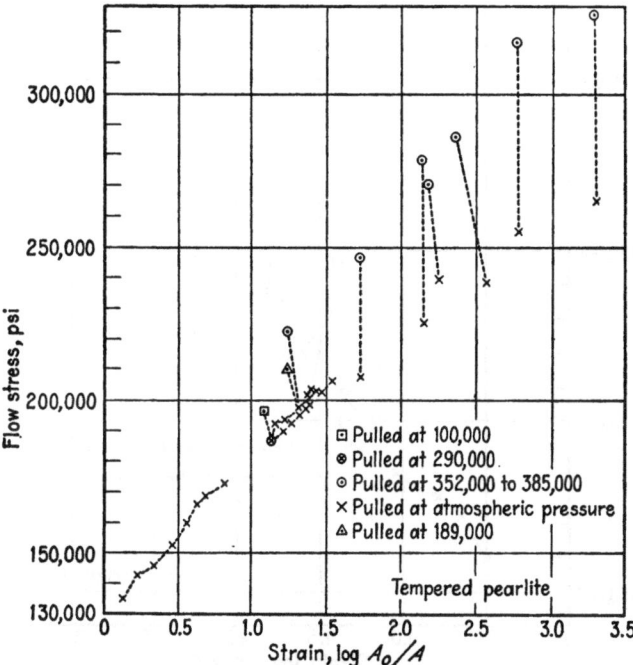

FIG. 23. The flow stress as a function of strain for tempered pearlite pulled at various pressures as indicated. The diagram also shows the results of second pullings at atmospheric pressure; these results will be further discussed in Chap. 17.

tion the strain-hardening curve is independent of the pressure of pulling but that when measurements are made with an attainable precision it is possible to establish an effect of pressure. This is shown in Figs. 23 and 24, which contain the most important results. These figures suggest that the idealized state of affairs, abstracting from the capricious variations which always obscure experiments on plastic flow, is as represented in Fig. 25 on an exaggerated scale. Here flow stress F is plotted against strain for various hydrostatic pressures of pulling. The curves for different pressures do not coincide. It would appear that the family of curves for different pressures is fan-shaped, and it is not improbable that the *percentage* effect of a given pressure on the flow stress is greater the greater the strain. Furthermore, the present experiments are good enough to indicate definitely the direction of deviation from linearity of the relation between stress and strain at any given pressure. Flow stress

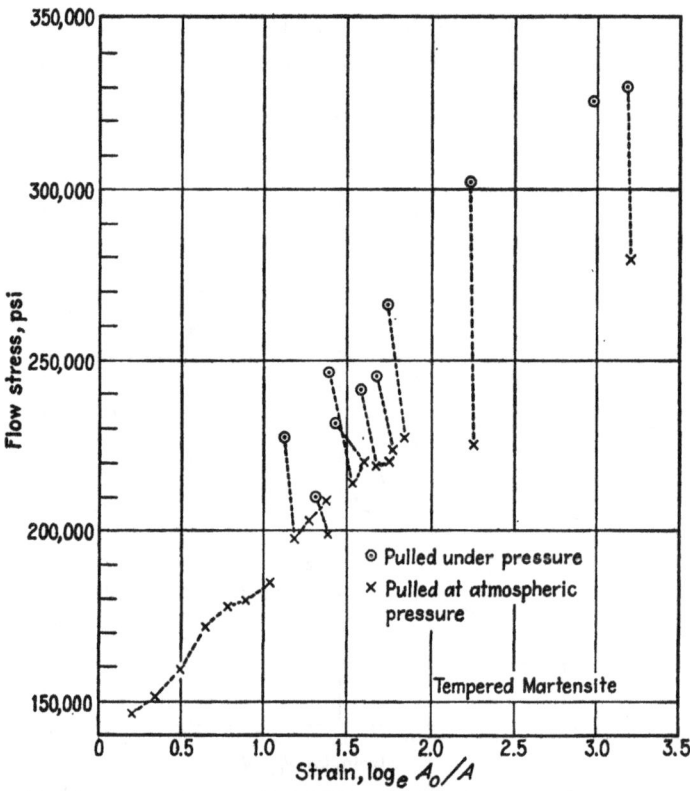

FIG. 24. The flow stress as a function of strain for tempered martensite pulled at various pressures which are given in detail in Table V. The diagram also shows the results of second pullings at atmospheric pressure; these results will be further discussed in Chap. 17.

against strain is concave toward the pressure axis, or, in other words, the rate of strain-hardening decreases at high strains. This would be expected on a priori grounds, since otherwise an indefinitely high strength would be attainable at high enough strains.

At any single pressure of pulling the curve terminates in rupture; this is indicated in the diagram. This effect will be discussed more in detail presently; in general, the strain at fracture increases approximately linearly with pressure. In Fig. 25 the curve for atmospheric pressure

(p_0) is shown prolonged beyond the fracture point. This prolonged curve applies to specimens which have been pulled initially at a higher pressure and then pulled a second time at atmospheric pressure. These experiments, to be described in detail later, show that, on the second pulling, to the next degree of approximation, the point representing F as a function of strain moves along a single curve independent of the pressure of the previous pulling.

Figure 23 for tempered pearlite indicates the general order of magnitude to be expected for the two second-order effects. (1) At low strains in the neighborhood of 0.2 where necking starts the flow stress is nearly independent of the pressure of pulling; at a strain of 3.0 the flow stress increases approximately 20 per cent for an increase of hydrostatic pressure from atmospheric to 400,000 psi. At intermediate points experiment justifies no other assumption than that the increase is linear both in strain and in pressure. (2) Deviation of the stress-strain at constant hydrostatic pressure from linearity is of such a general order of magnitude that at the mid-point of the strain range from 0.2 to 3.5 the flow stress may be taken to be 5 per cent higher than would be given by a straight line joining the extreme points. The points in Fig. 24 for tempered martensite are

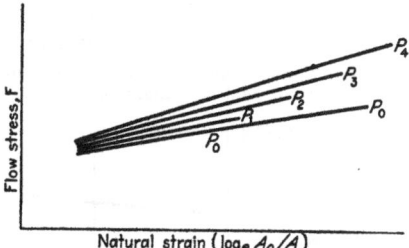

FIG. 25. Shows in an exaggerated form how the strain-hardening curve depends on the pressure of pulling.

somewhat more irregular than for pearlite, but within this irregularity there seems to be no difference either qualitative or quantitative.

In the derivation of the formulas for the distribution of stress across the neck of the tension specimen it was assumed that flow stress is independent of hydrostatic pressure, or in other words, that the von Mises condition is independent of pressure. This now appears not to be exact. However, a correction for this effect is probably hardly worth attempting in view of experimental irregularities. The direction of the correction arising from the dependence of flow stress on pressure is such as to decrease the difference between the mean stress and the calculated flow stress.

The primary interest of most of this experimental work has been in the region of large plastic deformations leading up to fracture. There is also of course an effect of pressure on the initial stages of plastic yield. The reason that this was not investigated more fully was that experimental accuracy is lower in the region of small deformations. However, if there are many experiments, capricious inaccuracies may be expected to cancel out, and certain definite trends seem to be established. The first exceeding of the elastic limit could usually be located with some definiteness. There seems no doubt that in general the tensile stress at the elastic limit is raised by hydrostatic pressure, and by an amount roughly proportional to the pressure. It is possible to establish with greater accuracy the effect of pressure on the maximum load. That there is an

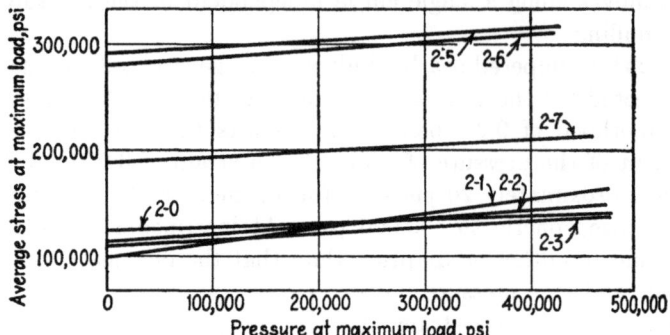

FIG. 26. The average stress at the maximum load against the pressure at the maximum load for various heat-treatments of steel 2.

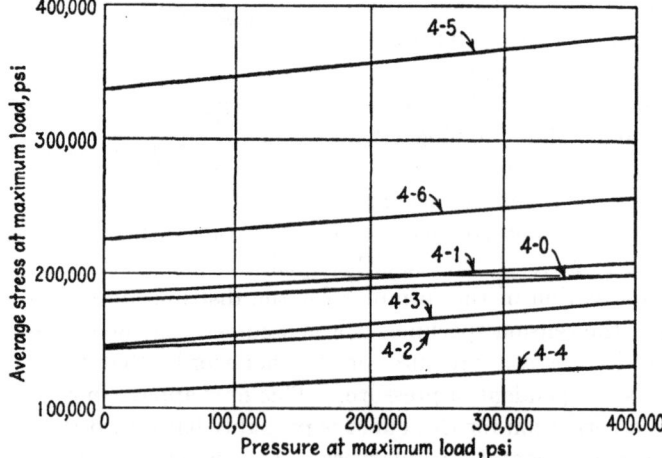

FIG. 27. The average stress at the maximum load against the pressure at the maximum load for various heat-treatments of steel 4.

effect of this sort has already been obvious from Figs. 18 and 19. These figures are typical of all the stress-strain curves taken. In general, the maximum load itself varies noticeably with pressure, but the total elongation at the maximum load, and therefore presumably the strain at the neck when necking starts, is sensibly independent of pressure. The data for this effect have been given in the fifth column of Table V. Figures 26 and 27 reproduce the results for steels 2 and 4, for which the data are by far the most complete. In the first place, the relation between pressure and maximum load, or the average stress at maximum load, which is proportional to it, is linear within experimental error. The individual points have not been plotted in Figs. 26 and 27 in order not to obscure the results. It is to be remarked, however, that these points fall more consistently on lines than they do for most of the other phenomena of tension. In general, it appears that the average stress at maximum tensile load rises with increasing pressure, and the slope is rather closely the same for all steels with one possible exception (4-1) which may be due to experimental error. The slope is roughly $1/15$, or an increase of 10,000 psi for an increase of hydrostatic pressure of 150,000 psi. Since the slope is constant independent of the absolute height of the lines, this means that

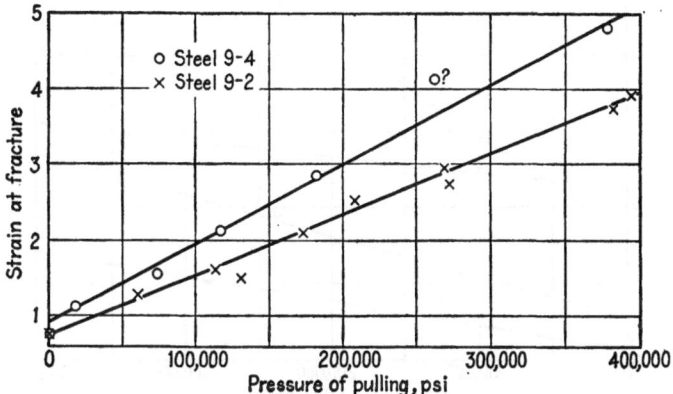

Fig. 28. The strain at fracture against the pressure of pulling (ductility curve) for steels 9-2 and 9-4.

the percentage effect of a given increment of pressure is less on the maximum load for the stronger steels than for the weaker steels. All the remarks just made apply strictly to the maximum load, and they apply to the "average stress at maximum load" only insofar as the assumption is justified that there is no change of strain of maximum load with pressure. If the strain at maximum load departs consistently with pressure from the constant value 0.2 assumed here (there is no experimental evidence for any departure), then there would be a consistent effect on the slope of the lines of strain at maximum load versus pressure.

Fracture. We next consider the second aspect of these tension tests, namely, the effect of pressure on fracture. The spectacular effect of pressure is to postpone the occurrence of fracture to much greater strains than those which accompany fracture at atmospheric pressure. We may express this by saying that the ductility is increased by pressure, and define ductility as the natural strain at which fracture occurs. To a first approximation this natural strain will be the average strain across the entire section of the neck; but to a second approximation, and we shall find that we have to take into account the second approximation, it will be the strain at the point at which fracture is initiated, that is, on the axis.

Table V gives extensive material for determining the effect of pressure on ductility. For any heat-treatment of any steel it is merely required to plot the third column of the table against the second for those specimens marked with an f. in the fourth column. Figure 28 shows the experimental points for heat-treatments 2 and 4 of steel 9, and Fig. 29

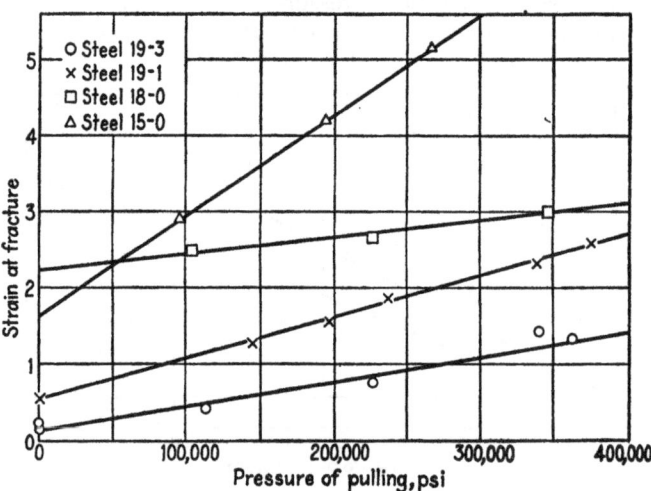

FIG. 29. The strain at fracture against the pressure of pulling (ductility curve) for steels 15-0, 18-0, 19-1, and 19-3.

the results for two of the stainless steels and for the high-carbon steel 19, treated to hardnesses above Rockwell C 48. The relation between pressure and ductility is obviously linear within experimental error for the steels of these two figures. This was true of all the other steels investigated, with the two exceptions of steels 4-3 and 4-6.

This leads to the second generalization with respect to the effect of pressure, namely, that ductility increases linearly with pressure. Assuming the linear relation, the equation of the ductility line may be written

$$P = \alpha + \beta S_f \tag{2-2}$$

where S_f is the natural strain at the neck at fracture. The parameters of the ductility line have been determined for all the steels for which the requisite data are given in Table V. These were determined graphically from plots similar to Figs. 28 and 29. The parameters are listed in Table VII, along with the parameters of the strain-hardening curves.

A remark is called for with regard to the accuracy of the pressure-ductility lines. It will be noticed that in general these lines are carried to higher strains than were the strain-hardening lines. The reason is that in general the strain at fracture can be determined with less error than the stresses at the upper end of the strain-hardening lines, since the latter involve the measurement of loads which become increasingly small because of the decrease of section. In fact, Figs. 28 and 29 indicate no greater error at the upper end of the ductility curves than at the lower end. There is at present no evidence, therefore, that the linear relation between pressure and ductility will cease to hold at pressures materially higher than those reached here. Thus if one wants to indulge in an extrapolation, it would appear that under a hydrostatic pressure of 100,000 atm (1,420,000 psi), which I have reached in a large number of laboratory measurements, the strain at fracture of steel 9-4 would be 15.3, or an elongation at the neck of 4.4×10^6-fold.

The physical significance of the parameters α and β is apparent. A large value of β means a small increase in the strain at fracture for a given increment of pressure, that is, a large β means a small pressure coefficient of ductility. A numerically large α (α is intrinsically negative, as appears in the table), means a large strain at fracture at atmospheric pressure, that is, a soft steel. Judging by the behavior of the parameters of the strain-hardening curves, one might perhaps expect the same sort of thing here and anticipate a correlation between α and β, large values of one going with large values of the other. To test this, α may be plotted against β for all the steels of the table. It will be found on making the plot that the points scatter much more than they did for the strain-hardening curve, so that it is hardly profitable to speak of a correlation at all. It hardly seems necessary to present a diagram to make this point.

There is, however, a correlation between the strain-hardening parameters and the ductility parameters. On plotting in Fig. 30 the b of Table VII against β, it will be found that one is roughly proportional to the other, with a degree of correlation evidently better than the correlation already remarked between a and b. Roughly, $b = 0.7\beta$; that is, as a rough average, the flow stress at fracture increases 0.7 times as fast as the corresponding increase of pressure. A steel may be made indefinitely strong by exerting enough pressure on it, but the increase of strength is not so great as the increase of pressure required to bring it about.

From the two linear relations of the strain-hardening and the ductility lines we may obtain a third linear relation by an elimination. Since in general the strain-hardening line is given by $F = a + bS$, we have in particular $F_f = a + bS_f$. The ductility line is given by $P = \alpha + \beta S_f$. Hence, eliminating S_f,

$$P = \left(\frac{\alpha}{\beta} - \frac{a}{b}\right) + \frac{\beta}{b} F_f \tag{2-3}$$

This gives the explicit relation between pressure and the maximum flow stress that can be attained at that pressure, or the strength of an unnecked specimen at that pressure.

If we want to indulge in extrapolation again, we find from the data of Table VII that at a pressure of 100,000 atm the strength of steel 9-4 is

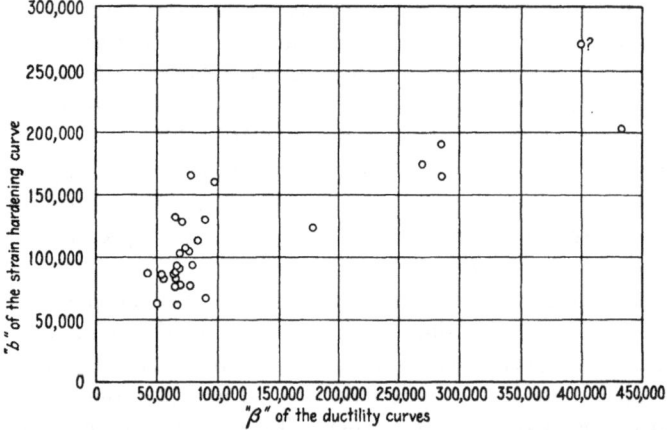

Fig. 30. The parameter b of the strain-hardening curve against the parameter β of the ductility curve for all the steels of Table VII.

1,180,000 psi. This steel, however, in order to attain this strength, would have to be drawn out to a wire 4.4×10^6 times the length of the original bar, or a bar initially 1 in. in diameter would have to be drawn down to a diameter less than 0.0005 in.

Physical Characteristics of Tensile Fracture under Pressure. Most of the steels listed in Table V exhibit a cup-cone type of tensile fracture at atmospheric pressure. A photograph of a section of a tension specimen through a typical such fracture is shown in Fig. 31. The break consists of a crater with usually burnished sides on which the failure is by shearing slip, and a flat bottom along which the failure is by tensile separation of the longitudinal fibers. The flat bottom is always situated accurately at the narrowest part of the neck; the sides of the crater emerge to the outer surface at some little distance from the neck. The sides of the crater do not always remain entirely on one or the other half of the tension specimen, but they may tear across, so that finally part of the periphery is attached to one half and the rest to the other half, as shown in Fig. 31. The line of separation between the flat bottom of the crater and the sides is usually so definite, and the outline is usually so closely circular, that measurements of considerable precision can be made on the diameter and so on the area of the flat bottom of the crater. The outside diameter of the narrowest part of the neck may similarly be measured, and thus the ratio of the area of the flat bottom, or of the tensile part of the break,

FIG. 31. Typical "cup-cone" fracture of a tension specimen at atmospheric pressure. This particular specimen was from an armor plate designated as C14 in the fourth NDRC report.

to the total neck area may be determined. This ratio is characteristic of the steel. It is small for the softer steels and becomes larger for the harder steels. Even a glass-hard steel always breaks in tension with a small shearing lip around the edge of the fracture.

The appearance of the normal cup-cone break changes drastically with pressure. At first the effect of pressure is to diminish the relative extent of the tensile part of the break. Eventually at higher pressures the flat bottom of the crater entirely disappears, and from here on the break is entirely shearing. This shearing break may go through a complicated course of evolution with increasing pressure, the exact details depending on the particular steel. Multiple shearing breaks with a dimple at the center are common. Sometimes several shearing vortexes appear at one place or another of the section, the appearance of one of the halves of the specimen being like that of the surface of a whirlpool in a liquid. Eventually, however, when the section at fracture becomes very much reduced, only a single shearing plane survives, and the final failure is by slip on this single plane entirely across the section, rotational symmetry having

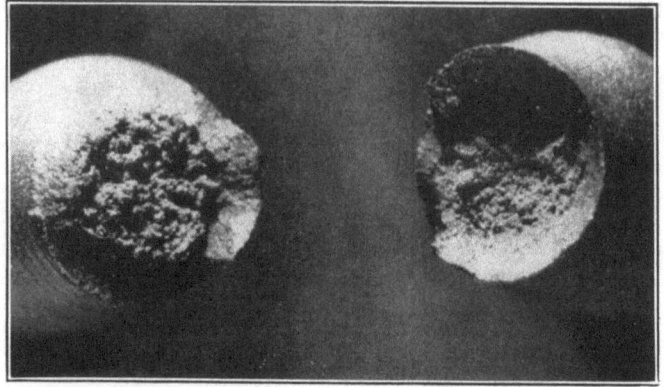

FIG. 32. The fracture of specimen 9-3-11, broken at atmospheric pressure.

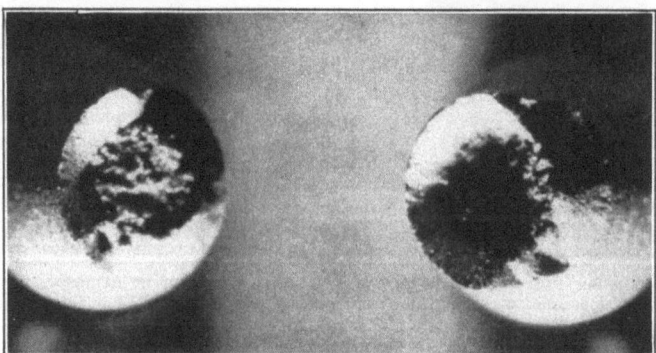

FIG. 33. The fracture of specimen 9-3-9, broken at 34,000 psi.

entirely disappeared in the final act of fracture. The shearing part of the break becomes more and more highly burnished as fracture takes place at higher pressures. The angle of the shearing plane, which initially is at approximately 45° to the axis, becomes oriented progressively more nearly at right angles to the axis.

Figures 32 to 37 show a series of fractures of steel 9-3 under progressively increasing pressure. In addition to the geometrical features just discussed, there is also a progressive change in the texture of the fracture as fracture occurs under progressively higher pressure. At first the

THE TENSION OF STEEL UNDER PRESSURE

fracture may be coarse-grained, as shown in the first of the series above, but it becomes progressively finer at higher pressures.

If the ratio of the area of the tensile part of the break to the total area is plotted as a function of pressure it will be found that the relation is roughly linear, the ratio vanishing at some finite pressure. Above this pressure the fracture is entirely shearing. Data are given in Table V for the ratio of the areas in those cases where it could be determined. A

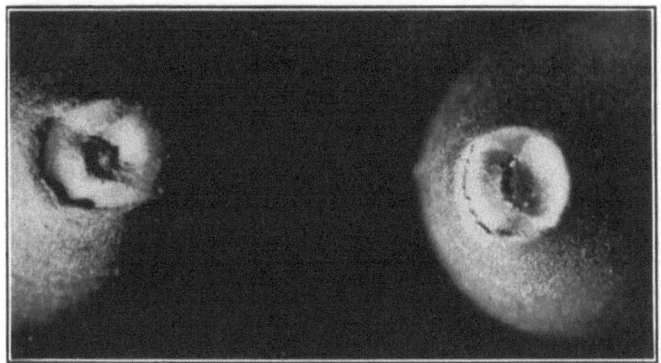

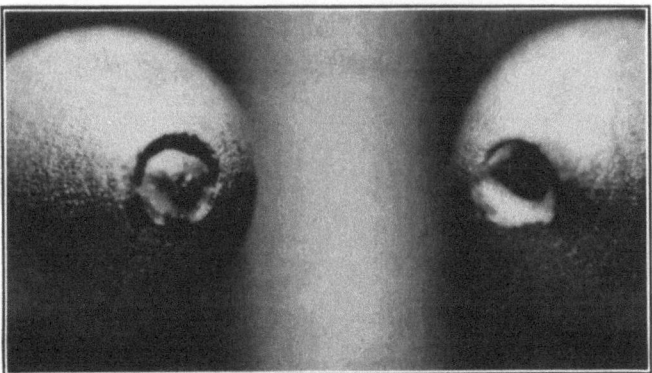

FIG. 35. The fracture of specimen 9-3-5, broken under 186,000 psi.

typical plot is shown in Fig. 38 for an armor-plate steel (C 14). In general, the pressure at which the tensile part of the break disappears is greater for the harder steels, as might be expected. For the hardest steel, in these experiments 19-5, the tensile break is not greatly affected by pressure. The relations for the hardest steels are shown in Fig. 39. As a rough average, the tensile break disappears for most of the steels of this work between 150,000 and 300,000 psi.

It is especially to be emphasized that the linear relation between hydrostatic pressure and strain at fracture continues smoothly without break

78   DEFORMATION UNDER HYDROSTATIC PRESSURE

through the point where the tensile part of the fracture vanishes. The justification for this statement is given by a detailed discussion of all the relevant data in the fifth Arsenal report, which is hardly worth reproducing here.

It might perhaps be thought that the disappearance of the tensile break is a function of the geometry of the specimen and not a legitimate

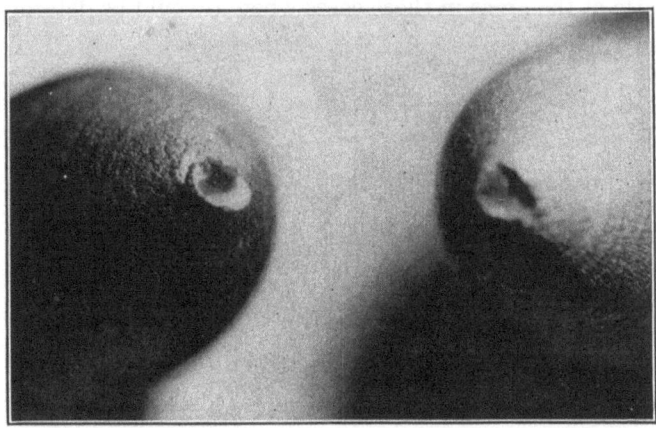

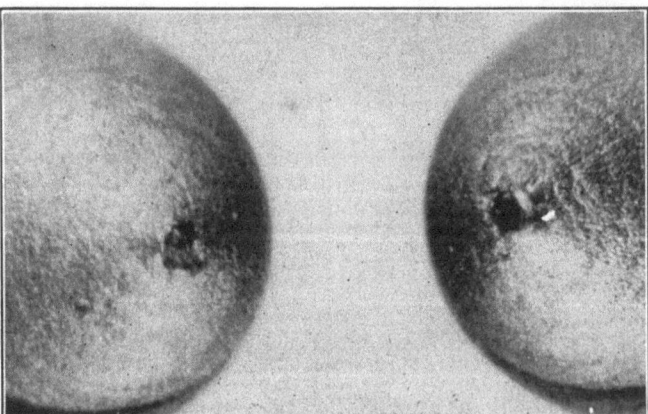

Fig. 37. The fracture of specimen 9-3-10, broken under 387,000 psi.

pressure effect for the reason that there is a progressive change in the geometry (increasing a/R) as fracture occurs at progressively higher pressures. Light can be thrown on this question by artificially changing the geometry. Two experiments were made with this question explicitly in mind. Both gave the same result. In one experiment a virgin specimen was machined to have such a value of a/R as corresponds to fracture under a hydrostatic pressure so high that the tensile part of the break has entirely disappeared. This specimen was fractured in tension in the normal way at atmospheric pressure, and in spite of the abnormal a/R value it fractured with a cup-cone break with the same value for the ratio of the areas, 0.43, as corresponds to normal fracture at atmospheric pressure of an initially straight specimen. In the second experiment a speci-

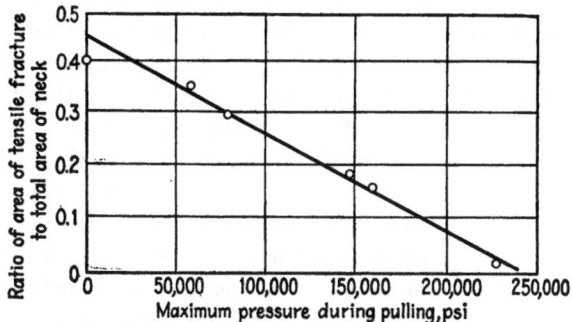

FIG. 38. The ratio of the area of the tensile part of the break to the total cross section of the neck as a function of the pressure of pulling for the same steel shown in Fig. 31.

men was pulled to a strain of 1.69 under 227,000 psi pressure. It was then removed from the pressure apparatus, machined in such a way as to reduce the value of a/R to nearly zero, and then reintroduced in the pressure apparatus and repulled to fracture at a pressure of 267,000 psi. The additional strain to fracture on the second pulling was 2.74, making the total strain at fracture 4.43. The a/R at fracture was now 1.0. The break was entirely shearing in character, as it would have been also if the fracture had been produced by a single straining without refiguring at 267,000 psi. On the other hand, if the specimen had been fractured at a pressure at which the normal a/R value was 1.0, the break would have been cup-cone, with a ratio of areas of 0.4.

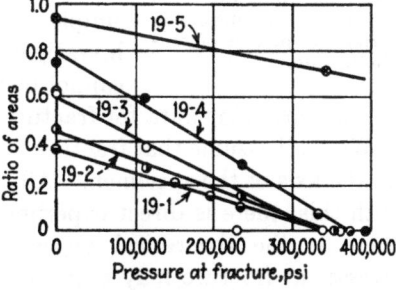

FIG. 39. The ratio of the area of the tensile part of the break to the total cross section of the neck as a function of the pressure of pulling for several of the hardest steels as indicated.

The two halves of a tension specimen almost never fit exactly together after fracture, but there is distortion in the act of fracture. What the nature of the distortion may be is suggested by Fig. 40. If the fracture takes place under a surrounding pressure at a pressure so low that the tensile part of the break has not yet disappeared, it is obvious that as the two halves are being pulled apart the external pressure must tend to force the walls of the crater into the vacant space on the axis and in so doing change the curvature of the external surface. In particular, this means that the value of a/R measured on the fractured halves is falsified as compared with the value which prevailed at the moment of fracture, which is the value pertinent to the calculation of the conditions of fracture. This agrees with the experimental finding that the values of a/R measured on fractured specimens were capricious and did not lie on the values extrapolated from lower stresses. Figure 40 also suggests the reason for the progressively more highly burnished appearance of the shearing parts of the break at higher pressures.

Not all varieties of steel fracture with the conventional cup-cone fracture. Exceptions were particularly common among the armor-plate steels which have not been described here in detail. Among the steels listed in Table V consistent deviations from the cup-cone fracture were exhibited by steels 9-7 and 9-8 (tempered pearlite and tempered martensite) when fractured by a second pulling at atmospheric pressure after a first pulling at high pressure. This matter will be discussed further on page 304.

The Stresses at Fracture. We now have the material completely in hand for a determination of the stresses at the tensile fractures under the simple conditions of Table V. A complete investigation of the stress conditions at fracture cannot be made until we have other data in hand, and in particular data for fracture after prestraining.

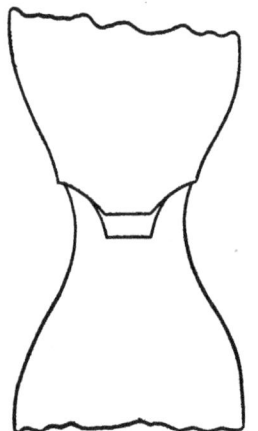

Fig. 40. Indicates the nature of the distortion during the process of tensile fracture.

We shall assume that fracture in the simple cases of Table V is initiated on the axis of the specimen. All the experimental evidence is consistent with this; there is direct experimental evidence by other observers, and there is the theoretical expectation as suggested on page 20. The stresses at fracture may therefore be taken from Eqs. (1-12) specialized for the axis, that is, by setting $r = 0$. This gives, for the actual stresses at the point of fracture,

$$\widehat{rr}_f = -P + F_f \log\left(1 + \frac{1}{2}\frac{a}{R}\right)$$
$$\widehat{\theta\theta}_f = -P + F_f \log\left(1 + \frac{1}{2}\frac{a}{R}\right) \quad\quad (2\text{-}4)$$
$$\widehat{zz}_f = -P + F_f \log\left(1 + \frac{1}{2}\frac{a}{R}\right) + F_f$$

We now inquire whether fracture is associated with characteristic values of these stresses or with characteristic combinations of them. It has been the tacit assumption in a great deal of engineering work that fracture will occur when some function of these stresses reaches a critical value, and a number of such criteria of rupture in terms of the stresses have been formulated and may be found in the literature. The simplest of these is that fracture occurs when the maximum tensile stress in the direction of the fracture reaches a critical value. This maximum tensile fiber stress is simply $\hat{zz}$ in the formulas above. Hence, if this criterion is correct, when $\hat{zz}$ at fracture is plotted against strain at fracture a horizontal straight line should be obtained. Another commonly proposed criterion is that fracture occurs when the maximum shearing stress reaches a critical value. The maximum shearing stress is one-half the difference between the maximum and minimum principal stresses, which from Eqs. (2-4) is $\frac{1}{2}F_f$. Hence, if this criterion is correct, $\frac{1}{2}F_f$, or F_f itself, when plotted against strain at fracture, should give a horizontal straight line. It will turn out that neither of these criteria is satisfactory.

In the search for other possible criteria we may consider other combinations of the stresses. Since rupture starts on the axis, a factor in determining fracture is obviously the hydrostatic tension that prevails on the axis in virtue of the necking of the specimen, that is, the component of hydrostatic tension on the axis dissociated from the impressed over-all hydrostatic pressure. We may abbreviate this hydrostatic tension by HT. Equations (2-4) give $\text{HT} = F_f \log [1 + \frac{1}{2}(a/R)]$. Or again, an argument might be made for the significance of the maximum tensile fiber stress undissociated from the impressed hydrostatic pressure. This is $F_f + \text{HT}$ and according to the formulas is $F_f \{1 + \log [1 + \frac{1}{2}(a/R)]\}$. Finally, the mean total hydrostatic pressure, or one-third the sum of the three principal stress components, which we abbreviate by Σ, may be thought to have a significance, because it represents the mean volume dilatation at the moment of fracture. Equations (2-4) give

$$\Sigma = -P + F_f\{\tfrac{1}{3} + \log [1 + \tfrac{1}{2}(a/R)]\}$$

These various possibly significant stress functions, and also F_f and the impressed hydrostatic pressure at fracture, the latter taken for convenience as positive, are plotted in Figs. 41 to 46 for six of the steels of Table V.

Inspection of the figures shows in the first place that the lines are all straight. This of course we already know should be the case for F_f and P as a function of strain at fracture. It is not at all obvious, however, that it should be linear for the others, because these involve HT, or $F_f \log [1 + \frac{1}{2}(a/R)]$, which involves a/R, which we already know to be a highly nonlinear function of the strains. The fact that HT is linear now emerges as a new and independent experimental fact. The meaning of the linearity is that the nonlinearity in a/R compensates for the nonlinearity in the logarithmic function. In fact, the process may be reversed, and by assuming the linearity of HT and F in the strain, the

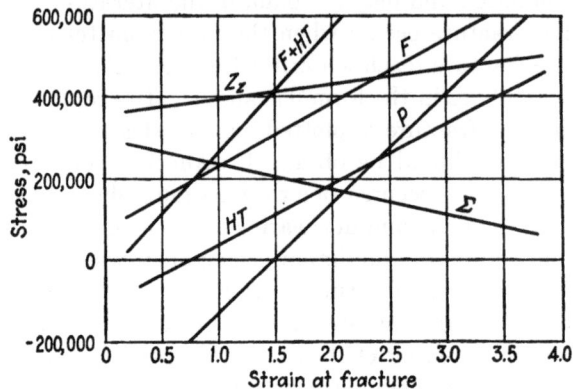

Fig. 41. Various possible significant stresses (see text for notation) as a function of the strain at fracture for steel 17.

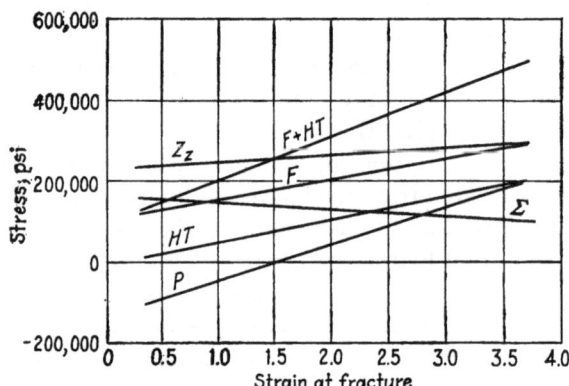

Fig. 42. Various possible significant stresses (see text for notation) as a function of the strain at fracture for steel 16.

functional connection between a/R and strain may be computed and compared with the connection found empirically. Agreement will be found within experimental error. The equations will be found worked out in detail in the fifth Arsenal report, pages 36$f\!f$., but need not be considered further here.

Returning now to the criterion of fracture, inspection of the figures shows that none of the lines consistently lies horizontal in all the figures and that therefore none of the suggested criteria can be universally valid.

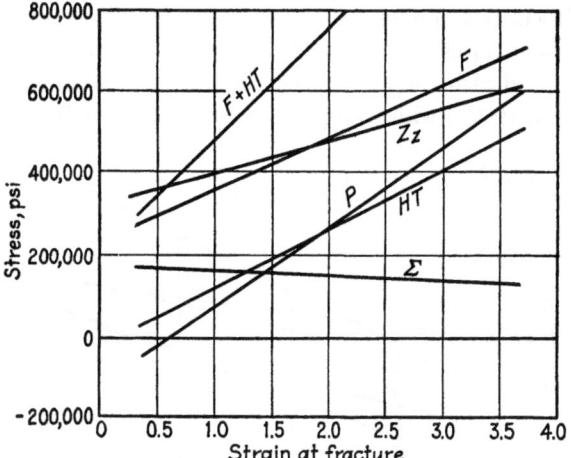

Fig. 43. Various possible significant stresses (see text for notation) as a function of the strain at fracture for steel 19-1.

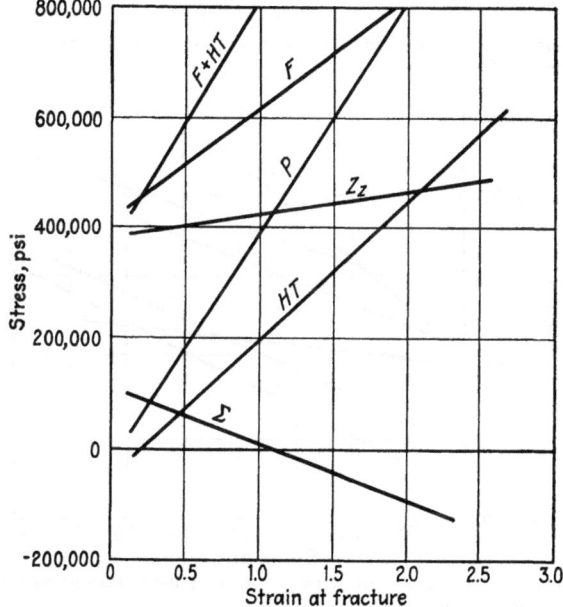

Fig. 44. Various possible significant stresses (see text for notation) as a function of the strain at fracture for steel 19-4.

Of the various possibilities, $\widehat{zz}$ and Σ come nearest to meeting the requirements, but these may vary by more than 2-fold in the range covered here. At the time the fifth Arsenal report was written it appeared that the criterion of constant Σ applied rather well to the results covered in that report, and indeed Σ is fairly constant in Figs. 45 and 46 for steels 9-4 and 9-2. The criterion does not stand up so well, however, for the wider range of steels covered here. Of all the criteria considered here, perhaps the simplest, and also the one which conversation with engineers shows to be the one with the widest appeal of intrinsic plausibility, is $\bar{z}\bar{z}$. This

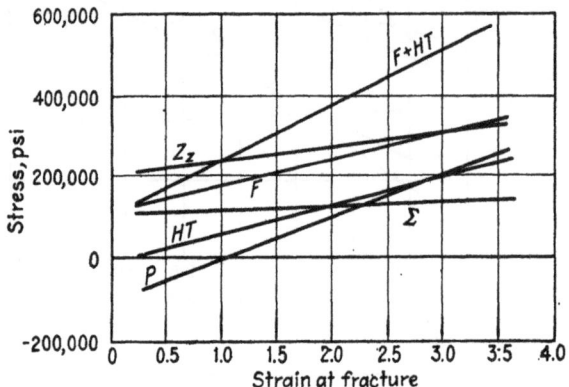

FIG. 45. Various possible significant stresses (see text for notation) as a function of the strain at fracture for steel 9-4.

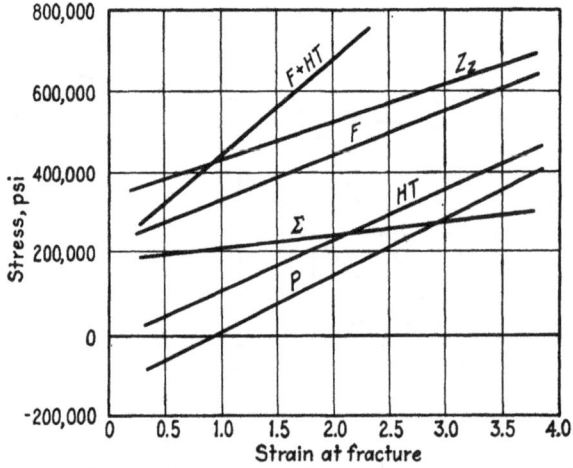

FIG. 46. Various possible significant stresses (see text for notation) as a function of the strain at fracture for steel 9-2.

criterion states that fracture occurs when the maximum fiber stress in any direction reaches a critical value, unaffected by the stress components in other directions. However, I think that even this loses its plausibility when it is considered that we are trying to cover a range of conditions so wide that the fracture changes its character from a tensile break to a break completely in shear.

It is of possible significance that for all the steels with the exception of stainless 17 the value of Σ extrapolated to zero strain is nearly the same, namely, 140,000 psi. Fracture at zero strain is a completely brittle fracture. It is brought about by subjecting the specimen to a negative hydrostatic pressure, the value of which can be found by extrapolating the line for P in the figures to zero strain. Whether the extrapolation of P to negative values has any more than a formal significance can hardly be known until some experimental method has been devised of applying high hydrostatic tensions.

It would appear that there is no universally valid criterion of fracture in terms of the stresses only. It will appear later that neither can there be a more general criterion in terms of both strains and stresses.

The NDRC Tests on Armor Plate. Brief reference will be made to these tests, which are not so informative as the tests already described because in general the composition of the steels was more complex, without systematic scheme of variation, and also because in many instances the material was not homogeneous. Tests were made on some 30 different plates of different compositions in addition to those already listed in Table V. Except for the greater variability of the results, largely due to the inferior grade of the material, the results were not different qualitatively from those already found. The same linear strain-hardening and ductility curves were found. Considerable attention was paid to possible difference of properties in different directions in the plate, and a number of tests were made on specimens cut in three mutually perpendicular directions. Except for obvious flaws, usually manifested by premature fractures, no difference with direction could be established, either for the strain-hardening curves or for the ductility curves. The parameters in the linear relations for strain-hardening and ductility, and in particular the pressure coefficient of strain-hardening and the pressure coefficient of ductility, would therefore appear to be functions of the material only. Lack of homogeneity was most often manifest by premature fracture of the specimens cut perpendicular to the face of the plate. The change in the character of the fracture with increasing pressure, that is, the change from cup-cone to purely shearing, was also found for these armor plates. The linear relation between pressure and ratio of area of tensile part of the break to total neck area was found here also. In fact, Fig. 31 has shown this for one of the armor plates. The appearance of the fracture was much more variable than for the steels of Table V. The fracture was often obviously affected by the presence of mechanical inclusions such as slag in the steel. These slag inclusions were comparatively unaffected by the pressure, retaining their hardness and not flowing under pressure. The surrounding steel, however, which under pressure becomes almost indefinitely plastic, flows around the flaws, and produces in some instances strange and unusual patterns of fracture, since the flaws are a much larger proportional part of the total area of the section for the fractures at high pressures.

In view of the nature of the material the results for the armor plates were not examined for any possible criterion of fracture in terms of the stresses.

CHAPTER 3

TWO-DIMENSIONAL TENSION UNDER PRESSURE[1]

Introduction. Two-dimensional tension was applied by subjecting thin tubes to tension along the axis. These tests were made in order to obtain information about the conditions of flow, strain-hardening, and fracture over a range of conditions of stress and strain different from those covered by the conventional tension test on cylindrical test pieces. In order to cover a wide range of strain, a number of the tests were made under high hydrostatic pressure, which was found to increase the ductility greatly for this type of strain as well as for the conventional cylindrical specimen.

Experimental Method. Because of the limitations of space imposed by pulling under hydrostatic pressure, the experimental difficulties were considerably greater than for the conventional test on a cylindrical specimen. The test specimen and the method of mounting it are shown in section drawn to scale in Fig. 47. A photograph of the exterior of the specimen before pulling is shown in the left-hand half of Fig. 48, and in the right-hand half a view of a different specimen of the same initial dimensions after pulling under pressure just to the verge of fracture. Tension is applied to the tube by means of a compressive force exerted by the ramrod as shown in Fig. 47. The assembly of Fig. 47 takes the place of the assembly for the solid test pieces described in Chap. 2, and it is mounted in the pressure vessel in exactly the same way as before. The total force on the specimen (compression in the ramrod, tension in the tube) is measured as before by the change of resistance of the calibrated grid.

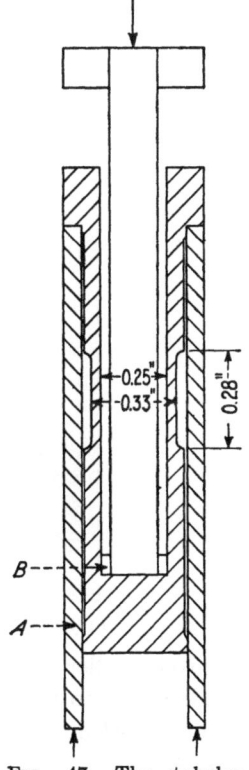

FIG. 47. The tubular specimen for two-dimensional tension and the method of mounting.

Care must be taken that the distortion and flow are symmetrical; otherwise, particularly in the later stages of flow, the tube thins down on

[1] This chapter is based on material contained in the sixth Watertown report.

87

one side more than another, and the average values are of diminished significance. In order to maintain axial alignment during flow, bearing surfaces, easily visible in the photographs, are provided at either end of the specimen, which fit accurately into the hardened and ground supporting sleeve A. The ramrod is centered in the lower end of the specimen by the brass washers B; the upper end of the ramrod is sufficiently centered by the walls of the pressure vessel. It is necessary to make the

FIG. 48. Tubular tension specimen, before and after pulling.

ramrod smaller than the hole by a generous amount in order to prevent binding due to inward radial flow of the tube during the initial stages. This inward flow is plainly shown in the section of the test specimen in Fig. 49.

Figure 49 is a longitudinal section through the specimen of Fig. 48 after fracture by the next application of tension. In order to specify completely the strain at any stage of flow it is necessary to know both internal and external contours. The external contour was measured optically. The specimen was mounted with two micrometer screws at right angles to each other so arranged that the readings of the two screws give the rectangular coordinates of the cross hairs of the reading microscope. The cross hairs were set on the external contour in groups of four orientations 90° apart. Combination of these readings gives the external diameter for two perpendicular orientations as a function of position along the axis. The wall thickness was measured with a specially constructed device in which the distance between two points, one brought just into contact with the internal surface and the other into contact with the external surface on the same radius, was measured with an Ames 1/10,000-in. jeweled gauge. Provision was made for making these readings at different positions along the axis and in different orientations. The wall thickness is the most important dimension in determining the strain, and measurements were regularly made at eight equispaced orientations. In the neighborhood of the neck, groups of readings in eight orientations were spaced along the axis at intervals of 0.005 in.; spacing as close as this was necessary in order to get good values for the radius of curvature of the contour of the neck. The thickness was repeatable and consistent to about 0.00003 in. By combining the measurements of the external contour with the thickness, the internal contour may be obtained, and in this way the radius of curvature of both external and internal contours at the neck. In the early stages of strain all irregularities

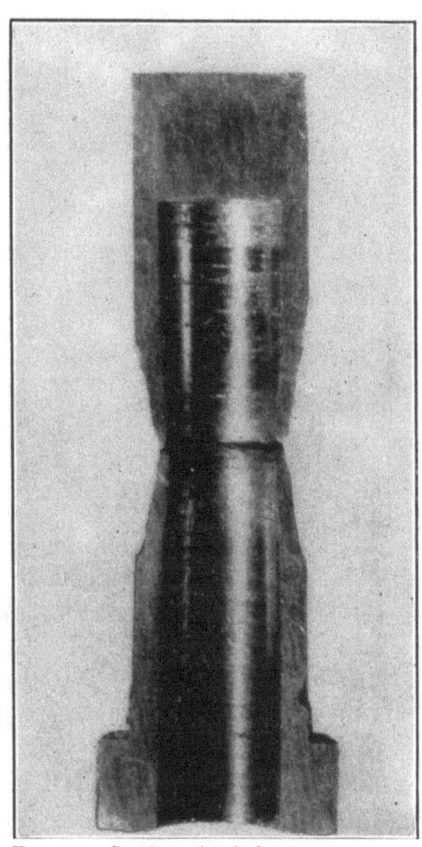

FIG. 49. Section of tubular tension specimen on the verge of fracture. The inward flow at the inside before necking starts is plainly visible.

of strain were neglected and the average of the eight readings of wall thickness combined with the average external diameter at the same axial position. The contours in the neighborhood of the neck were plotted on a much enlarged scale, and the radius of curvature at the neck found either graphically, by a method of trial and error with a compass, or, for the larger radii, by computation from the coordinates.

The experimental procedure was to pull each specimen to progressively larger strains, releasing the tension and pressure to zero and measuring the dimensions after the successive pullings. Two specimens were pulled at atmospheric pressure, one at a pressure of 150,000 psi and three at roughly 350,000 psi. These were all eventually pulled to fracture. The pullings before fracture give points on the stress-strain curve which may be accepted with considerable confidence. The parameters at fracture,

FIG. 50. Enlarged view of the tubular tension specimen of Fig. 48. At the right-hand edge is a place where the circle of necking has deviated from the equator.

however, were not so unambiguous. Even for solid rods it has been found that the fracture parameters show considerable irregularity when fracture occurs under such high pressure as to result in great ductility. The difficulties were exaggerated for these tubular specimens. In the last stages of pulling, just before fracture, geometrical irregularities become accentuated; the necking may become nonuniform around the circumference, the line of necking may wander from the equatorial plane, and the contour may in places lose its smooth curvature, and reentrant angles may appear. This is suggested by Fig. 50, showing an enlarged view of the specimen of Fig. 48. This was pulled at 350,000 psi and is just on the verge of fracture. When fracture does occur it is likely to start at a single point and travel around the circumference. Fairly satisfactory values for the fracture parameters were obtained only for the specimens fractured at atmospheric pressure, where the ductility is so small that geometrical instabilities do not appear before fracture occurs. In general it is much more difficult to control the phenomena in this two-dimensional work than when working with solid rods. Fracture occurs much more abruptly, with less total elongation, and the process of fracture itself is likely to be unstable, there being enough energy stored as elastic energy in other parts of the specimen and the ramrod to carry fracture through to completion, once it has started, without further displacement imposed on the ramrod from outside. The instability in the moment of fracture was somewhat lessened by making the ramrod of carboloy, the elastic constants of which are three times greater than those of steel. Even with the carboloy ramrod, the instability before fracture was so high that a further measure of control was supplied to all except the first two specimens by the use of stops to limit the motion. These stops themselves, however, permitted a small amount of elastic deformation, so that when the stop was only partially engaged it was not possible to know what fraction of the load was carried by the stop and what by the specimen. As a result, the load on the specimen corresponding to the maximum strain when the stop was used was indeterminate. The dimensions of the stops were so adjusted that the maximum load was reached before the stop came into action at all. By assuming that the maximum load was reached without further plastic flow beyond the final strain of the previous run, points on the stress-strain curve could be obtained by correlating the maximum load of one run with the final strain of the previous one. This procedure has the disadvantage of possible error from time effects, such as hardening on resting, but the consistency of the results was such as to indicate that any errors from this effect were probably small.

Slight geometrical imperfections in the original specimen may have an unusually large effect in the neighborhood of fracture. In order to minimize such effects the outside of the specimens was polished, and the hole in the last two specimens was lapped to a smooth finish. The initiation of necking at the center of the length was ensured by diminishing the external diameter initially at this locality by a fraction of 0.001 in.

The plastic deformation has three components e_z, e_r, and e_θ. The constancy of volume demands $e_z + e_r + e_\theta = 0$. In the first stages of longitudinal extension there is considerable transverse contraction of the tube, which sucks in, as shown by Fig. 49. If the tube were infinitely long with no end effects the transverse strain would be symmetrical in all dimensions, as if the tube were part of a solid cylindrical test specimen, and we would have $e_r = e_\theta = \frac{1}{2}e_z$. This sort of flow would continue up to the point where necking starts. In the actual case of a tube of finite length the ends are constrained from flowing, a component of stress is generated in a direction to oppose radial contraction, and the decrease of diameter is not so great as it would be if the specimen were infinitely long.

When longitudinal stretch is pushed beyond a certain point necking starts. This necking is unsymmetrical, running in further from the outside surface, where the curvature of the contour is in the same direction as the curvature generated by the necking, than from the inside surface, where the direction of curvature of the contour is opposite to that of the necking. This means that a/R is greater at the outer surface than at the inner surface; in fact, in the early stages of necking a/R at the inner surface may be negative. In applying the mathematical formulas for the corrections in terms of a/R, a mean value of R was used, calculated by the expression

$$\frac{1}{R_{\text{mean}}} = \frac{1}{2}\left[\frac{1}{R_{\text{outer}}} + \frac{1}{R_{\text{inner}}}\right] \tag{3-1}$$

As soon as necking begins, the decrease of diameter of the tube is much retarded, the internal diameter presently ceases to decrease, and beyond a certain degree of necking the internal diameter at the neck increases to meet the decreasing external diameter. In this final stage e_θ is practically constant, and $\Delta e_z = -\Delta e_r$. In the final stage the curvatures at outer and inner surfaces approach each other, and the neck approaches the symmetry assumed in the mathematical analysis for the necking of a plane sheet so clamped as to prevent change of width.

The strains were calculated as follows. A_0 is the initial total cross-sectional area $[A_0 = \pi(r_0^2 - r_i^2)]$ and A the cross-sectional area at the neck after pulling $[A = \pi(r_0'^2 - r_i'^2)]$. Here r_0 and r_i are the initial external and internal radii and r_0' and r_i' the external and internal radii after deformation. Then $e_z = \log_e (A_0/A)$ and $e_r = \log (t'/t)$ where t is the initial wall thickness and t' the thickness after deformation. e_θ is then determined by the condition on the volume.

Experimental Details. The steel was the 1045 steel designated as 9-4 in Table VI. The numerical results are collected in Table VIII. Each line in the table, except those giving the initial dimensions, corresponds to a complete run, with load starting from zero. The three components of strain tabulated are the strains corresponding to the final points of the runs, obtained in the manner already described from the dimensions in the second, third, and fourth columns. The dimensions were measured on release of load and removal from the pressure apparatus so that the strains given are the components of permanent plastic deformation and do not involve any elastic components produced by the tensile load or the hydrostatic pressure. The stresses tabulated in each line after the strains are not always the stresses corresponding to the strains in that line. Unless care is taken this may lead to some confusion in plotting the results. The necessity for this undesirable procedure arose, as already explained, from the fact that in many cases it was necessary to limit the motion with a stop in order to prevent the unstable incidence of fracture and in those cases the fraction of the total load supported by the specimen could not be determined. In such cases, the stress tabulated is that corresponding to the maximum load, which was reached before the stop came into play. It was assumed that at the maximum load there was

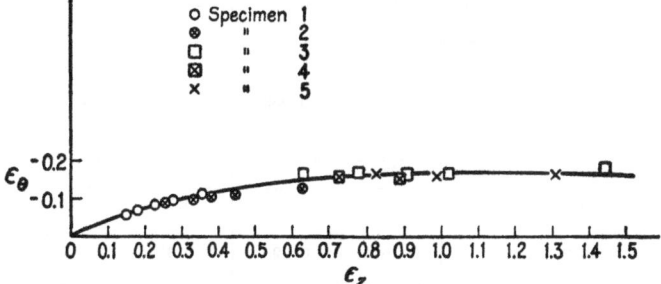

FIG. 51. The circumferential strain against the longitudinal strain at the neck for two-dimensional tension.

no further plastic flow beyond the flow reached in the previous application, and the strain of the previous application was used in calculating the stress from the load. There is doubtless some error introduced by this procedure because of time effects. These were irregular, so that correction for them would be difficult, and in any event it is probable that the error is small. In some cases two sets of stresses are given, one corresponding to the maximum load and the other to the final strain. In these cases the goodness of the approximation may be checked by comparing the stresses for the same strains obtained by the two methods. In the column of remarks will be found a statement as to which procedure is to be used. The first example of the uncanonical procedure is in the third line for specimen 2. Here will be found the value 117,500 psi for longitudinal stress; this should be plotted against the final strain of the preceding run given in the second line.

It is particularly to be noted that the "stresses" of columns 11, 12, and 13 do not include the hydrostatic pressure in the ambient liquid. Thus the total principal stress in the z direction in the surface of the neck,

94 DEFORMATION UNDER HYDROSTATIC PRESSURE

TABLE

Max pressure of pulling, psi (1)	Outside diameter, cm (2)	Inside diameter, cm (3)	Area, cm² (4)	Radius of curvature at outside, cm (5)	Radius of curvature at inside, cm (6)	Mean $\frac{a}{\bar{R}}$ (7)	$\log_e \frac{A_0}{A}$ $(= e_z)$ (8)
Specimen 1	0.8406	0.6384	0.2349	......	....	......	......
Atmos.	0.7863	0.6012	0.2017	......	....	......	0.153
Atmos.	0.7778	0.5971	0.1952	......	....	......	0.185
Atmos.	0.7674	0.5919	0.1873	......	....	......	0.227
Atmos.	0.7531	0.5848	0.1768	......	....	......	0.282
Atmos.	0.7371	0.5789	0.1642	......	....	......	0.358
Atmos.	......	......	......	0.22±	....	0.1±	......
Specimen 2	0.8410	0.6380	0.2358	......	....	......	......
Atmos.	0.7617	0.5905	0.1818	0.833	−2.0	0.015	0.259
Atmos.	0.7484	0.5881	0.1683	0.45	−2.05	0.034	0.336
Atmos.	0.7429	0.5888	0.1613	0.31	−1.92	0.052	0.379
Atmos.	0.7836	0.5893	0.1498	0.145	−0.35	0.073	0.452
Atmos.	0.7139	0.5917	0.1253	0.063	−0.31	0.193	0.633
Specimen 3	0.8384	0.6380	0.2323	......	....	......	......
410,000	0.6935	0.5687	0.1237	0.11	0.76	0.15	0.631
360,000	0.6812	0.5724	0.1071	0.090	0.31	0.19	0.775
350,000	0.6751	0.5795	0.0942	0.065	0.057	0.39	0.904
360,000	0.6696	0.5839	0.0843	0.045	0.037	0.53	1.015
365,000	0.6461	0.5892	0.0552	0.020	0.037	0.54	1.437
Specimen 4	0.8397	0.6375	0.2346	......	....	......	......
345,000	0.6904	0.5764	0.1134	0.121	0.109	0.25	0.727
350,000	0.6792	0.5817	0.0966	0.048	0.105	0.37	0.888
Specimen 5	0.8427	0.6375	0.2385	......	....	......	......
350,000	0.6844	0.5780	0.1055	0.073	0.091	0.33	0.815
355,000	0.6757	0.5856	0.0893	0.0048	0.067	0.40	0.978
360,000	0.6623	0.5961	0.0654	0.027	0.009(?)	0.60(?)	1.295
365,000	......	......	......	......	....	......	......
Specimen 6	0.8419	0.6387	0.2363	......	....	......	......
155,000	Fracture on first pulling			......	....	......	0.82(?)

TWO-DIMENSIONAL TENSION UNDER PRESSURE

VIII

e_r (9)	e_θ (10)	Average longitudinal "stress," psi (11)	Longitudinal "stress" at surface of neck, psi (12)	Hydrostatic "tension" at center of neck, psi (13)	Remarks (14)
....	....	....	....	....	Initial dimensions
−0.091	−0.062	101,000			
−0.111	−0.074	106,000			
−0.143	−0.084	111,000			
−0.185	−0.097	127,000			
−0.245	−0.113	128,000			
....	....	....	....	....	On sixth stretching ruptures with little further yield at slightly less than previous load
....	....	....	....	....	Initial dimensions
−0.170	−0.089	88,000 114,000	113,500	850	At max, on original area At final strain
−0.236	−0.100	118,000	117,500	880	
−0.275	−0.104	125,000	124,000	2,100	Stress from max load and previous strain
−0.341	−0.111	131,000	129,000	3,400	
−0.507	−0.126	137,000 156,000	134,000	4,900	Stress from max load and previous strain Specimen cracked. Fracture stress extrapolated
....	....	....	....	....	Initial dimensions
−0.472	−0.159	108,000 187,000	177,000	13,000	At max load, on original area At final strain
−0.610	−0.165	178,000 193,000	169,000 182,000	12,000 17,000	From max load, and previous strain At final strain
−0.738	−0.166	180,000 195,000	170,000 174,000	16,000 31,000	From max load and previous strain At final strain
−0.850	−0.165	204,000	176,000	41,000	Stresses in much doubt because of uncertain correction for partial contact with stop
−1.261	−0.176	208,000	179,000	43,000	Breaks on next application
....	....	....	....	....	Initial dimensions
−0.574	−0.153	102,000 182,000	169,000	20,000	At max load, on original area At final strain
−0.730	−0.158	181,000 187,000	168,000 168,000	20,000 28,000	From max load and previous strain At final strain Breaks with very slight yield on next application at 182,000 psi calculated from max load and previous strain
....	....	....	....	....	Initial dimensions
−0.656	−0.159	101,000 173,000	157,000	23,000	At max load, on original area At final strain
−0.824	−0.153	190,000 210,000	172,000 187,000	26,000 34,000	From max load and previous strain At final strain
−1.135	−0.160	205,000 230,000	183,000 196,000(?)	33,000 51,000(?)	From max load and previous strain At final strain
....	....	220,000	187,000(?)	49,000(?)	From max load and previous strain. Breaks with no perceptible yield
....	....	....	....	....	Initial dimensions
....	....	93,000 160,000(?)	....	....	At max load on original area. Roughly estimated stress and strain at fracture

$\widehat{zz}_a$, is column 12 minus column 1, and the total $\widehat{rr}$ at the center of the neck is column 13 minus column 1. It will be convenient to designate by a prime the part of the total stress system which does not include the pressure in the liquid; thus column 12 may be designated by $\widehat{zz}_a'$.

Comment has already been made on the tendency for the sucking in of the tube to approach an asymptotic value with increasing longitudinal strain. This is shown graphically in Fig. 51 where e_θ, which measures roughly the decrease of external diameter on sucking in, is plotted against e_z. The rise to an asymptote is clearly shown; points for all specimens lie on the same curve within error, whether or not the specimen was pulled under pressure. Rise of e_θ to a horizontal asymptote means that in the

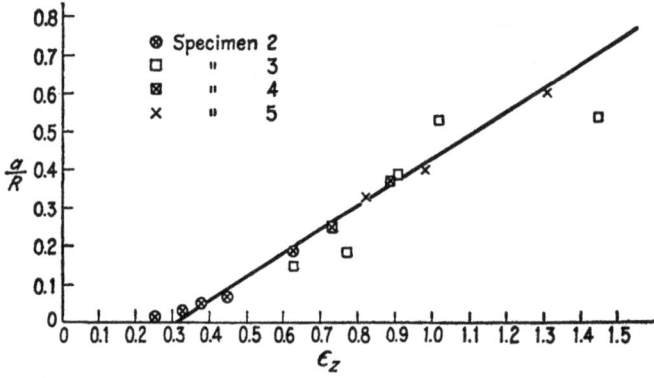

Fig. 52. a/R in two-dimensional tension against longitudinal strain.

limit e_r increases numerically at the same rate as e_z, its absolute value being less than that of e_z by a constant amount. It is hardly worth making a separate plot to illustrate this point.

Radius of Curvature at the Neck. The shape of the contour is specified by columns 5, 6, and 7, giving radius of curvature at outside and inside and the mean value of a/R. The initial reverse curvature for the inner surface is shown by the negative values of R for specimen 2, pulled at atmospheric pressure. The parameter a/R in column 7 was calculated from the half-wall thickness at the neck and the mean R determined by Eq. (3-1). In general, a/R would be expected to increase smoothly with the strain e_z. In Fig. 52 the a/R of column 7 of Table VIII is plotted against the e_z of column 8. Over the range of these measurements, that is, up to a strain of approximately 1.5, a/R increases roughly linearly with the strain. There is, however, considerable scattering, markedly greater than for the tension tests on solid specimens. The R used here is calculated from the mean of four measurements of external contour at 90° intervals around the circumference. The scattering would be much greater if the individual values of R at the 90° intervals were used. In some cases R was measured at various points on the circumference by the technique used for solid rods, that is, by sliding a cone over the contour until the best fit was found. Thus in one case, specimen 3 after the third pulling, values of R were obtained by this procedure varying from 0.052 to 0.97, with a mean of 0.73, against an R of 0.065 obtained from the averaged micrometer measurements of external and internal diameter. In virtue of such large local variations, considerable irregularity is to be expected in the averaged results and in the stress distribution calculated from the averaged results. Another source of uncertainty in the calculations of the stress distribution is the approximation involved in applying the formulas for plane sheets to this case with unsymmetric distortion from the two sides. In view, however, of the experimental irregularities, the approximation is doubtless good enough to give a qualitatively adequate picture, particularly in view of the smallness of the effects in comparison with the corresponding effects for solid rods. It is to be remembered that the maximum strain in these experiments is less than 1.5 against strains of 4 or more for the solid rods and that the hydrostatic tension at the center in these experiments seldom exceeds one-fifth of the flow stress, whereas for solid rods it may rise to 0.7 of the flow stress. There is another consideration which at first thought might appear to contribute to the uncertainty, but which probably does not, namely, that the exact relation of the stress $\widehat{\theta\theta}'$ to $\widehat{zz}'$ is in some doubt. But it is to be remarked that the "correction factor" has been found to be the same in cases 1 and 2, that is, it is the same in the two limiting cases of $\widehat{\theta\theta}'_a = \frac{1}{2}zz'_a$ and $\widehat{\theta\theta}'_a = 0$. The correction factor is therefore probably applicable in the intermediate cases.

It is interesting to compare a/R as a function of strain as shown in Fig. 52 with the corresponding diagram for solid rods. The place at which the line crosses the axis marks the strain at which necking begins; this point for solid rods occurs at a strain of about 0.2. A linear extension of the line in Fig. 52 would lead to the expectation that on the average necking for the tube would begin at a strain of around 0.3 instead of 0.2. The number of experimental points is, however, probably too small to justify a final conclusion. A further difference between solid and tubular specimens is that, for the solid, a/R increases more rapidly with the strain; its value, for example, at a strain of 1.0 is 0.8 for the solid against 0.4 for the tube. A consequence of this is that, although the correction for the same value of a/R is greater for the tube than for the solid, the correction for the same strain is less.

The Complete Stress System and the Conditions of Flow. In Fig. 53 the "average longitudinal stress" of column 11 is plotted against longitudinal strain at the neck. In Fig. 54 the "longitudinal stress at the surface of the neck," or zz'_a, which is given in column 12, is plotted against longitudinal strain. This is merely the average stress of Fig. 53 corrected according to Eq. (1-29) for the effect of necking. This stress is the

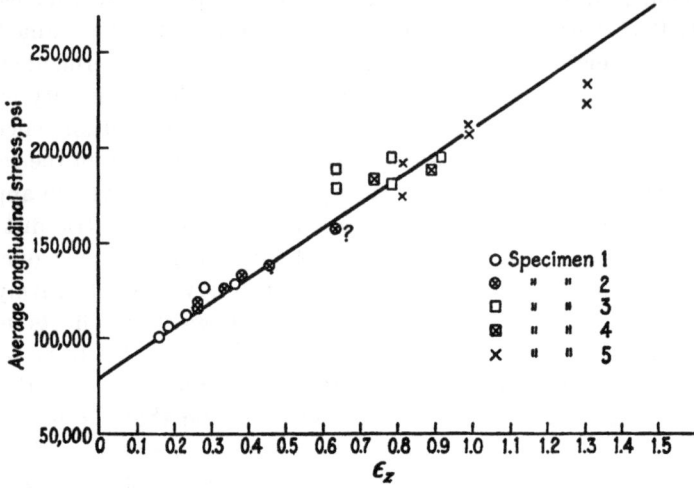

FIG. 53. The average longitudinal stress against longitudinal strain at the neck for two-dimensional tension.

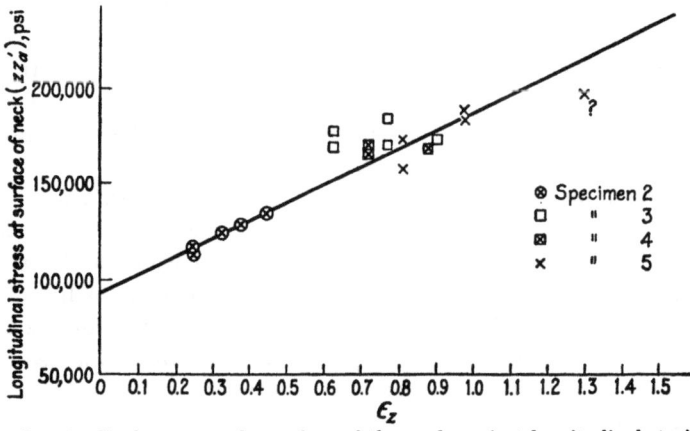

FIG. 54. Longitudinal stress at the surface of the neck against longitudinal strain at the neck for two-dimensional tension.

analogue of the "flow stress" used in discussing the results for solid rods. In the case of the solid rod the flow stress was the only component of the primed stress system, the two other components being zero. In the present case of the tube the solution is not so simple. One of the other principal components vanishes, that is, the component perpendicular to the surface, or $\widehat{rr}'_a$, but the third principal component, $\widehat{\theta\theta}'_a$, does not vanish. In the later stages of flow we may expect to approach the state of affairs when $e_\theta = 0$, which means that $\widehat{\theta\theta}'_a$ approaches $\frac{1}{2}\widehat{zz}'_a$, but in the early stages of flow, when there is appreciable sucking in of the tube, the relation will be different. In fact, it may be deduced from Fig. 51 that in the early stages of flow e_θ is approximately $-0.4e_z$, and correspondingly e_r is $-0.6e_z$.

Taking the flow stress corrected for necking of Fig. 4 as a first approximation, it is possible to obtain an approximate expression for the circumferential stress $\widehat{\theta\theta}'_a$. We assume the equations of plastic flow in their integrated form. This approximation should be good enough, in view of the numerical relations, to give a fairly adequate picture.

$$\frac{e_r}{\widehat{rr}'_a - \frac{1}{2}(\widehat{\theta\theta}'_a + \widehat{zz}'_a)} = \frac{e_\theta}{\widehat{\theta\theta}'_a - \frac{1}{2}(\widehat{zz}'_a + \widehat{rr}'_a)} = \frac{e_z}{\widehat{zz}'_a - \frac{1}{2}(\widehat{rr}'_a + \widehat{\theta\theta}'_a)} \quad (3-2)$$

At the free surface we have put $rr'_a = 0$. We use the empirical relation between e_θ and e_z shown in Fig. 51 and approximate to the curve by splitting up into three linear ranges as follows:

Range 1: $0 < e_z \leq 0.2$ $e_\theta = -0.4e_z$ $e_r = -0.6e_z$
Range 2: $0.2 < e_z \leq 1.0$ $e_\theta = -0.06 - 0.1e_z$ $e_r = +0.06 - 0.9e_z$ (3-3)
Range 3: $1.0 < e_z$ $e_\theta = -0.16$ $e_r = +0.16 - e_z$

Substituting these strains into the flow equations gives

Range 1: $\quad\quad\quad\quad\quad \theta\theta'_a = \frac{1}{8}\widehat{zz}'_a$

Range 2: $\quad\quad\quad\quad\quad \widehat{\theta\theta}'_a = \frac{0.4e_z - 0.06}{0.95e_z - 0.03}$ $\quad\quad$ (3-4)

Range 3: $\quad\quad\quad\quad\quad \widehat{\theta\theta}'_a = \frac{0.5e_z - 0.16}{e_z - 0.08}$

These stresses and strains may now be used to check various criteria of flow. Perhaps the most widely used flow relation at the present time assumes that the "octahedral" strain is the same function of the "octahedral" stress for all possible systems of plastic flow. We already have data for tensile tests on solid rods of the same steel as these tubes, so that by comparing the results for solid rods and tubes we have a check on the assumed relation.

Since the strains in this work are large, a preliminary examination must be made of the form of the relation for large strains. For large strains we use, of course, the natural or logarithmic strains instead of the conventional strains. The octahedral strain γ_n is not simple for large natural strains. Nadai[1] gives a differential definition, subject to the restriction that during the entire shearing process there be no rotation of the axes of principal stress. Fortunately these conditions apply here. Nadai's definition is

$$d\gamma_n = \tfrac{2}{3}[(de_1 - de_2)^2 + (de_2 - de_3)^2 + (de_3 - de_1)^2]^{1/2} \quad (3\text{-}5)$$

where the de's are the increments of natural strain. In our case the integration has to be performed separately for each of the three ranges. The integration is straightforward and simple and yields

Range 1: $\quad \gamma_n = 1.424 e_z$
Range 2: $\quad \gamma_n = -0.026 + 1.556 e_z \quad (3\text{-}6)$
Range 3: $\quad \gamma_n = -0.098 + 1.628 e_z$

It is to be remarked that, if we abandon the differential definition of γ_n and arbitrarily define a new quantity,

$$\gamma'_n = \tfrac{2}{3}[(e_1 - e_2)^2 + (e_2 - e_3)^2 + (e_3 - e_1)^2]^{1/2} \quad (3\text{-}7)$$

the numerical difference is not large even in the present comparatively large range of strain. The analytical expressions are

Range 1: $\quad \gamma'_n = 1.424 e_z$
Range 2: $\quad \gamma'_n = \tfrac{2}{3}(5.46 e_z^2 - 0.288 e_z + 0.0216)^{1/2} \quad (3\text{-}8)$
Range 3: $\quad \gamma'_n = \tfrac{2}{3}(6.0 e_z^2 - 0.96 e_z + 0.1536)^{1/2}$

In the first range the expressions for γ_n and γ'_n are identical. At the end of the second range ($e_z = 1$), $\gamma_n = 1.530$ and $\gamma'_n = 1.519$. In the third range, at the point $e_z = 2$, $\gamma_n = 3.17$ and $\gamma'_n = 3.00$. The divergence between γ_n and γ'_n becomes less at larger strains.

It has recently been proposed[2] as a hypothesis that γ'_n is a universal function of the octahedral shearing stress. Although the theoretical justification offered for this relation seems perhaps questionable, nevertheless it makes little numerical difference whether γ_n or γ'_n is used.

For the octahedral shearing stress we have the general formula

$$T_n = \tfrac{1}{3}[(\widehat{rr}' - \widehat{\theta\theta}')^2 + (\widehat{\theta\theta}' - \widehat{zz}')^2 + (\widehat{zz}' - \widehat{rr}')^2]^{1/2} \quad (3\text{-}9)$$

Using the relations already given we obtain T_n as a function of zz'_a with a different formal expression for each of the three ranges. The deviations from linearity are so slight that it will be sufficient to make the numerical calculations for the end points of the ranges.

[1] A. Nadai, *J. Applied Phys.*, **8**, 205–213, 1937.
[2] J. E. Dorn and E. G. Thomsen, Report to the NDRC on The Effect of Combined Stresses on the Ductility of Metals, OSRD No. 3218, Serial No. M-213, Feb. 2, 1944.

γ_n obtained in this way as a function of T_n for the tubular specimen is to be compared with γ_n as a function of T_n for the solid specimen. For the solid we have

$$e_r = e_\theta = -\tfrac{1}{2}e_z \qquad \widehat{rr}' = \widehat{\theta\theta}' = 0 \qquad (3\text{-}10)$$

Substitution in the general formulas gives

$$\left.\begin{array}{l} \gamma_n = 1.414 e_z \\ T_n = 2.12\widehat{zz} \end{array}\right\} \text{solid} \qquad (3\text{-}11)$$

The results are now collected in Table IX and are plotted in Fig. 55. The relation between γ_n and T_n for the tube is very nearly linear over the

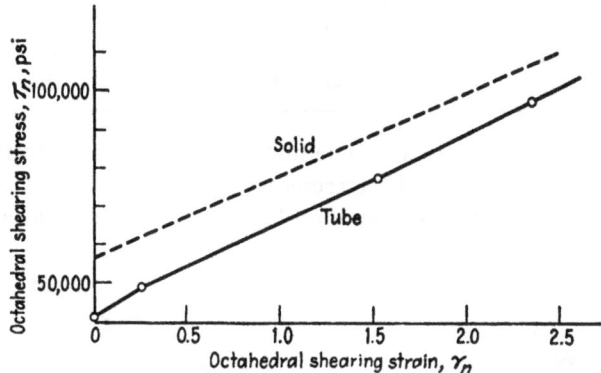

Fig. 55. The "octahedral" shearing stress against "octahedral" shearing strain for ordinary solid and for tubular tension specimens. The two lines should coincide if there is a universal relation between these two parameters.

entire range in spite of the fact that three different formulas are concerned in three ranges. Furthermore the slope of the line of γ_n against T_n for

TABLE IX

	Tubular specimen				Solid specimen		
e_z	$\widehat{zz}'_a$, psi	γ_n	T_n, psi	e_z	$\widehat{zz}'_a$, psi	γ_n	T_n, psi
0	92,000	0	40,900	0	120,000	0	56,000
0.2	110,000	0.285	49,000				
1.0	187,000	1.53	77,300		Relation linear		
1.5	236,000	2.34	96,500	1.5	215,000	2.12	100,900

the tube is not much different from that for the solid. The absolute value of T_n for the solid is, however, from 35 to 10 per cent greater than for the tube.

It is interesting to compare other criteria of flow. One of the oldest of these, which has very recently been used by Koehler and Seitz[1] in a theoretical discussion because of the simplicity of its mathematical formulation, is that the maximum shearing stress is a universal function of

TABLE X

Tubular specimen				Solid specimen			
e_z	$\widehat{zz}'_a$, psi	Max shearing strain	Max shearing stress	e_z	$\widehat{zz}'_a$, psi	Max shearing strain	Max shearing stress
0.0	92,000	0.00	46,000	0	120,000	0	60,000
0.2	110,000	0.16	55,000				
1.0	187,000	0.92	93,500		Relation linear		
1.5	236,000	1.42	118,000	1.5	215,000	1.125	107,500

the maximum shearing strain. The numerical results are given in Table X and plotted in Fig. 56. The agreement for solid and tubular specimens is on the whole somewhat better than for the octahedral components.

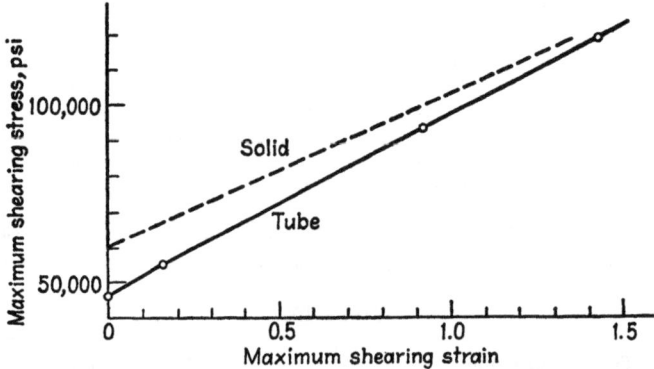

FIG. 56. The maximum shearing stress against maximum shearing strain for ordinary solid and for tubular tension specimens. The two lines should coincide if there is a universal relation between these two parameters.

Another simple criterion is that the maximum principal "reduced" stress is a universal function of the maximum principal strain. The "reduced" stress is the total actual stress diminished by the mean hydrostatic component. In this case the maximum principal reduced stress is $\widehat{zz}'_a - \frac{1}{3}(\widehat{zz}'_a + \widehat{\theta\theta}'_a)$. Figure 57 shows maximum principal strain against maximum principal reduced stress. The slopes of the lines for tubular

[1] James S. Koehler and Frederick Seitz, Report to the NDRC on The Stress Waves Produced in a Plate by a Plane Pressure Pulse, OSRD No. 3230, February, 1944.

and solid specimens are very nearly the same, but the difference in absolute values is somewhat greater than for the other two criteria.

On the whole, there is not a great deal to choose among these three criteria under these conditions; it is perhaps surprising that all three work as well as they do over such a wide range of strain. It is to be noted that all three criteria give lines for the tube lying below those for the solid. Since in general the corrections for necking are less for the tube than for the solid, the discrepancy between the various criteria applied to tube and solid would have been greater if the uncorrected stresses had been used.

The difference between the flow criteria for solid and tube is doubtless connected with a striking physical difference in the two cases which has

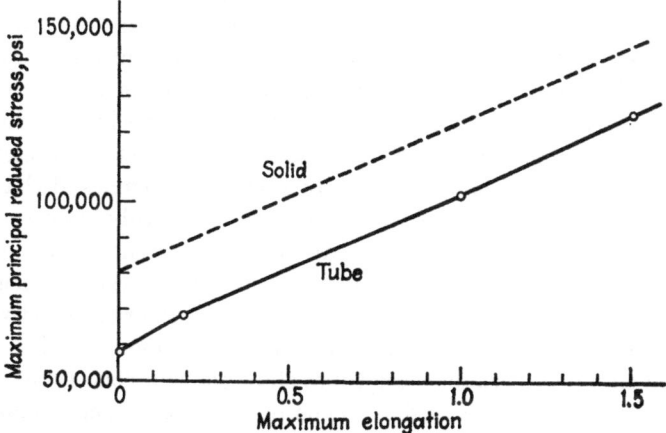

FIG. 57. The maximum "principal reduced stress" against maximum elongation for ordinary solid and for tubular tension specimens. The two lines should coincide if there is a universal relation between these two parameters.

not yet been explicitly mentioned, although it is obvious enough on contemplation of the results. This is that in the two-dimensional case in most instances the pressure in the liquid is greater than the stress arising from the tensile load, so that the total stress $\bar{z}\bar{z}$ is negative in most cases; that is, in most of these two-dimensional cases deformation by extension takes place in a direction in which the total stress component is compressive. Paradoxical as it may seem, this is of course the sort of thing that must eventually occur for any substance at sufficiently high pressures. In Figs. 32 to 37, for example, it is shown by the curves for P rising more rapidly and eventually crossing the lines for $\bar{z}\bar{z}$. In the two-dimensional case this occurs much earlier than for the solid.

Rupture. It is to be remarked first of all that the incidence of rupture for the tubes is much more irregular and capricious than for solid rods.

Thus the average strain at fracture of one of the specimens fractured at atmospheric pressure was approximately 0.36 and that of the other 0.63. The strain at fracture of the three specimens subjected to approximately 350,000 psi was about 1.44, 0.89, and 1.30, respectively. In the case of solid specimens, the strain at fracture was regular enough so that curves could be drawn, and in fact it was found that the strain at fracture was a linear function of the pressure. Obviously no such relationship can hold here. In partial explanation of the greater capriciousness it is to be remarked that we are here concerned with a two-dimensional instability instead of with a one-dimensional instability, corresponding to necking along the radius and along the circumference.

In general, it would seem safe to say that rupture under the conditions imposed on the tubes, that is, where the constraints force one of the components of strain to vanish, takes place at smaller strains than in the conventional solid specimens where there are three nonvanishing strain components. At atmospheric pressure the highest strain at fracture of a tubular specimen was 0.63, whereas all the solid specimens broke at a strain in the neighborhood of 1.0. At high hydrostatic pressure the results are more capricious, but it is at least to be remarked that at the highest pressure the average strain at fracture of the tubes never exceeded 1.5, whereas for the solid specimens at the same pressure it was over 4.0.

A careful examination of the appearance of all the fractures leaves the impression that there are important geometrical differences between the fracture of these tubes and that of the conventional solid specimen. Under no circumstances do these tubes show the cup-cone type of break. The "separation" or "extension" type of fracture perpendicular to the line of maximum principal tension does not appear, but the surfaces of fracture, when they can be identified, are shearing surfaces, inclined at roughly 45° to the tube axis. In none of the specimens is there any appearance to suggest that the fracture may have started at the mid-point of the tube wall. The fracture surface is usually a single shearing plane, running all the way across the wall of the tube from outside to inside. There are often, however, limited sectors of the circumference in which there are two shearing planes at approximately 90°, running across each other in such a way as to leave a sharp dihedral angle. This sort of thing is to be expected, because it would seem to be a matter of accident whether the shearing starts on a plane in the first or the fourth quadrant; and, if fracture starts at more than one point, effects like this are to be expected when the two fractures meet. In the specimens broken at atmospheric pressure, the shearing surface of the fracture has a coarsely granular structure, with no evidence of burnishing. There was no particularly striking difference in the appearance of the two atmospheric specimens.

In the specimen broken at a strain of 0.63 the dihedral part of the shearing fracture extended around about 90° of the circumference, and in the one broken at a strain of 0.36 around 45°. Neither was there any direct correlation between the appearance of the fracture of the specimens broken under high pressure and the strain at fracture. In the specimen broken at a strain of 1.30 the fractured part was pulled out to a knife-edge around all but 20° of the circumference; in this 20° there was a region of transition from one level to another, the course of the fracture around most of the circumference being along a helix of low pitch. The fracture on most of the circumference of the specimen which broke with a strain of 0.89 was along a definitely recognizable narrow shear plane, only slightly burnished. The fracture of the specimen which broke with a strain of 1.44 was of a somewhat intermediate type; for most of the circumference a surface of shearing slip could be recognized, but it was not sharply defined and seemed to be multiple and in places merged gradually into a dragged-out well-defined knife-edge. In this specimen it was difficult to decide what was the original external surface and where new external surface had been created by the exposure of the ends of bands of internal slip.

In seeking for possible reasons for the geometrical differences between the fracture of tubes and of solid rods it is to be considered that the stress systems in the two cases are essentially different. The stress system for the solid rod is biaxial, with two equal components of principal stress. In the case of the tube, however, the stress is irreducibly triaxial, one component of principal stress being the arithmetic mean of the other two. The geometrical appearance of the fracture, would seem, therefore, to be sensitive to the type of stress.

CHAPTER 4

TENSION TESTS UNDER PRESSURE ON MATERIALS OTHER THAN STEEL[1]

Comparatively little work has been done in this field, which at the date of writing invites extensive further work. I am just beginning a more extensive examination of some of the common metals.

MATERIALS NORMALLY DUCTILE

Aluminum. A specimen of ordinary commercial rod fractured at atmospheric pressure at a natural strain of 1.74 (82.5 per cent reduction of area). The "true stress" at maximum tensile load was 19,000 psi. The stress at fracture was not determined. Under a hydrostatic pressure of 410,000 psi the same material was stretched to a natural strain of 2.87 (reduction of area 94.4 per cent) without fracture, although when the specimen was removed from the pressure apparatus it appeared on the point of fracture. The "tensile strength" under pressure was 21,800 psi, which is greater than that at atmospheric pressure by about the same amount that would be suggested by the experiments on steel. The "true stress" at the maximum strain under pressure was 63,000 psi. It is evident, therefore, that qualitatively aluminum shows the same sort of strain-hardening and increase of ductility under pressure as steel. More measurements would be necessary to obtain numerical values. My impression is that probably the effects are not so great as for steel.

A technical difficulty appeared which is going to interfere with carrying the strains as high as for steel in that the circular figure of the cross section is lost at a much lower strain than for steel. In the experiment just quoted the section had become roughly square in section, and gave the impression of a single crystal. It is possible that the original grain size in this specimen was large, and the possibility must not be lost sight of that there was recrystallization during pulling under pressure.

Copper. A specimen of commercial copper rod fractured at atmospheric pressure at a strain of 0.95, having shown a "tensile strength" of 49,000 psi. The "true stress" at the moment of fracture was 86,000 psi. A similar specimen was pulled under 410,000 psi to a natural strain of 2.93 with no sign of fracture. The reduction of area was so large that the load at the final strain could not be determined with sufficient accu-

[1] This chapter is based on material contained in B20, B24, and B26.

racy, so that there are no figures for the strain-hardening. The "tensile strength" under pressure was 57,000 psi. The cross section lost its circular shape to an extent definitely greater than for steel at the same strain, but not nearly so much as for aluminum. Qualitatively, the behavior seems to be in the same class as that of steel.

Brass. A specimen of brass rod from the stock of the laboratory machine shop fractured at atmospheric pressure at a strain of 0.72, with a "tensile strength" of 64,000 psi and a "true stress" at fracture of 120,000 psi. The same brass under a pressure of 390,000 psi supported a strain of 1.31 with no sign of fracture, with a "true stress" of 194,000 psi at the greatest extension, and a "tensile strength" of 73,000 psi. This material exhibits a much greater elongation than steel before necking starts, and it was accordingly more difficult to accommodate it in the range of the present apparatus. Except for the deferred appearance of necking the behavior is qualitatively not unlike that of steel.

Bronze. Partial measurements have been made on a welding bronze. The results were incomplete because of the development of leak during the pressure run. The only significant result obtained under pressure was that the "tensile strength" under a pressure of 225,000 psi is 77,000 psi against a value 71,000 at atmospheric pressure. At atmospheric pressure this bronze fractured with a "true stress" of 170,000 psi at a strain of 1.08.

MATERIALS NORMALLY BRITTLE

Introduction. The tensile testing of materials normally brittle demands a radical change in technique. If conventional tension specimens such as those of Fig. 17 are used, break invariably occurs at one or another of the corners because of the well-known stress-concentration effects. The effects may be minimized by rounding the contours at the corners, but even with the harder steels this demanded so great a change in the geometrical figure that it seemed of prohibitive difficulty to attempt to prepare tension specimens of this kind for substances as brittle as glass, for example, to say nothing of the mechanical difficulty of fashioning the specimens. Fortunately a very simple technique proved applicable, which makes the tensile testing of brittle specimens under pressure even simpler than that of materials like steel. This new technique suffers only from the disadvantage that it cannot be applied at atmospheric pressure. This is not a great disadvantage, however, particularly when it is considered that in many cases the chief interest in studying brittle substances under pressure is to find whether pressure imparts ductility.

The experiments on tension to be described in the following are in a certain sense an extension of experiments made long ago on what I called the "pinching-off" effect.[1] In these experiments rods of glass or other brittle substances passed completely through the pressure vessel, emerging through stuffing boxes, the part of the rod within the vessel being exposed to hydrostatic pressure. On raising pressure high enough, the glass rod fractured by a clean tensile break at some point within the pressure vessel. Those experiments were mostly qualitative in nature, and no attempt was made to get consistent values for the pressure which produced fracture. The chief result of the experiments was to emphasize that it is possible to have a tensile fracture across a plane across which there is no component of stress. This is, from the point of view of some of the traditional conceptions of fracture, a highly paradoxical thing. The conclusions which I drew from those experiments had, therefore, understandably met with considerable skepticism in some quarters and with the attempt to explain away the phenomenon as due to various secondary effects. For a long time I had had it in mind to make a more exact study of the "pinching-off" effect. A few years ago I found, in some unpublished experiments, that the precise pressure at which the pinching-off fracture is produced in glass is extraordinarily sensitive to surface conditions. This indicated the necessity for a more elaborate investigation of the whole effect. With these new experiments, this extended examination is now possible. We have now two degrees of freedom in the stress system, whereas formerly there was only one. Formerly the setup was such that the longitudinal component of stress was continually zero, and the two lateral components were increased until fracture occurred. Now the longitudinal component and the two equal lateral components may be independently varied, so that fracture may be studied over a much wider range of conditions. The paradox that characterized the pinching-off effect is now carried much further, for now tensile fractures can be produced across planes on which the *compressive stress* is so high as to be beyond any possible experimental error. Thus, to mention only a single example, tensile fracture has been produced in glass under a mean hydrostatic pressure of 370,000 psi with a superposed tension of 156,000 psi, so that the net stress across the plane on which the tensile break occurred was a compressive stress of 114,000 psi.

Method. The method and the general nature of the problems encountered may be indicated by means of a highly idealized mental experiment. Figure 58 represents a compound tension specimen consisting of a central cylinder A of the brittle substance to be tested, and of two "pull" pieces B with shoulders and the same small diameter as the specimen, mounted together in the thin cylindrical sheath C which is attached by solder or

[1] *Phil. Mag.*, July, 1912, p. 63.

otherwise to the pieces B at D in such a way as to be liquid tight. The whole compound specimen is immersed completely in a liquid, to which any desired hydrostatic pressure is applied, and, superposed on the hydrostatic pressure, a tensile force is then applied to the pull pieces B. The situation is idealized by supposing that the pull pieces are infinitely strong, that the lateral contraction of A and B under the complete stress system is the same, so that their diameters continue to be the same, and that the sheath C has negligible longitudinal strength, offers zero friction on its inside surface, and is infinitely stiff so that it cannot be pushed by the external pressure into any cracks which might be formed on the surface of A or B. We now inquire what will happen if the pressure is first raised to some high value, at which it is kept constant, and the tension is then increased indefinitely. It is evident in the first place that the compound specimen cannot separate at the surface of separation between A and B as long as the tension is less numerically than the pressure. For if incipient separation should start at this surface, the surfaces of separation would be pushed back into contact by the pressure, which is greater than the tension. All the more, if separation cannot occur at surfaces already free between A and B, it cannot occur in the material of the solid rod A. That is, under a hydrostatic pressure P, the tensile strength of the material A is, under these conditions, at least as high as the pressure P. Ideally, it seems impossible to place any limit to this behavior, so that by increasing indefinitely the hydrostatic pressure the tensile strength may also be indefinitely increased. This means that the numerical value of the pressure coefficient of tensile strength approaches unity under these conditions. This in itself is highly anomalous. It is a matter of experience that the pressure coefficient of most physical properties of

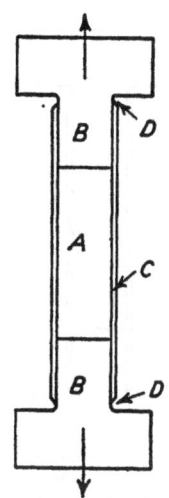

FIG. 58. Shows the method for applying a tensile pull to a straight cylindrical specimen A through the medium of the steel pull pieces B when immersed in a fluid carrying hydrostatic pressure.

substances like glass is of the order of magnitude of perhaps 10^{-5}, if the unit of pressure is taken as the kilogram per square centimeter. Yet here we have the necessity for a coefficient 10^5 times as high. The situation is also anomalous when considered from the point of view of strain. Under combined tension and pressure the specimen A elongates, and we have apparently the necessity for an indefinitely great elongation with no fracture. To say that beyond a certain point the specimen will receive plastic set on release of stress does not relieve the paradox of the situation while the stress is in action.

This idealized experiment, leading to such paradoxical conclusions, may be made the basis of the actual experiment, and also suggests the direction in which the experimental results may be expected to go, which is indeed the direction of paradox. In practice the pull pieces B may be made of heat-treated steel, and the specimen A made of the brittle substance under examination, such as glass or brittle alloy or what not. The sheath C may be made of copper, a few thousandths of an inch thick. We now see at once that we have a method of applying tension to a brittle specimen without shoulders or regions of stress concentration in the specimen itself. The pull pieces B may be reused from experiment to experiment, so that the preparation of the specimen could not well be simpler. In the simple form shown in Fig. 58 any tension can be applied up to the magnitude of the pressure itself, assuming that there are no limiting complications in practice. These complications do, in fact, occur, since the copper is not infinitely rigid, and would eventually blow into the surface of separation between A and B before tension reaches its theoretical upper limit. This difficulty may be avoided by artificially fashioning a neck on the tension specimen, as shown in Fig. 59. The neck may be easily made of such proportions that there is no harmful stress concentration in the neighborhood. By making the neck of suitable dimensions, any tensile stress may be applied in this manner, without grips, to any material up to the fracture point, whether that is above or below the surrounding hydrostatic pressure. This arrangement has been used, for example, in testing the tensile strength of carboloy under pressure, the tensile strength being much higher than the highest hydrostatic pressure that was applied.

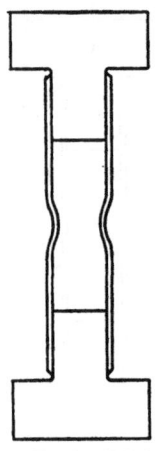

FIG. 59. Modification of the arrangement of Fig. 58, in which, by necking the specimen, it is possible to apply to it a tension in excess of the pressure in the surrounding liquid.

The material of the sheath is a vital factor in the result, for not only may it be blown between the surface of separation of A and B, but it may also be blown into any incipient cracks which appear in the surface of the specimen A. One would anticipate, therefore, that the measured tensile strength would be a strong function of the surface conditions, as exemplified by the material of the sheath. This did indeed prove to be the case. Various materials were tried. In practice the sheath sometimes was made compound, still keeping copper over the surface of separation of A and B, but replacing the central part by some other substance such as lead or rubber, which must naturally be attached in such a way as to prevent access of liquid. It would have been desirable if conditions could have been pushed to the limit by permitting contact between the liquid and the specimen all over its lateral surface, while preventing entry of the liquid into the interface between A and B, but no feasible way of doing this presented itself. As it was, a sufficiently wide range of external conditions was attained by the use of copper, lead, and rubber to indicate the general nature of what might be expected.

The arrangement indicated in Figs. 58 and 59 has points of resemblance to an arrangement probably first used by Richart, Brandtzaeg, and Brown[1] in studies of the effect of compound stresses on concrete and later by Balsley[2] working with Griggs on minerals. Both these investigators used straight cylindrical steel pull pieces, attached to the specimen by a sheath. The pull pieces were taken out of the pressure chamber through stuffing boxes and on the outside were subjected to an independent compressive stress, which, as it varied from greater to less than the internal hydrostatic pressure, was capable of exposing the specimen to a superposed stress of either compression or tension. There are technical difficulties with this arrangement, however. The stuffing boxes have to be accurately aligned and are not capable of as wide a pressure range as that used here. The maximum pressure in the Illinois experiments was 4,300 psi, and in those of Balsley 142,000 psi. Neither of these experimenters used necked specimens in order to permit raising the tension to values exceeding the hydrostatic pressure, but there is no reason why necked specimens should not be used with externally projecting pull pieces.

During the war Gurney and Rowe[3] in England made a large number of measurements of the pinching-off effect (which they describe as fracture by "radial" pressure) on glass and several plastics, and in a number of instances used necked specimens, thus permitting any desired tensile stress depending on the dimensions of the neck. These experiments were limited, however, because of the extension of the specimen itself through stuffing boxes, to hydrostatic pressures not greater than the "pinching-off" pressure of the straight cylindrical specimen.

The various other details of the experiment, such as the methods of applying and measuring the force, were the same as in the tension experiments on steel, and the details have already been sufficiently described.

The Experiments. *Glass.* The experiments on glass were made on pyrex, all specimens being taken from the same batch. Five experiments

---

[1] F. E. Richart, A. Brandtzaeg, and R. L. Brown, *Univ. Illinois Eng. Expt. Sta. Bull.* 185, 1928.

[2] J. R. Balsley, *Trans. Am. Geophys. Union*, Part II, p. 519, 1941.

[3] C. Gurney and P. W. Rowe, Royal Aircraft Establishment, Farnsborough, Report No. Mat. 5, May, 1945; and Rep. No. Mat. 8, November, 1945.

were made in all. In the first a copper sheath was used 0.016 in. thick over the junction between steel and glass, turned down to a thickness of 0.005 in. over the central part of the glass. No extravagant precautions were taken with the glass, which was 0.44 in. long and 0.12 in. in diameter. The lateral surface was the natural surface of the rod as provided commercially. The ends were lapped by hand with a fine carborundum stone perpendicular to the axis. Care was taken not to chip the edges. This specimen was exposed to a pressure of 385,000 psi, and then a superposed tension added up to 270,000 psi, when separation occurred between glass and one steel pull piece. The glass was undamaged, as was found by dissolving off the copper sheath in acid. The pull pieces were found to be too soft, and under this tension received a permanent elongation of 1.5 per cent. This flow of the steel may have been responsible for the separation. The glass received no measurable permanent change of dimensions.

In the next experiment, copper ferrules were used over the surface of separation of glass and steel, but the central part of the glass was covered with a soft neoprene tube, extending over the copper ferrules. At a hydrostatic pressure of 355,000 psi and a superposed tension of 84,000 psi, making the net compressive stress in the glass 271,000 psi, this failed by a tensile break at the edge of the copper ferrule. On removing the ferrules from the glass, the glass fell apart into several disks separated from each other by clean tensile breaks. In this experiment the edges of the ferrules were comparatively thick, and it seemed probable that the fracture was in some way connected with a discontinuity of stress at the edge. The experiment was accordingly repeated, but with the edges of the copper ferrules feathered down to vanishing thickness. At a hydrostatic pressure of 360,000 psi and a superposed tension of 161,000 psi failure occurred at the interface. The entire central portion of the glass, where it had been exposed to the neoprene, was separated into thin disks, still hanging lightly together. In the fourth experiment, copper ferrules again guarded the surface of separation of glass and pull pieces, and the central part of the glass was covered with a heavy lead sheath, soldered to the copper. The central part of the glass was necked down by grinding and polishing from 0.120 to 0.106 in. in order to concentrate the tensile stress at the center and locate the fracture. Under a hydrostatic pressure of 396,000 psi this broke in clean tensile fracture across the smallest diameter of the neck under an average superposed tensile stress at the neck of 220,000 psi. Tensile fracture, therefore, occurred here against a net compressive stress of 176,000 psi. The lead was melted off, and the copper dissolved off in aqua regia. An additional tensile break appeared on thus stripping the specimen, and the entire mass of the remaining glass was copiously fissured with fine disking cracks, the glass still hanging lightly together. In the fifth experiment, the glass was necked down at the center from 0.120 to 0.080 in. by grinding, polishing, and fire polishing to remove the last trace of mechanical cracks. The glass, together with the pull pieces, was mounted in a heavy straight copper sheath 0.21 in. in diameter, which was then collapsed tightly around the neck in the glass by a preliminary exposure to 110,000 psi, and then turned down to a thickness of 0.005 in. over the neck. Under a hydrostatic pressure of 380,000 psi, this failed by a clean tensile break across the neck at an average superposed tension at the neck, corrected for the strength of the copper, of 345,000 psi. This makes fracture to have occurred at a net compressive stress of 35,000 psi.

Carboloy. Tests were made on carboloy of grade 999, which is the grade containing the smallest quantity of binder furnished by the General Electric Co. To increase the stress the specimen was necked by grinding with diamond. Two unsuccessful preliminary experiments were made in which the strength of the carboloy was not estimated high enough and the neck diameter accordingly was not made small enough. In two final experiments the area at the neck was made somewhat less than one-quarter the area at the ends. In the two final experiments hydrostatic pressure was raised to 380,000 psi. In the first of the two, there was a leak at the maximum load, suddenly throwing an uncontrolled and unmeasured increment of tension on the sample, under which it fractured. The best estimate of the superposed tensile load at the neck at fracture was somewhere between 800,000 and 850,000 psi. In the second, there was no mishap and tensile fracture occurred at a superposed tension at the neck of 770,000 psi. In both cases the break was a clean fine-grained tensile break square across the narrowest part of the neck, with no evidence whatever of any plastic flow in the neck, and no trace of the little shearing rim around the edge of the break that is almost always found in hardened steel.

The tensile strength under 380,000 psi is between two and three times as high as any value that has ever been recorded for any grade of carboloy at atmospheric pressure by the method of the bending of slender rods. It was a surprise that there was no evidence of plasticity. Under one-sided compressive stress this grade of carboloy premits plastic shortening up to at least 10 per cent under this pressure.

Beryllium. Beryllium as ordinarily obtained commercially, even in the purest form, breaks at atmospheric pressure in tension with no appreciable elongation and with tensile strength not more than 28,000 psi. Being a crystalline material it was anticipated that under hydrostatic pressure it might exhibit a degree of plasticity. The first experiment with beryllium, which was kindly furnished by Professor John

Chipman of MIT, employed steel pull pieces, by the same technique as used for glass and carboloy. The specimen was necked to 0.7 of the full diameter, and mounted in a copper sheath, collapsed around it as in the last experiment on pyrex. Under a pressure of 390,000 psi this broke at the neck with a further reduction of area at the neck of 48 per cent, thus showing the anticipated ductility under the action of pressure. The neck at the break was very jagged, and the contour was finely corrugated around the periphery in an unusual way. The superposed tensile stress at fracture was 110,000 psi, on the original area of the neck. Two other experiments were made with beryllium, in which the conventional shaped tensile specimens were used, like Fig. 58, entirely fashioned out of a single bar. Mr. Charles Chase skillfully found how to fashion the beryllium rod with a carboloy tool. In these two experiments the beryllium was directly exposed to the action of the pressure-transmitting liquid, a mixture of isopentane and commercial "pentane." In the first experiment the specimen pulled apart at the shoulder under a hydrostatic pressure of 325,000 psi. This was evidently the effect of stress concentration. There was, however, measurable extension in the body of the specimen. In the second specimen, the fillet at the shoulder was given a larger radius and the center of the specimen was further necked down to localize the break. Under a pressure of 250,000 psi, this broke across the center of the neck with a coarse granular break, with a 20.6 per cent reduction of area, and a maximum superposed tensile stress of 76,000 psi calculated on the original area.

There would seem to be no reason to think that the transmitting liquid exercises any effect in the case of this material. It would appear that increasing hydrostatic pressure progressively raises the ductility and the tensile strength.

Phosphor Bronze. For commercial welding a bronze is used consisting of 93 per cent copper and 7 per cent phosphorus. Under atmospheric conditions this material as ordinarily supplied in the cast condition is coarsely crystalline and completely brittle. It was of interest to find whether ductility is imparted by pressure. Three experiments were made. The specimens were made in one piece from the commercial cast rod and were exposed to direct contact with the pressure liquid. The first two of these were not pulled to fracture, the ductility proving to be greater than anticipated, so that the dimensions originally given did not permit extension to fracture. The first of these was pulled under 310,000 psi to a reduction of area of 12.6 per cent, and the second under 410,000 psi to 18.5 per cent reduction. The elongation of both these specimens was uniform, necking not having yet started. The maximum superposed tensile load on the second specimen was 123,000 psi. In the third experiment the original specimen was turned down farther over the central portion, to locate the break. Under a hydrostatic pressure of 420,000 psi this broke with an 80.2 per cent reduction of area. The break was a burnished shearing break at approximately 45° across the narrowest part of the neck. The load at fracture was not obtained.

A control run on a similarly shaped specimen at atmospheric pressure yielded, as expected, a completely brittle fracture at the bottom of one fillet.

Al_2O_3 (*Synthetic Sapphire*). The Linde Air Products Co. has developed a method of making single-crystal rods of Al_2O_3, and I am indebted to them for providing me with a number of specimens for my tests. The material as it ordinarily comes is not so perfect as the crystals formed in nature, but usually there are imperfections visible to a low-power glass in the form of bubbles scattered through the interior. This sort of material exhibits capricious behavior. Furthermore, the behavior is sensitive to the surface conditions. Two specimens enclosed in a neoprene sheath under a pressure of 350,000 psi or more broke with a completely brittle fracture at superposed tensile stresses of 70,000 and 130,000 psi. The first of these specimens was artificially necked for the experiment; the break did not occur at the neck, where the stress was 145,000 psi, showing the inhomogeneity of the material. When enclosed in a copper sheath, this material did not break at all, but failed by slip along well-defined slip planes, without loss of cohesion. The slip was produced by a superposed tension of 125,000 and 140,000 psi for the two specimens.

The behavior of material carefully selected for freedom from internal bubbles is quite different. Two specimens were tested in tension, in a copper sheath. Both of these withstood the maximum tensile pull possible, which was raised to the value of the hydrostatic pressure, when the specimen separated on the surface between sapphire and steel as in the method for calibrating the grid on page 42. In one of these experiments the angle between the crystal axis and the length was 0°, and the hydrostatic pressure and accompanying tension 400,000 psi. In the second example the angle between the crystal axis and the length was 85° and the pressure and accompanying tension 380,000 psi. In neither of these cases was there any evidence of slip or measurable permanent change of dimensions. It would be desirable to continue these experiments on necked specimens instead of straight specimens in order to realize higher tensile stresses. However, in view of the sensitiveness to surface conditions, the fashioning of the neck would probably offer difficulties.

These experiments on Al_2O_3 are here described for the first time. They offer a somewhat different picture from the one that would be drawn from the published experiments in the *Journal of Applied Physics* (**18**, 246, 1947), which were performed on imperfect material.

Pipestone (*Catlinite*). This material, an iron-rich clay stone, has a typical composition: SiO_2 57.4, Al_2O_3 25.9, Fe_2O_3 8.7, H_2O 7.4, MgO and CrO, trace. Mechanically it seems to be an exceedingly fine cemented aggregate. I have used it extensively for the insulation of leads into the pressure chamber, and in this usage it is capable of supporting high compressive and shearing loads without appreciable flow. When it does yield in this usage it disintegrates to a fine powder.

Two experiments were made on pipestone. In the first, described in the *Journal of Applied Physics* (**18**, 246, 1947), a completely brittle fracture was found in a copper sheath under a pressure of 395,000 psi at a superposed tensile load of only 6,000 psi. This, if the argument on page 109 is valid, is an impossible state of affairs if the copper sheath retains its integrity. However, the appearance of the sheath did not suggest failure, and the results were described for what they were worth. The experiment was later repeated, and is here reported for the first time. The expected sort of behavior was now found. Under a hydrostatic pressure of 425,000 psi a piece of pipestone in a copper sheath was pulled without fracture to a 17.5 per cent reduction of area with a superposed tensile stress of 170,000 psi. This normally so brittle and weak material therefore under pressure acquires ductility and considerable tensile strength. The results on the first specimen must have been due either to an undetected surface flaw, which permitted the intrusion of the copper, or else to a failure of the copper sheath in spite of its appearance.

NaCl (*Rock Salt*). Under normal conditions at atmospheric pressure rock salt breaks brittlely in tension on one of its numerous cleavage planes. It is well known, however, that under special conditions, as when properly supported or when in aqueous solution, it may be made to support plastic deformation. Two tests were made of the effect of hydrostatic pressure. Both specimens were cut from the same natural crystal, with one of the natural cleavage planes perpendicular to the length. They were used in the form of straight cylinders, mounted between steel pull pieces. The first was used with a copper sheath 0.007 in. thick over the central part, and the second with copper ferrules and lead over the central portion. The first was pulled under a hydrostatic pressure of 420,000 psi to a reduction of area of 20 per cent under a superposed "true" tension as calculated on the final diameter, of 7,300 psi. At the necked part there was no loss of optical homogeneity—no evident slip planes or other evidence of crystal structure, and the cross section remained round. The second specimen, with lead sheath, was pulled under 410,000 psi pressure. The experiment was terminated by leak, which prevented the measurement of the tension, but the specimen necked down to a reduction of area of 14.6 per cent. Again the specimen retained its perfectly homogeneous appearance.

Voigt[1] experimented on the effect of gas pressures up to 850 psi exerted by CO_2 on the tensile strength of NaCl. He found that up to this pressure the superposed tensile load required to fracture was independent of pressure and that the fracture continued to be a clean brittle break with no trace of plastic flow. Break occurred on the cleavage planes, and the superposed stress to cause fracture was 810 psi. Evidently his pressures were not high enough to bring about the plasticity which is exhibited at higher pressures. It is to be noted that Voigt's tensile strength was only one-tenth of the stress which at high pressure produced only flow without fracture.

Solenhofen Limestone. This material is normally completely brittle. Measurements were made on only a single specimen. Balsley[2] has already reported the plastic deformation of limestone in tension under hydrostatic pressure and has made a number of measurements, but at a considerably lower pressure.

A single straight specimen, 0.438 in. long and 0.103 in. in diameter, was pulled in the conventional way between steel pull pieces, in a copper sheath. Pulled under a pressure of 400,000 psi, this necked down before fracture to a 53 per cent reduction of area (natural strain 0.76). The fracture was on a single shear plane completely across the specimen inclined at 57° to the length. The true stress at the maximum load (beginning of necking) was 210,000 psi. The load at fracture was not obtained, fracture being without warning. Necking started at a reduction of area of 21.5 per cent.

Cast Iron. I have already published the statement in several places that in several experiments cast iron remained completely brittle under pressure. These experiments were all made on naked specimens directly exposed to the pressure liquid. If, however, the cast iron is sheathed in copper to prevent contact with the liquid, the result is entirely different. Two experiments were made on straight cylindrical specimens of gray cast iron from the stock of the laboratory machine shop. The first of these was pulled without fracture under a pressure of 420,000 psi to a strain of 0.60 (45 per cent reduction of area) with a "true stress" at the neck of 203,000 psi. The second specimen was pulled at a pressure of 425,000 psi to fracture at a strain of 1.77 (83 per cent reduction of area). Fracture occurred on a single shear plane across the neck, inclined at approximately 45° to the length. The "true stress" at the neck at fracture was 480,000 psi.

[1] W. Voigt, *Göttinger Nachrichten*, 521, 1893.
[2] J. R. Balsley, *Trans. Am. Geophys. Union*, Part II, p. 519, 1941.

CHAPTER 5

SIMPLE COMPRESSION UNDER HYDROSTATIC PRESSURE[1]

Ductile Materials. The only ductile material for which experiments of this type have been made is steel, and this was only a single specimen of an armor-plate steel. The work was done under the NDRC contract and was started toward the end of the war. The contract was terminated after the apparatus and method had been developed when measurements on only one steel had been completed.

The method is simpler than for the tension tests. Because of the geometrical differences between tension and compression, there being no concentration of strain by necking in compression, the strains accessible with the form of apparatus used were much less for compression than for tension. The specimen is in the form of a cylindrical block 0.15 in. in diameter and 0.20 in. long. It is mounted in a holder between carboloy blocks, the holder occupying the same position in the pressure vessel as did the tensile samples. Compression is applied to the specimen by direct contact through spacer blocks with the piston by which the hydrostatic pressure is generated, in the same way as in the tension experiments. The numerical value of the one-sided compression is given by the excess of the change of resistance of the grid over that due to hydrostatic pressure only, exactly as in the tension experiments. The amount of distortion of the specimen during compression is given by the motion of the piston. The various corrections to the motion of the piston are determined by blank runs and by comparison with the total permanent shortening measured after removal of the specimen from the pressure vessel. The specimen was mounted in an easily deformable lead centering device. The ends were lubricated with colloidal graphite and a sheet of lead foil 0.002 in. thick where they come in contact with the carboloy. The lubrication was effective in avoiding barreling and keeping the diameter constant during the upsetting process within a fraction of 1 per cent for a shortening of 10 per cent.

The armor-plate steel, on which the single series of measurements was made, was designated as APL 22. It has a Brinell hardness of 234. At atmospheric pressure it fractures in tension with a reduction of area of 58 per cent and under 350,000 psi pressure with 86 per cent reduction.

[1] This chapter is based on material contained in the sixth NDRC report and B13, B24, and B26.

The flow stress in tension increases by a factor of 2.17 over the pressure range. The "tensile strength" at atmospheric pressure is 132,000 psi, and the "true stress" at fracture under pressure 315,000 psi.

Under simple compression it was found that, in order to produce a plastic shortening of 10 per cent under a pressure of 170,000 psi, an increase in the stress of simple compression of something of the order of 5 per cent was required in comparison with the simple compressive stress required to produce the same shortening at atmospheric pressure. This 5 per cent excess of stress was approximately independent of the amount of shortening in the range between 0 and 10 per cent. The increase is thus of the same general order as the increase in the elastic limit or the tensile strength in tension.

Brittle Materials. Many more measurements were made on brittle materials, the results having an ulterior interest in the early days of my experiments to 100,000 atm in the search for the best piston material.

Glass (Pyrex). Two experiments were made on glass of the same origin as that of the tension tests. The first specimen was 0.235 in. long and 0.125 in. in diameter. It was compressed in the regular way between carboloy platens, after soldering into a lead sheath to prevent contact with the pressure liquid. Compressive load was applied at a hydrostatic pressure of 400,000 psi. At an additional compressive load of 270,000 psi the capacity of the grid to measure compressive stress was reached, and

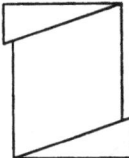

FIG. 60. Manner of failure of a cylinder of pyrex glass in simple compression when supported by hydrostatic pressure.

the experiment had to be discontinued. The glass was found undamaged except for a slight chipping around one edge. In the second experiment, the specimen was made smaller, 0.131 in. long and 0.069 in. in diameter. The method of mounting was as before. Compressive load was applied under a hydrostatic pressure of 400,000 psi. The specimen failed under a superposed compressive load of 670,000 psi, calculated on the original area. The lead sheath maintained the fragments in approximate position; when the sheath was removed the specimen was found with the appearance suggested in Fig. 60. Two wedge-shaped regions on the ends had slipped sidewise on planes at an angle with the axis of about 65°; the material in these wedges was completely comminuted to a fine powder, opaque and white. The central portion was filled with conchoidal fracture surfaces; the fragments were much larger than at the ends and were transparent.

It is to be noticed that the effect of hydrostatic pressure is markedly greater in increasing the compressive strength of glass than in increasing its tensile strength.

Al₂O₃ (*Synthetic Sapphire*). As in the case of the tension tests the results depended to a great degree on the nature of the material. I had found a number of years ago[1] that a specimen of sapphire cut from a large boule made in the standard way exhibited a crushing strength of 1,000,000 psi when immersed in a liquid at 350,000 psi pressure. Later, experiments were made on a number of single crystals from small rods prepared by the Linde Air Products Co. by their method. The specimens which were not specially selected gave, as before, capricious results. One of these speci-

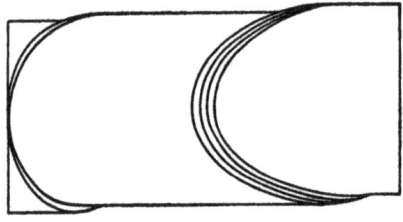

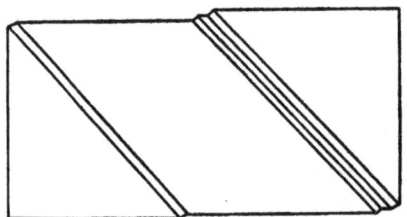

FIG. 61. The slip planes developed in a single crystal of Al₂O₃ under simple compressive stress when supported by hydrostatic pressure. There were minute internal flaws in this material before exposure to stress.

mens showed slow creep under compression, with slip along definite planes, not the same as the normal cleavage plane, as indicated in Fig. 61, at a pressure of 340,000 psi. Four specimens of the Linde single crystal, carefully selected for freedom from internal bubbles, gave more consistent and higher results. Two specimens tested at atmospheric pressure failed with a completely brittle fracture. One, with crystal axis at 84° to the length, failed at 300,000 psi simple compressive load. Another, with crystal axis at 7° to the length, failed at 600,000 psi compressive load. Another specimen with crystal axis at 87° to the length, when subjected to a hydrostatic pressure of 350,000 psi failed by twinning at a compressive stress of 750,000 psi. A photograph of the specimen distorted by twinning is shown in Fig. 62. Even after distortion by twinning a total compressive load of 500,000 psi was supported. A second specimen, with crystal axis at 0° to the length, when supported by a medium with pressure of 390,000 psi, failed under a simple compressive stress of 1,300,000 psi. The failure was by slip along a shear plane inclined at 36.5° to the longitudinal axis, as indicated in Fig. 63. Except for the sheared-off wedge, the rest of the specimen was coherent. It was an opaque white, with many irregular fracture surfaces at haphazard angles, superposed on many regularly spaced disking cleavages. In these two last experiments direct contact between the sapphire and the pressure liquid was prevented by embedding the specimen in a lead sheath.

[1] *J. Applied Phys.*, **12**, 469, 1941.

Sintered Carbides. A number of observations have been made on these materials incidentally in the course of explorations for materials best suited for the construction of high-pressure vessels. It has already been stated that carboloy (tungsten carbide cemented with cobalt) will support plastic shortening in simple compression when supported by high hydro-

FIG. 62. Twinning distortion induced by simple compressive stress in a single crystal of Al_2O_3 when supported by hydrostatic pressure. There were no obvious internal flaws in the virgin material.

static pressure. The pistons of the piezometers used in measuring compressions to 100,000 atm are made of carboloy, and because of this, considerable work has been done in examining the behavior of the piston material. In general the grades of carboloy containing small percentages of cobalt are best suited for pistons. I have used the grades known as 905, and lately exclusively 999. Shortenings in simple compression up to 10 per cent have been observed at hydrostatic pressures around 425,000 psi. Carboloy does not work-harden when it yields plastically in compression, but the shortening continues proportionally to time until terminated by fracture. Compressive stresses as high as 2,100,000 psi have been obtained in carboloy for a short time. The rate of yield under these conditions is, however, so high as to make the use for pistons infeasible. My pistons in the compression measurements have seldom been carried above 1,560,000 psi, and more usually are not carried above 1,490,000 psi. Under these stresses the total plastic shortening of the pistons is something of the order of one-quarter of 1 per cent for a duration of maximum stress of 15 min. There is considerable variation in piston material of ostensibly the same composition, and it pays to make a preliminary examination in order to pick out the best pieces.

The results for carboloy appear to be the same whether it is directly exposed to the pressure liquid or sheathed with lead, copper, or bismuth.

It is not inconceivable that materials other than carboloy would be better adapted for standing high compressive stress when supported by hydrostatic pressure, even though the material might not be so well adapted for ordinary conditions of use. The natural direction in which to seek for improvement is in diminishing the amount of binder, since it would appear that the plastic flow takes place in the binder. A cursory examination was made of some of the more obvious or easy possibilities, but with no improvement. In fact, none of the materials tried approached carboloy in performance; doubtless we went too far in the direction of minimizing the binder. In this exploration I was fortunate to obtain the cooperation of Mr. deWald, who has specialized in the subject, and who prepared the specimens for me at MIT, where he enjoyed the facilities of the Department of Metallurgy. The following materials were tried with the following results.

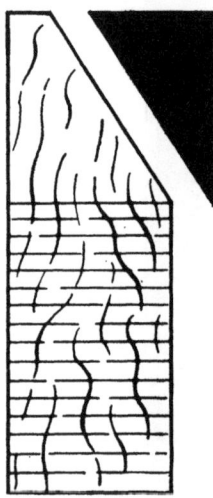

FIG. 63. Manner of failure under a simple compressive stress of 1,300,000 psi of another specimen of single crystal Al_2O_3 when supported by a pressure of 390,000 psi.

1. A mixture of 97 per cent TaC, 2 per cent VaC, and 1 per cent Mo_2C with no binder. Under a pressure of 325,000 psi this failed with a completely brittle fracture under a superposed compressive stress of 510,000 psi.

2. WC with 0.25 per cent Co as binder. Under 350,000 psi this failed brittlely under 675,000 psi superposed compressive stress.

3. TaC with 0.25 per cent Ni as binder. Under 350,000 psi hydrostatic

pressure this failed brittlely under 435,000 psi superposed compressive stress.

There is room for much more work on this subject. In particular, systematic investigation should be made of the effect of varying the grain size and the time and temperature of sintering.

Possible Piston Material. We may now complete this discussion of compressive strength under hydrostatic pressure by describing various exploratory investigations made for the purpose of finding the best material for the apparatus for 100,000 atm, an apparatus which functions immersed in a liquid at 425,000 psi.

Steel was naturally investigated first of all for piston material. There is an increase of strength under hydrostatic pressure, but not enough to give any promise. Thus a piston of glass-hard Teton steel was upset and permanently shortened by 9.2 per cent under 680,000 psi one-sided compressive stress under a hydrostatic pressure of 360,000 psi, against a normal strength of approximately 425,000 psi. The increase is by less than the confining pressure. The possibility of plastic deformation is noticeably enhanced by the confining pressure. However, under normal conditions this steel even in the glass-hard condition will tolerate a permanent shortening of from 2 to 3 per cent before fracture. Another grade of steel gave on a single test figures perhaps 150,000 psi higher than the Teton steel, but there were experimental uncertainties. Other arrangements were used with short bosses of steel, but there was never sufficient increase of strength to suggest that any steel would become a serious competitor of carboloy.

Quartz was naturally tried because of the great increase of strength under confining pressure first found by Griggs and Bell,[1] and stated in a first note of mine to have been verified by me. Measurements under improved conditions did not give nearly such high figures as at first reported. In explanation of the original high values it is to be remarked that the experiments of Griggs were made under the same conditions as mine, that is, with a small apparatus in which stress was exerted entirely by a soft metal with no true liquid. Six new tests were now made with quartz confined with a true liquid; in four of these the stress was measured with the grid. Specimens of a single crystal of quartz, cut parallel to the c axis, gave for the compressive strength 570,000 psi under 360,000 psi hydrostatic pressure, 550,000 psi strength with 330,000 pressure, and 470,000 strength with 230,000 pressure. The fracture in all cases was complete, with reduction to a fine powder and no evidence of plastic deformation. Sosman in International Critical Tables reports the strength at atmospheric pressure to be 340,000. Again it is evident that

[1] D. Griggs and J. B. Bell, *Bull. Geol. Soc. Am.*, **49**, 1723–1746, 1938.

there is no such exponential increase of strength with hydrostatic pressure as has been suggested; in fact the increase seems to be linear within the error of the measurements.

Quartz glass does not behave markedly differently from the crystal; the compressive strength of one sample under 360,000 psi hydrostatic pressure was 560,000 psi.

Other minerals are known to be considerably stronger than quartz under normal conditions, and accordingly small cylinders of the most promising of these were tested to destruction. Under a hydrostatic pressure of 360,000 the compressive strength of a spinel was found to be 580,000 psi, of sapphire 1,000,000 psi, and of tourmaline 1,260,000. In all cases fracture was complete, to a fine powder. Thus none of these minerals is equal to carboloy.

Finally, in order to proceed to the logical conclusion and be sure that nothing had been missed, a small cylinder of clear diamond, 0.060 in. in diameter and 0.090 in. long, was fractured in compression. The crushing strength under a hydrostatic pressure of 330,000 psi was 1,830,000 psi. Failure was complete, to a fine powder. The carboloy anvils, with which the compressive stress was transmitted to the diamond, were marked with fine lines, indicating cracks in the diamond just before rupture, but there was no appreciable general indentation, showing that the plastic flow of the carboloy in the experiments with carboloy bosses does not begin noticeably at pressures below 1,800,000. It would thus appear that diamond may be superior to carboloy as piston material under hydrostatic pressure, but the superiority is too slight to justify the expense except perhaps for very special purposes.

General Discussion of Brittle Fracture. Perhaps this is as good a place as any to present certain general considerations with regard to the fracture of brittle substances, although it must be admitted that it interrupts somewhat the continuity of the exposition.

It seems to me that the most important result of the experiments on both the fracture of brittle substances in simple tension and in simple compression is to emphasize the reality of a condition of fracture which may often be disregarded. This is: fracture cannot occur unless the process of fracture is an energy-releasing process. The reason that in the idealized experiment with which we started our exposition tensile fracture could not occur until the superposed tension reached a value at least equal to the ambient hydrostatic pressure was merely that fracture under these conditions was not an energy-releasing process. For, if fracture were to occur, the work done against the hydrostatic pressure would be greater than the work received from the force producing the tension. The fact that fracture occurs in the actual experiment under tensions less than the hydrostatic pressure, that is, against a net compressive stress, is a consequence of the difference between the ideal and the actual case. This difference is in the lateral conditions. In the ideal case, there was an infinitely rigid sheath shielding the lateral surface from entry by the pressure liquid if a crack should appear. In the actual case, the sheath is not infinitely rigid, and if the external pressure is high enough, the sheath may be forced into incipient cracks, and this is an energy-releasing process. According to this view, tensile fracture of a brittle substance in a medium carrying hydrostatic pressure starts on the external surface. This view is confirmed by the extreme sensitiveness of tensile fracture under these conditions to the surface conditions. Thus in the case of glass, we may have fracture by the "pinching-off" effect at a pressure of 21,000 psi if the external medium is water, and at a pressure of 42,000 or 57,000 if the medium is oil, or otherwise expressed, fracture at superposed tensions of 21,000 and 42,000 or 57,000, respectively. But, if the external medium is neoprene or lead or copper, we have seen that tensile fracture occurs at superposed tensions of 84,000, 220,000, and 350,000 psi, respectively. There is a specific effect of hydrostatic pressure concealed in this comparison, but this effect does not obscure the main result. The surface conditions at glass are doubtless complicated, and at least in the case of water involve a chemical factor as well as a purely mechanical factor, since it is known that water is forced by pressure into the surface layers of glass. Brittle substances thus differ essentially from the substances which flow plastically before yielding in tension, which have been made the subject of previous study. In the case of these ductile substances, fracture starts in the interior on the axis, where the tensile stress is a maximum, because of the stress redistribution produced by the necking itself, and always at values of the tensile stress materially higher than the ambient hydrostatic pressure. Fracture under these circumstances is again an energy-releasing process, because although work is done against the external pressure when a cavity appears in the inside, nevertheless the work received from the tensile stresses more than compensates because of their greater intensity. The fact that fracture starts at the inside explains the consistency of the results on fracture of ductile substances as contrasted with the capriciousness of the results for a brittle substance like glass.

From the point of view of energy release it is easy to understand the very much greater strength of glass to compressive stresses when supported by hydrostatic pressure than to tensile stresses. In the experiments above under a hydrostatic pressure of 380,000 or 400,000 psi the maximum superposed tensile stress supported was 350,000 psi (copper sheath), whereas a superposed compressive stress of 670,000 was observed.

When fracture occurs under tensile stress combined with hydrostatic pressure two processes occur, both energy-releasing. The force producing the tension does work, and the pressure does work, because when tension is released the volume of the specimen decreases. But when fracture occurs under one-sided compressive stress, only one of the processes is energy-releasing. The force producing the compressive stress again does work during the fracture, but, since the volume increases on release of compressive stress, work is done against the pressure by the fracture.

From the point of view of energy release there is no particular significance as to whether the superposed tensile stress under which fracture occurs is greater or less than the hydrostatic pressure, that is, whether the tensile fracture occurs against a net tensile or compressive stress. For in either case the process of fracture may be resolved into two components, both energy-releasing. The significant thing is the lateral conditions, and the extent to which the surrounding hydrostatic pressure may follow through into any crack which appears, doing work in the process. In the case of ductile materials, when a crack appears on the surface the propagation is by a process of slip, which is not energy-releasing for the pressure, so that the fracture does not spread.

Both the strong dependence on the surface conditions in brittle fracture under hydrostatic pressure and the condition of energy release emphasize that there can be no single criterion of fracture in terms of stress and strain at a point only.

The actual experimental conditions differ from those of the idealized experiment in another important respect than the rigidity of the sheath, namely, in the molecular structure of the specimen under test. In the idealized experiment, the material was thought of as structureless, and capable of indefinite and microscopically homogeneous distortion, as in the mathematical equations of elasticity theory. Actually, the material is composed of molecules, which have a certain impenetrability preventing indefinite approach, so that when strain is pushed too far individual molecules must be forced out of their positions by a series of discrete operations. It is doubtless something of this sort that is responsible for the remarkable separation of the glass tension specimens into thin disks, often, but not invariably, observed after simultaneous exposure of glass to pressure and tension. As tension increases, lateral contraction increases, and there must ultimately come a time when some of the molecules are brought into such close contact that they are forced out of their normal positions into neighboring spaces that have been opened by the simultaneous longitudinal extension. This process of molecular transfer is irreversible. When stress is released these displaced molecules act like internal wedges, distending the structure, which eventually fractures as in tension when stress is reduced far enough. A similar sort of thing has been found in other situations; fracture on release of stress would appear not to be uncommon. A somewhat similar state of affairs may be brought about in glass if foreign molecules are forced into the interior at the surface. Thus disking was almost always found in glass after a failure by the pinching-off effect in water as the transmitting medium.

The question has been much discussed of the great discrepancy between the experimental values of fracture stress and those calculated theoretically from various points of view. These considerations may have some bearing on this question. Strength in tension for a brittle substance we have seen to be an indeterminate matter, depending on the surface conditions. Under hydrostatic pressure breaking strengths in tension may be expected, with proper surface conditions, up to at least the magnitudes of the hydrostatic pressure. With pressures of the magnitude reached in these experiments, this is bringing the breaking stress up within reach of the theoretical values. In compression, dependence on the surface conditions is not to be anticipated, because the surface action that would take place at fracture is not energy-releasing. Very large values of compressive strength can be reached, as, for example, 620,000 psi in pyrex glass against some 25,000 under ordinary conditions, or an increase of some 25-fold, again getting within sight of theoretical expectations.

It seems probable that the increases in compressive strength brought about by hydrostatic pressure are much greater for glasslike substances than for crystals. This is caused by the interlocking of the molecules in a glass, which is accentuated by pressure, making relative displacement more difficult. The effect is analogous to the very large increase of viscosity under pressure observed in some liquids with complicated molecules. In a crystal, on the other hand, the regular structure provides the possibility of internal slip. The plastic distortion of the crystal of Al_2O_3 is a case in point. The behavior of a crystal under compression is doubtless a highly specific property of the kind of crystal; I have never been able to produce measurable flow in crystalline quartz, although experiments on the collapsing of negative crystals show that some flow must be present.

Certain of the points emphasized here recall the point of view of Poncelet in a paper in Vol. VI of Alexander's "Colloidal Chemistry." Poncelet regards brittle fracture in tension as orginating at the surface and also emphasizes the role played by the "particulate" structure of ordinary matter as distinguished from the homogeneous isotropic structure assumed in the mathematical theory of elasticity.

Rowe and Gurney[1] have concluded from their experiments on the fracture of glass under "radial" pressure (my "pinching-off" effect) that Griffith's theory of fracture is essentially correct. According to this theory an actual specimen of glass is filled with minute crevices, with corners so sharp that under ordinary tensile load there may be stress concentrations at the corners by as much as a factor of 600. The breaking strength of geometrically perfect glass would thus be of the order of 6,000,000 psi, which is of the order of the theoretical value, whereas the strength of an actual piece of glass is of the order of only 10,000 psi. Under radial pressure fracture starts at the surface cracks. Rowe and Gurney have shown that, assuming stress concentrations of 600, failure under radial pressure would be expected to occur at a pressure very approximately equal numerically to the ordinary breaking stress in tension, and this does in fact approximately agree with their experimental results. I think it must be admitted that qualitatively the Griffith point of view has considerable to recommend it. The strong dependence of fracture on surface conditions, for example, is consistent with its picture.

Qualitatively, Griffith's picture would lead to the expectation of pressure coefficients of tensile strength of completely sheathed brittle material of the order of magnitude of unity, and this also agrees with the experimental results above, and again indicates a real value in Griffith's point of view. It would not be easy to make a more precise calculation of the exact coefficient to be expected for completely sheathed specimens, however, because of the complicated nature of the surface conditions, and various other oversimplifications discussed in the next paragraph.

I question, however, whether other considerations do not play a vital role and whether the Griffith's picture is not much oversimplified. For instance, the surface conditions are in fact complicated and involve chemical as well as purely mechanical factors, as shown by the very great effect of pressure media such as water or alcohol as contrasted with the oil used by Rowe and Gurney. Again, the ordinary concepts of stress and strain are carried down to a fineness of scale where there are stress concentrations of 600, whereas stress and strain are properly macroscopic concepts. At this scale of magnitude, the conditions of fracture are applied in the form that fracture occurs when the maximum tension reaches a critical value, irrespective of the other components of stress and independent of the strains. This condition of fracture does not seem to me to have any greater plausibility on the microscopic scale than on the macroscopic scale. The assumption of such a criterion amounts, among other things, to postulating that a pure hydrostatic pressure exerts

[1] C. Gurney and P. W. Rowe, Royal Aircraft Establishment, Farnsborough, Report No. Mat. 8, November, 1945.

no specific effect on the fracture properties of a substance. This is certainly not the case for ductile substances, as shown by my experiments on the enormous increase of the ductility of steel under pressure. The same thing is shown by the experiments of Rowe and Gurney themselves on plastics, which under their conditions of radial pressure tolerated an elongation before fracture 18 times greater than under normal conditions. These results can properly be described as an increase of ductility under pressure. If plastic materials have their properties thus drastically altered by hydrostatic pressure it would seem to be only reasonable to expect some effect also on brittle substances. This is strongly suggested by the great increase of strength to simple compressive stresses found above for glass under high pressure.

The high tensile strength acquired by carboloy under pressure is worthy of notice. The tensile stress at fracture is much higher than the surrounding hydrostatic pressure in the case of this material, so that the net stress at fracture is a tension. It is not necessary in this, or similar cases, therefore, to suppose that surface conditions play any role in the process, and the increase of strength may be a specific effect of pressure. The absolute value of the tensile strength, between 700,000 and 850,000 psi, is not far from that which has been observed in steel, pulled to the limit of ductility and strain-hardened under pressure. A possible explanation of the effect in carboloy is that it is associated with the cementing film of cobalt on the surface of each grain of carbide; the cobalt becomes ductile under pressure and strain-hardens in a way similar to steel.

A question that naturally presents itself in connection with these experiments is under what conditions does a substance lose its brittleness and become ductile under pressure? In order to deal with this question it is necessary to consider the concept of brittleness itself. It is not uncommon to speak of a brittle fracture, and in fact this usage has been followed in this book. Such usage, however, employs a specialization and extension of meaning, because "brittle" properly does not refer to the process of fracture, but to what comes before the fracture. If the substance fractures before it receives permanent set, as shown by a permanent change of dimensions on release of stress, then the substance is brittle; if the substance fractures after it has received permanent set, then it is plastic or ductile. So far as the act of fracture itself is concerned it may be characterized as a shearing fracture or a tensile fracture, but not as a brittle fracture. When a brittle fracture is spoken of, what is meant is the fracture of a brittle substance. But strictly, from this point of view, all fractures might be spoken of as brittle, for a plastic substance which has been work-hardened to the limit and then fractures is, when at the point of fracture, a brittle substance. There is thus a considerable verbal element in the concept of brittle. Furthermore, the concept is not sharp, because if measurements are increased in sensitiveness plastic flow may be discovered before fracture where formerly with rougher measurements none appeared. In practice all that can be meant is fracture with no noticeable preliminary distortion.

In answer to our question as to what substances lose their brittleness under pressure, it may now be said that as a rough qualitative matter a crystalline substance, particularly if it is cubic, is more likely to become measurably plastic under pressure than an amorphous substance like glass. The limits are not sharp, however, and there is enormous variation in the numerical magnitude of the effect, as illustrated by the variation from soft steel to beryllium. In fact, even in the case of quartz crystal or glass an infinitesimal amount of flow must exist in principle, as shown by the spontaneous breakage of these materials under some circumstances a long time after exposure to high pressure.

CHAPTER 6

BRINELL HARDNESS UNDER HYDROSTATIC PRESSURE[1]

The "Brinell" hardness is obtained from the diameter of the impression made by pressing a hardened steel ball with a known force against the flat surface of the material to be tested. The formula for calculating the hardness is

$$B = \frac{P}{\frac{1}{2}\pi D(D - \sqrt{D^2 - d^2})} \qquad (6\text{-}1)$$

where B is the Brinell hardness number, P is the load in kilograms, and D and d are the diameters of the sphere and of the indentation, respectively, in millimeters.

The Brinell hardness does not express directly any of the fundamental properties of the material, but it is of considerable technical interest in indicating the practical behavior of the material under certain conditions. It is apparently by far the most important single parameter determining the ballistic behavior of an armor plate under attack by a projectile. Because of this, the effect of pressure on the Brinell hardness of a number of armor-plate steels was measured at the beginning of the NDRC program.

The procedure was very similar to that for the tensile tests. The indentations were made with hardened-steel balls 0.250 in. in diameter. The samples for indentation were disks $7/16$ in. in diameter and 0.25 in. thick. They were cut from rods sawed out of the plates, cut to length, and finished by surface grinding on both plane faces. They were piled in a stack separated by the balls, which were centered in a suitable holder, and kept in line with an outer tube. A stack of the specimens was mounted in the pressure vessel, and one-sided compression, in addition to the hydrostatic pressure, was applied and measured on the grid exactly as in the tension experiments. The compressive load was increased to a predetermined maximum. Readings of the piston displacement were made as a matter of routine but were not used in the final calculations. After the termination of the pressure part of the experiment, the diameters of the impressions were measured and the hardness calculated according to the formula.

[1] This chapter is based on material contained in the first and second NDRC reports.

Comparison runs were made at atmospheric pressure. Although the dimensions of the specimens were not such as to give perfectly satisfactory values for the absolute hardness, they should allow sufficiently good comparison of the hardness under high pressure with that at atmospheric. Care was taken to keep the total load in the pressure run as nearly as possible equal to that in the comparison run at atmospheric pressure—a load in the neighborhood of 1,500 kg. At first, four or five different grades of plate were used in the same stack. Control of the conditions in the different runs was obtained by making the disks at the ends of the stacks of drill rod, all cut from the same piece. Agreement of the values of the hardness of the drill rod from run to run would furnish a check. The check so obtained was fairly good, but presently inconsistencies appeared which could mean only that the plates themselves were not entirely homogeneous. This was checked by measurements at atmospheric pressure of the Rockwell C hardness of a number of plates at points well scattered through the mass of the plate. Considerable variation was sometimes found. In order to avoid error from fluctuations within a single plate the course was later adopted of making the stacks of a number of disks taken from different parts of the same plate. In the earlier runs with many disks from the same plate the comparison disks of drill rod at the ends were dispensed with, and only four disks used altogether. With four disks, six impressions were obtained, two on the interior disks and one on each of the exterior disks. In the later experiments the use of the tool steel controls was resumed.

The experimental results are summarized in Table XI.

The hardnesses were corrected by multiplying the hardnesses as given directly from the diameters of the impressions by that factor which would reduce the hardness of the drill-rod controls to the average of the drill rod for all the pressure runs. The figures were improved on the average by applying this correction, but in at least two cases—APL 18 and CA 11—it would appear that there must have been sufficient inhomogeneity in the drill-rod controls to vitiate the correction. Of course, error from inhomogeneity would also affect the results based on only two measurements on each plate, so that too much significance must not be attached to the details of the variation of the pressure effect from plate to plate. In parentheses in the second column of Table XI are given the Brinell hardnesses of several of the plates as determined at the Navy with a regular Brinell machine. The differences are a composite effect of lack of homogeneity, small size of the present samples, and difference in the size of ball. The differences are probably not large enough to indicate significant errors.

The percentage increase of hardness produced by an increase of pres-

TABLE XI. SUMMARY OF MEASUREMENTS OF BRINELL HARDNESS OF ARMOR PLATE UNDER HYDROSTATIC PRESSURE

Designation of plate	Hardness at atmospheric pressure	Hardness under pressure	Pressure, psi	Total increase of hardness, per cent	Per cent increase for 100,000 psi
87207	249	276	357,000	10.9	3.1
APL 15	246	285	385,000	15.8	4.1
APL 18	258	268	352,000	3.8	1.1
APL 22	234	256	352.000	9.4	2.7
APL 32	243	264	370,000	8.7	2.4
APL 36	250	297	410,000	21.8	5.3
APL 42	289	331	387,000	14.4	3.7
CA 5	264 (262)	308	387,000	16.6	4.3
CA 7	280 (293)	319	343,000	14.0	4.1
CA 9	295 (286)	337	343,000	14.2	4.2
CA 10	295 (286)	355	341,000	20.3	5.9
CA 11	235 (263)	296	420,000	26.0	6.2
B 5	382 (375)	442	341,000	15.7	4.6
C 14	389 (387)	435	420,000	11.7	2.8
GG 125	241 (241)	280	365,000	16.2	4.4
58-7404	242 (255)	289	365,000	19.4	5.3

sure of 100,000 psi varies between 2.4 and 6.2, except for one low figure. The average for all 16 plates is 4.0. No significant correlation appears between the increase of hardness produced by pressure and the effect of pressure on the tensile properties as previously determined. Some sort of correlation might be expected. The number of observations on any one grade of plate is probably too small, and, furthermore, the samples are too small to lead to the expectation of any very consistent values for the hardness.

The conclusion would seem to be that, for grades of armor plate of the general hardness range measured here, and presumably for other comparable steels, the Brinell hardness may be expected to increase under a pressure of 100,000 psi by roughly 4 per cent. This increase is not large compared with the effect of pressure on ductility or strain-hardening. The increase is, however, of the same order of magnitude as the increase in the elastic limit or the "tensile strength" produced by pressure. Since the diameter of the Brinell impression is obviously connected with the strength of the material in the comparatively early stages of plastic flow, the general magnitude of the effect is what might be expected.

CHAPTER 7

PUNCHING UNDER PRESSURE[1]

Introduction. The punching process is obviously complicated, perhaps too complicated to justify study in order to obtain fundamental information about plastic flow. It is, however, of great practical interest, and particularly plays a large part in the penetration of armor by a projectile, the final stage in penetration often being the expulsion of a punching from the back of the plate. It therefore suggested itself that a study of the effect of hydrostatic pressure on punching might throw light on the problem of armor penetration. Accordingly a program of measurements was initiated for the NDRC and measurements completed on the punching of three heat-treatments of steels 2 and 4 at the single pressure of 215,000 psi, before termination of the contract. It appeared during the course of these measurements that they might have more than an immediately practical interest, because the deformation accompanying punching is predominantly an intense shearing in the narrow annular zone between punch and die. Punching under pressure therefore offers information about the effect of pressure on resistance to shearing deformation isolated from the other effects with which it is connected in pure tension. It would have been desirable if the shearing deformation could have been produced under more well-defined conditions, as perhaps by conducting torsion experiments under pressure. No feasible method of doing this presented itself, however. In particular, there would have been great technical difficulties in applying and measuring forces of torque through angles of many complete revolutions within the pressure chamber. A couple of attempts were made to develop other methods, but these did not appear promising, and the punching experiments were therefore continued. These should be of at least orienting values. The continuation comprised an extension of the measurements on the NDRC steels to higher pressures and also on the 9-0 steel of the Watertown Arsenal.

The Method. The apparatus is shown in Fig. 64. The specimen to be punched, shown shaded, is compressed tightly between an upper and a lower annular clamp in order to prevent warping when the punch penetrates. The disk to be punched is 0.050 or 0.030 in. thick, depending on its hardness. The two surfaces were ground to parallelism in a sur-

---

[1] This chapter is based on material contained in the sixth NDRC report and B22.

face grinder. The punch was of Teton tool steel, left glass-hard. The edge of the punch was sharp and the face flat. The cylindrical side was full diameter for a length of 0.019 in. and then relieved. The disk through which the punching is expelled was of the same Teton steel, with sharp edges, and tapered at a double angle of 1° to eliminate friction in the die on the expelled punching. The face diameter of the hole in the die was 0.1906 in. The cylindrical holder for punch, specimen, and die was of a Cr Mo steel, heat-treated to a Rockwell C hardness of 30. All parts were ground and fitted accurately together. The specimen in the holder was mounted in the pressure vessel in the same position as that taken previously by the tension specimens. The force driving the punch was measured with the same "grid" that was used to measure tensile load, the displacement of the punch was measured in the same way as the elongation of the tension specimens, and the whole technique of the experiment was very similar. Since the total motion of the punch from initial contact to complete penetration was less than the total elongation of the tensile specimens, the total change of hydrostatic pressure during the course of the punching was less than in the corresponding tension experiment, which is an advantage.

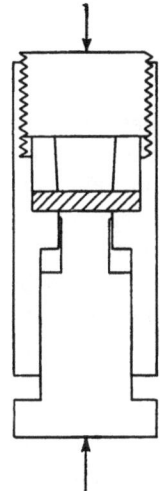

FIG. 64. The punch with specimen, shaded.

The measurements consist in simultaneous measurements of hydrostatic pressure, the force on the punch, and the displacement of the piston, the latter being measured externally from the displacement of the piston producing the hydrostatic pressure. The latter displacement includes, in addition to the actual penetration of the punch, the elastic distortions of all intermediate parts, including the punch itself, the transmitting parts within the pressure chamber, and the external piston. These elastic distortions were corrected out of the final result by making a sufficient number of readings before plastic flow began to establish the slope of the line of elastic distortion, and subtracting from the total displacement after plastic flow began an amount proportional to the total load and the slope.

The total load on the punch was converted into shearing stress per unit original area by dividing the total load, after correction for friction on the punch, by the area of the sheared region (product of thickness of plate and circumference of punch). This shearing stress on the original area was then converted to "true" shearing stress by multiplying by the ratio of the original thickness to the running thickness (initial thickness minus punch penetration). The "true" shearing stress was finally plotted against the penetration of the punch. The friction on the punch was computed from the readings at the end of the punching process, when the face of the punch had been driven completely through the plate, but the unrelieved sides of the punch were still in contact with the plate. This contact persisted for a displacement of 0.019 in.; during contact the driving force dropped proportionally to the distance still to go. The maximum correction for the friction of the punch was about 10 per cent.

It would have been desirable if the "true" shearing stress could have been plotted against the strain, so as to get a strain-hardening curve for pure shear. The conditions of plastic flow were not, however, well enough defined to permit any clean-cut evaluation of the strain. Ideally, the strain might be localized in the thin cylindrical shell included between the diameter of the punch and the diameter of the die as suggested in Fig. 65. Actually, however, the strain spread out into the surrounding region from this cylindrical shell in a way that would be most difficult to compute. The question was investigated by using punches of different diameters, thus varying the thickness of the cylindrical shell. If a true strain-hardening curve of stress against shearing strain could be obtained by the idealized calculation, then there should be a simple connection between the results obtained with the different punches, the true shearing stress for a certain penetration with one punch being the same as that at a penetration twice as great with a punch of twice the clearance, for example. Three punches were used, with clearances varying from 0.0040 to 0.0008 in. At the same penetration of the punch the force with the largest clearance was only 3 or 4 per cent less than with the smallest clearance. This shows that the plastic flow must be pretty well spread outside the idealized cylindrical shell and frustrates the attempt to express the results in terms of absolute strains. For this reason these experiments on punching have a more qualitative significance than the tension experiments.

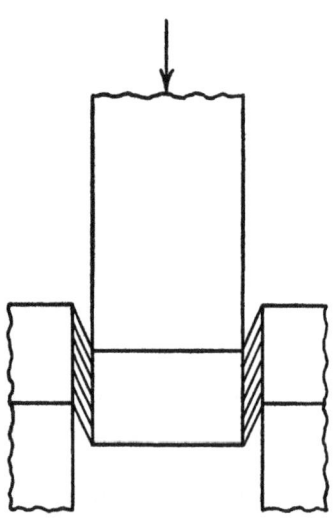

FIG. 65. Exaggerated and idealized zone of shear around the punch in process of penetration.

The Measurements. All the final measurements were made with the largest punch, with a clearance of 0.0008 in. Measurements were made on five grades of steel: 2-3, 2-6, 4-3, 4-6, and 9-0.

The qualitative nature of the effects is shown in Fig. 66, in which is plotted for steel 4-3 the shearing stress on the original area, which is proportional to the total driving force on the punch, against the corrected

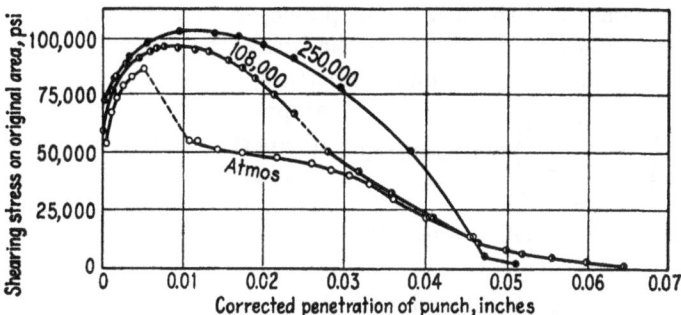

FIG. 66. The shearing stress on the initial peripheral area against penetration of the punch for steel 4-3, at several different pressures as indicated on the curves.

penetration of the punch. In this instance the material "breaks" at atmospheric pressure after a penetration of the punch to one-tenth the thickness of the plate. This is indicated by the dotted line in Fig. 66. The force jumps down, and the punch jumps in, the intermediate configurations being unstable and controlled by elastic distortion in the punch itself and the transmitting members. After the break, a force comparable with the maximum is still required to advance the punch, the force gradually dropping off as penetration becomes complete. The first effect of hydrostatic pressure is to suppress the "break," displacing it at the same time toward higher penetrations. In Fig. 66, at the next pressure above atmospheric, 110,000 psi, only a vestige of the break remains, and it has been displaced from a penetration of 0.005 to 0.025 in., the latter being one-half the thickness of the plate. At the next higher pressure, 250,000 psi, no vestige of the break remains.

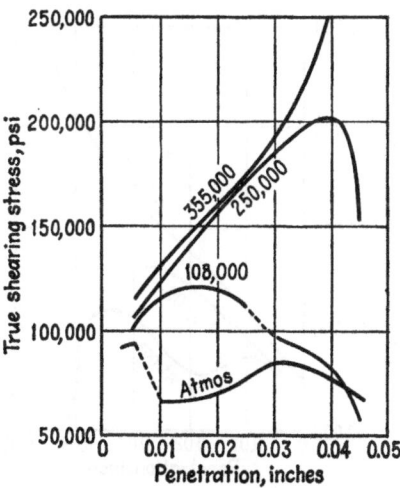

FIG. 67. The data of Fig. 66, reduced to "true" shearing stress, against penetration of the punch. The curve for the pressure 355,000 psi was not shown in Fig. 66, in order to avoid confusion.

On converting total loads into true shearing stresses and correcting for friction, Fig. 66 is converted into Fig. 67. The secondary rise in true stress at atmospheric pressure after passing the break is explained by the character of the punching, shown in Fig. 68. The punching is not truly cylindrical, but the exit diameter, in contact with the die, is slightly less than the diameter of the face of the punch, so that the punching has the shape of a truncated cone. After the break, the larger base of the cone has to be forced through a smaller aperture, like forcing a conical rubber stopper the "wrong" way through an aperture, thus accounting for the increase in force. The drop of force when expulsion nears completion is caused by distortion in the thin lip of the plate. With increase of hydrostatic pressure the truncated cone becomes a true cylinder, brightly burnished over its entire external surface, instead of being burnished only over the ring corresponding to the penetration of the punch before the break. The transition between the two conditions is gradual. In Fig. 67 the true stress at the first pressure above atmospheric, 110,000

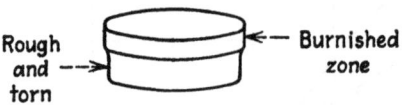

FIG. 68. Shows the general character of the punching.

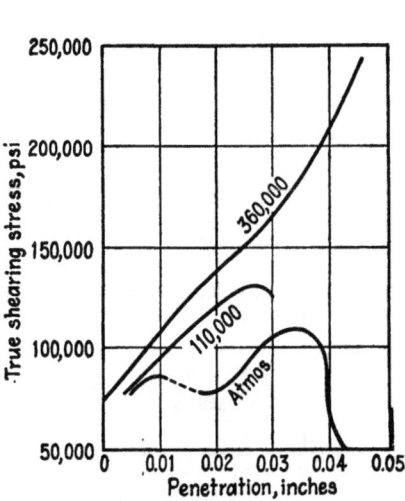

FIG. 69. "True" shearing stress against penetration of the punch at several pressures for the 9-0 steel. The heavy line at 0.05 in. indicates the total thickness of the plate.

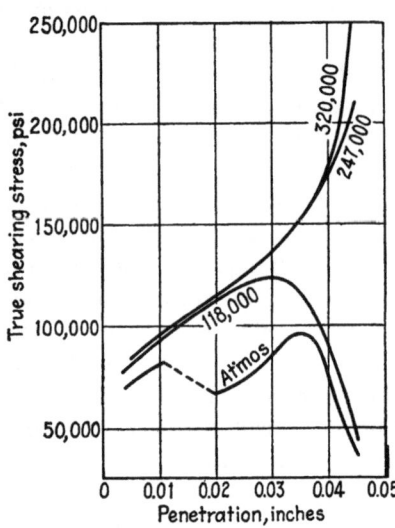

FIG. 70. "True" shearing stress against penetration of the punch at several pressures for steel 2-3.

psi, rises to a maximum and then drops slightly before the vestigial break occurs; that is, there has been some deterioration of the steel after a certain degree of penetration, but no wide-open break. At the next pressure, 250,000 psi, strain-hardening is approximately linear with the penetration up to a penetration to 0.8 the initial thickness. The drop beyond 0.8

represents deterioration, without actual fracture. The upper curve in Fig. 67 is for a pressure of 360,000 psi; the corresponding curve was not drawn in Fig. 66. At this pressure any deterioration is displaced so nearly to complete penetration as to be beyond experimental reach. The upward curvature at the end of this curve is probably real and corresponds to the fact that the average effective strain increases at an accelerated rate as penetration nears completion.

The 9-0 steel and the 2-3 steel are much similar. In Fig. 69 is shown true stress versus penetration for the 9-0 steel, and in Fig. 70 for steel

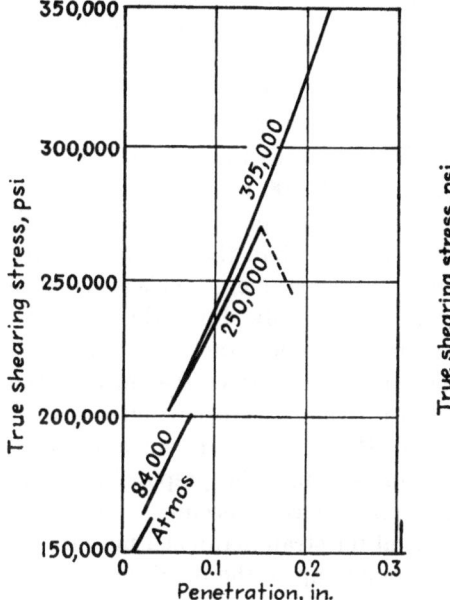

FIG. 71. "True" shearing stress against penetration of the punch at several pressures for steel 2-6. Plate thickness 0.0302 in.

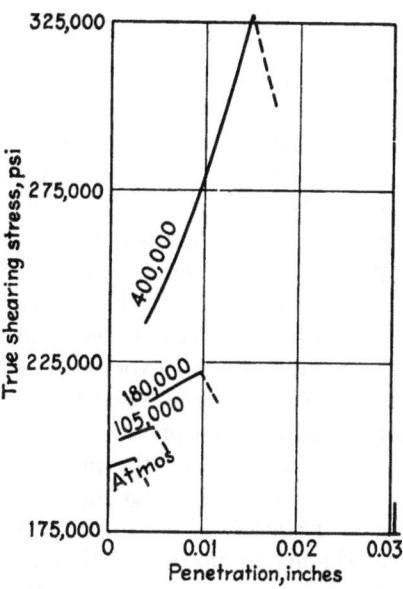

FIG. 72. "True" shearing stress against penetration of the punch at several pressures for steel 4-5. The initial thickness of the plate was 0.0302 in.

2-3. These steels are both softer than 4, and the curves for both show upward curvature at the deepest penetrations and the highest pressure. Toward the end of penetration the strain is increasing more rapidly than in the initial stages, because it has less chance to spread out from the idealized cylindrical shell, and the steels are soft enough not to show the deterioration at high penetrations of 4.

Figures 71 and 72 show the results for steels 2-6 and 4-6. The heat-treatment of these two specimens gives a much harder product, with higher resistance to punching and much less plastic yield when the break does occur. The effects of pressure are here in the same direction qualitatively as for the softer steels, but they do not go nearly as far. The fracture phenomena are not suppressed in the pressure range shown in the figures; however, the penetration of the punch before the break occurs is much increased by pressure. This agrees with the results previously found for tension, namely, that the phenomena for the harder steels are qualitatively the same as for the softer steels, but the scale of the effects is changed, a much higher pressure being required for the same plastic flow without fracture in harder steel as compared with a softer steel.

In order to demonstrate that the punching may be moved bodily through the surrounding metal under pressure without fracture, an experiment was made on the 9-0 steel in which the punch was driven 0.9 through the plate at a pressure of 325,000 psi, and then the punching operation was completed at atmospheric pressure. For comparison, another plate was punched at atmospheric pressure, which had been machined so as to have in it a depression of the same dimensions as that which had been driven into the plate by the punch under 325,000 psi. At atmospheric pressure the maximum force required to start the punch moving through the prepunched plate was 80 per cent greater than the maximum force supported by the geometrically similar virgin specimen. This is evidence of a considerable strain-hardening and incidentally shows that the prepunched plate could not have had many wide-open fractures, if any.

In another similar experiment the punch was pushed 0.4 through the plate at a pressure of 370,000 psi, and the punching operation was then completed at atmospheric pressure. In this case 42 per cent more force was required to start the punch moving through the prepunched plate than was needed for a geometrically similar virgin specimen. The prepunched plate behaved in the way usual for strain-hardened material in that the maximum penetration of the punch up to the break (calculated from the new zero) was only 0.38 of that for the virgin specimen. The true shearing stress at the maximum penetration under pressure was 126,000 psi; it required 104,000 true stress to make flow resume at atmospheric pressure. Some of the difference between the two figures may be caused by a loss of strain-hardening by an annealing effect on resting. However, if the total difference between these two figures is ascribed to a specific effect of hydrostatic pressure, it would mean a raising of the plastic flow stress by only 5.5 per cent per 100,000 psi pressure. This is of the same order of magnitude as the raising of the elastic limit or Brinell hardness by pressure.

Discussion. The qualitative resemblance between the effects of pressure on plastic flow in tension and in shear is evident. The greatest increase of shearing stress observed is that for the 9-0 steel, which shows a maximum increase under 370,000 psi pressure of something more than

3-fold. In tension, a 3-fold raising of the flow stress of this steel is produced by a strain of 3.75, which means an elongation of 43-fold.

In addition to this rough qualitative resemblance, there is one significant qualitative difference between the strain-hardening curves for tension and for pure shear. The strain-hardening curve for tension rises linearly with increasing strain until the point of fracture is reached, which occurs abruptly, with no previous warning that could be detected within the sensitiveness of these experiments. The difference between a strain-hardening curve at one pressure and another higher pressure is merely, to a first approximation, that rupture on the second curve is not reached until a higher strain than on the first. In shear, however, the strain-hardening curves for different pressures do not coincide over their entire extent, but, especially at the lower pressures, the approaching fracture announces itself by a dropping away from the strain-hardening curve for higher pressures. In a certain range of pressure the strain-hardening curve may show a maximum. This is evident in Figs. 67, 68, and 69 on the curves for the first pressure above atmospheric. This would indicate a deterioration of the steel before complete fracture.

The physical explanation of the difference between shearing and tension is to be sought in the difference of the stability relations. If an incipient fracture occurs in tension, the adjacent parts are immediately removed from the region of atomic interaction and the fracture is catastrophic. On the other hand, when tangential slip of one part on another occurs, the slipping parts are not necessarily removed from atomic contact and cohesion remains to an extent depending on the degree of destruction of the lattice structure. In other words, in tangential slip the phenomenon of self-healing can occur, and failure is not catastrophic, as in tension. But the self-healing need not be complete, and to the extent that it is not, we have deterioration in the quality of the steel, or a bending over of the stress-strain curve. Hydrostatic pressure keeps the slipping parts in more intimate contact, so that the self-healing is more complete the higher the pressure, and the bending over of the stress-strain curve is displaced to higher strains or is completely suppressed.

CHAPTER 8

THE COLLAPSE OF THICK HOLLOW CYLINDERS OF STEEL UNDER EXTERNAL PRESSURE[1]

Introduction. Experiments were made on this subject as part of the Arsenal contract and are described in the fourth report to the Arsenal. After the war, Professor MacGregor of MIT experimented on the collapse of thick hollow cylinders of aluminum by external pressure and analyzed the results by a mathematical analysis similar to mine. Professor MacGregor was not acquainted at the time with this work of mine, buried in a government report, but after he had sent his manuscript to the *Journal of Applied Physics* his attention was called to it. At his suggestion and with the encouragement of the editor, I wrote a summary of my results which appeared in the same issue of the *Journal of Applied Physics* as his article. In this summary I made a comparative analysis of the mathematics of Professor MacGregor and myself and summarized my principal experimental findings. There is, however, nothing in this *Journal of Applied Physics* article of mine not contained either explicitly or implicitly in the Arsenal report, and I shall not refer to it further. The Arsenal report now follows, practically entire.

If a heavy-walled hollow cylinder with closed ends is exposed to an external hydrostatic pressure sufficiently high to make it flow plastically, the material flows radially symmetrically toward the center, the outside and inside surfaces remaining circular in section. For any given external pressure flow ceases at some definite ratio of internal to external diameter. Flow is terminated by a combination of the geometrical effect of a relative thickening of the walls and the strain-hardening accompanying the flow.

A number of years ago I studied the collapse of cylinders of copper and steel under external pressures up to 170,000 psi. This pressure was sufficient to close the hole in the copper cylinders apparently completely, thus giving theoretically infinite strains at the center. The steel cylinders were not completely collapsed, but an extrapolation of the measurements, using the empirical result that the internal diameter was a linear function of the external pressure, indicated that collapse would be complete at pressures between 280,000 and 350,000 psi. This complete closing of the hole is a most surprising result, for theory would indicate that, if

[1] This chapter is based on material contained in the fourth Watertown report and B1, B2, and B25.

there is no strain-hardening at all, as expressed by either the von Mises or the maximum stress difference criterion for plastic flow (the two criteria become identical under these conditions), the logarithm of the ratio of external to internal diameter should become infinite as pressure becomes infinite, or in other words, the hole should never completely vanish no matter how high the pressure. If then the hole does vanish at some finite pressure, there must be strain-softening instead of strain-hardening when strain is pushed sufficiently high. Such an effect has not been indicated by any of the ordinary methods of testing, so that it becomes important to check the existence of such an effect. The pressures now available, up to 450,000 psi, are higher than the pressures indicated by the early

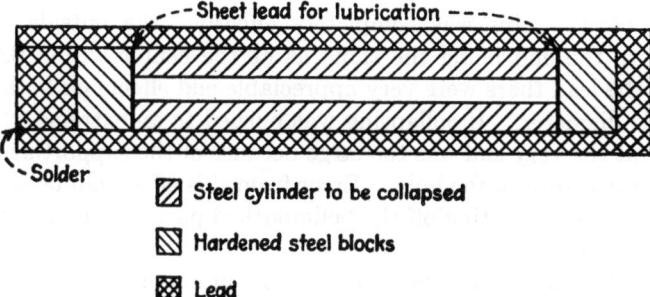

FIG. 73. The cylinder and method of mounting for collapsing tests under external pressure.

experiments as the pressure of complete collapse of the steel, so that it was thought to be worth while to reexamine the matter with the new range of pressure. Other information about plastic flow should also come out of the measurements.

Experimental Results. In the following, results are presented for collapsing tests on 27 different hollow cylinders. Twenty-two of them were made of a 1045 steel, in groups of 7, 5, 5, and 5, with four different heat-treatments, which were given at the Arsenal. This steel and the heat-treatments have been designated in Table VI as 9-2, 9-3, 9-4, and 9-6. The remaining group of five specimens were made of a 1020 steel, previously designated as steel 14, specially annealed at the Arsenal to give maximum softness. This heat-treatment consisted of 1 hr at 1650°F and 40 hr at 1320°F. The cylinders were all of approximately the same initial dimensions: length 1.813 in., OD 0.3125 in., and ID 0.0998 in. The setup is indicated in Fig. 73. The hollow cylinder was first drilled and reamed, then turned to outside dimensions between centers, and finally ground flat on the ends. The ends were closed with flat blocks of hardened steel. The cylinder and the closing end blocks were mounted in a sheath of about $3/32$ in. wall thickness, machined from solid stock, and sealed by soldering a closure block of lead in the open end through which the cylinder was inserted. The lead sheath with its contents was placed in the cylinder of the regular pressure apparatus, and pressure was applied to the entire external surface by the pressure liquid. Pressure was measured as always by a manganin gauge immersed in the liquid. The resistance to collapse offered by the lead was assumed to be negligible, and in the following the collapsing pressure on the outside of the steel cylinder was taken to be the pressure in the liquid.

Under these conditions the steel cylinder collapses uniformly, except at the very ends, where there may be end effects due to frictional resistance to radial flow offered by the steel blocks. In order to minimize this friction, a sheet of lead 0.002 in. thick was interposed between cylinder and end block. This was completely effective, the outside diameter right up to the end being the same as at the center within 0.0001 in. Without the lead there were very appreciable end effects on both outside and inside diameter. If collapsing was pushed to an extreme, the inside diameter at the very end was too large because of the supporting effect of the lead squeezed into the hole. Error from this effect can be eliminated when necessary by cutting off the bellmouthed part before measuring the inside diameter.

The usual procedure was to apply pressure at a uniform rate of increase, reaching the maximum in about 10 min, maintain at the maximum for 10 min, and release to zero in 2 or 3 min. Time effects under these circumstances are unimportant, as shown by one experiment in which pressure was immediately released after reaching the maximum (see sixth entry in Table XII), and by another experiment in which pressure was maintained at the maximum for 2 hr (see seventh entry in Table XII). These two special procedures gave the same distortion as the regular procedure. On release of pressure the lead sheath was cut away, the outside dimensions, length, and diameter at a number of points measured with micrometers to 0.0001 in., and a rough measurement made of the internal diameter by finding the drill which was the closest fit for the hole. The average internal diameter was then measured more accurately by grinding the ends flat, cementing a glass plate to one end, filling with mercury under vacuum, flattening off the meniscus in order that the filling should be complete by pressing another glass plate against the mensicus and wiping off the excess mercury, and weighing the mercury filling the hole. Agreement with the rough drill measurements, except for a couple of cases to be mentioned in detail later, was within the limits of error of the rough measurements, indicating approximately uniform collapse of the hole.

The first generalization to come out of the measurements is that the change of length after collapse is small. The extreme change of length varies from +0.0020 to −0.0035 in., that is, from +0.11 to −0.19 per cent. If the changes of length of all 27 specimens are plotted in a single diagram against the pressure of collapse, a distinct trend is evident, in spite of scattering of individual points over the entire range of the effect, the increments of length tending to be positive at low collapsing pressures and negative at high pressures. It is probable that these small changes of length, although real, are incidental effects due to the finite length and would vanish on a percentage basis if the cylinders were infinitely long.

There is some warping of the initially plane ends into a configuration like a ball race, as indicated in Fig. 74. This was disclosed on the second grinding of the ends; the depression averaged something of the order of 0.001 in. It is obvious that a warping like this must be an end effect, for considerations of symmetry show that warping must vanish in an infinitely long cylinder. Part of the warping is doubtless due to a trapping of the lead lubrication between cylinder and end block. A similar effect has been found in straight compression tests on solid cylindrical blocks with lead for lubrication. In view of these considerations the change of length will be assumed to be exactly zero in the mathematical analysis to be given presently, which applies only to infinitely long cylinders and neglects end effects.

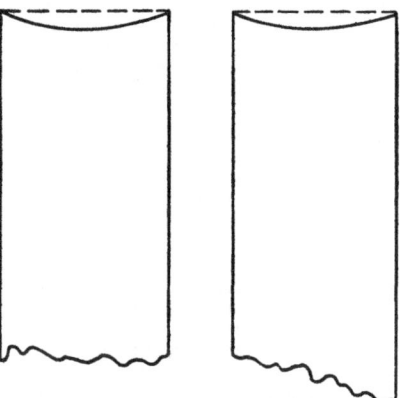

Fig. 74. Shows the manner of warping of the ends of the cylinders.

The numerical results are collected in Table XII. In Fig. 75 is shown $\log_e (r_0/r_i)$ plotted against the collapsing pressure, and in Fig. 76, $P/\log_e (r_0/r_i)$ against collapsing pressure. The points of Fig. 75 lie on straight lines with a high degree of regularity. If this result can be assumed to be general over an indefinitely wide range of pressure, the question proposed at the beginning is at once answered, namely, the hole is not squeezed shut by any pressure, no matter how high, thus contradicting the conclusion reached in the early work. How is the new result to be reconciled with the old one? In the first place it is to be noticed that there is no entry in the table for the cylinder of soft 1020 steel exposed to the highest pressure. The reason is that no measurements could be made, the hole apparently having been squeezed tight shut. The specimen that had been exposed to 355,000 psi still had a trace of the

TABLE XII. SUMMARY OF COLLAPSING TESTS

Designation of steel	Collapsing pressure, psi	Initial dimensions, in.		Final dimensions, in.			$\dfrac{r_0}{r_i}$	$\log_e \dfrac{r_0}{r_i}$	$\dfrac{P}{\log_e (r_0/r_i)}$ psi	e_r	
		Length	OD	Length	OD	ID				$\log_e \left(\dfrac{a}{r_i}\right)$	$\log_e \dfrac{b}{r_0}$
9-4	142,000	1.8123	0.3122	1.8141	0.3051	0.0731	4.18	1.430	99,400	0.315	0.0244
	213,000	1.8126	0.3128	1.8091	0.3005	0.0476	6.32	1.844	116,000	0.742	0.0410
	284,000	1.8123	0.3127	1.8120	0.2983	0.0312	9.68	2.270	126,000	1.163	0.0485
	355,000	1.8124	0.3125	1.8113	0.2972	0.0213	13.95	2.634	135,000	1.545	0.0516
	412,000	1.8130	0.3119	1.8096	0.2962	0.0144	20.58	3.022	136,000	1.937	0.0529
	355,000	1.8131	0.3115	1.8127	0.2962	0.0216	13.73	2.620	135,000	1.533	0.0517
	355,000	1.8127	0.3127	1.8120	0.2969	0.0209	14.20	2.654	134,000	1.564	0.0514
9-3	142,000	1.8125	0.3126	1.8145	0.3081	0.0847	3.64	1.292	110,000		
	214,000	1.8108	0.3127	1.8121	0.3024	0.0592	5.11	1.631	132,000		
	284,000	1.8150	0.3126	1.8161	0.2989	0.0402	7.44	2.006	142,000		
	355,000	1.8131	0.3126	1.8132	0.2978	0.0278	10.70	2.371	150,000		
	412,000	1.8128	0.3126	1.8126	0.2970	0.0209	14.22	2.656	155,000		
9-2	142,000	1.8132	0.3128	1.8134	0.3125	0.0998	3.13	1.138	125,000		
	213,000	1.8132	0.3124	1.8133	0.3120	0.0990	3.15	1.147	186,000		
	284,000	1.8130	0.3125	1.8125	0.3084	0.0876	3.52	1.258	226,000		
	355,000	1.8133	0.3126	1.8115	0.3030	0.0667	4.54	1.513	235,000		
	412,000	1.8132	0.3125	1.8112	0.2992	0.0525	5.70	1.740	238,000		
9-6	142,000	1.8133	0.3125	1.8146	0.3114	0.0962	3.23	1.172	121,000		
	215,000	1.8116	0.3123	1.8115	0.3043	0.0705	4.32	1.463	147,000		
	284,000	1.8122	0.3127	1.8110	0.3007	0.0501	6.00	1.792	159,000		
	355,000	1.8129	0.3122	1.8106	0.2982	0.0358	8.33	2.120	167,000		
	412,000	1.8130	0.3123	1.8098	0.2976	0.0282	10.56	2.358	175,000		
14	71,000	1.8134	0.3117	1.8135	0.3079	0.0869	3.54	1.264	56,000		
	170,000	1.8131	0.3126	1.8134	0.2989	0.0365	8.19	2.103	81,000		
	256,000	1.8132	0.3126	1.8128	0.2971	0.0171	17.37	2.855	89,000		
	355,000	1.8139	0.3124	1.8133	0.2964	0.0085*	35.0	3.555	100,000		
	412,000	1.8131	0.3126	1.8130	0.2963						

* From microscopic examination at the Arsenal.

cavity left that could be detected with a machinists' eight-power glass. However, the standard procedure for determining the average diameter failed because mercury could not be made to enter the hole. This specimen was further examined at the Arsenal. It was cut across at two places, the sections polished according to metallographic procedure, and photographs of the sections made with a high power. One of these is reproduced in Fig. 77. The mean internal diameters at the sections determined in this way were 0.21 and 0.23 mm; the mean is listed in the table. Similar examination at the Arsenal of the specimen collapsed at 412,000 psi showed under 200 magnification only slight traces of the original cavity. This is shown in Fig. 78. In both cases the final hole is smaller than the original grain size. The cylinder of 1020 steel which had been exposed to 265,000 psi, the next lower pressure, was examined after it had been cut across the axis at the center of the length. At this point the hole had the outline of a right triangle, instead of a circle. Furthermore, the cylinder of the 9-4 steel which had been exposed to the highest pressure was split longitudinally along the axis, and the hole found to be

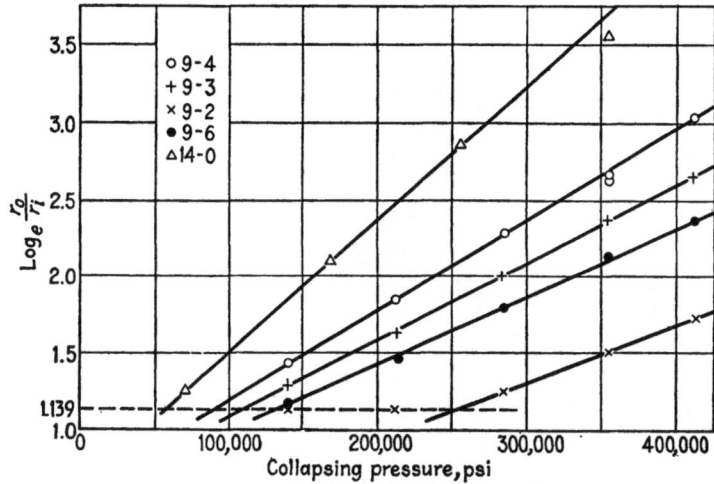

FIG. 75. $\log_e (r_0/r_i)$ against collapsing pressure for several steels.

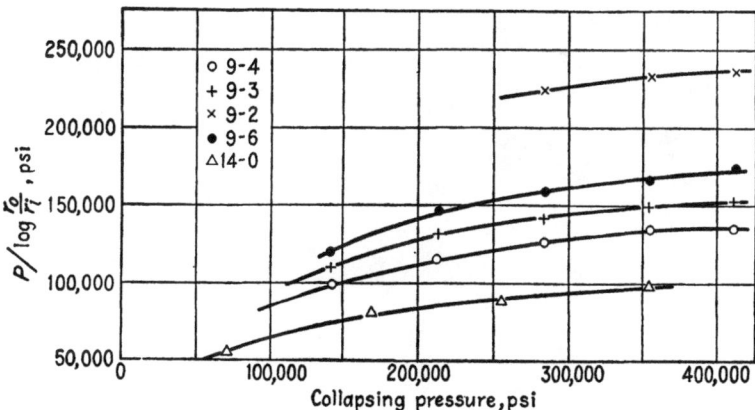

FIG. 76. $P/\log_e (r_0/r_i)$ against collapsing pressure for several steels.

far from regular, the walls being roughened by the apparent extrusion of individual grains. Because of the method of measurement, the diameters of the holes listed in the table are obviously average diameters. What apparently happens is that as the hole gets smaller it departs more and more from its circular figure, the departure arising from the finite size of the individual grains, within which slip takes place on planes. Somewhat similar effects are shown by tension specimens pulled out to extreme elongations under high pressure where there is great ductility, for the surface of such specimens becomes increasingly rough. In the case of tension specimens of copper, brass, and aluminum, it has already been

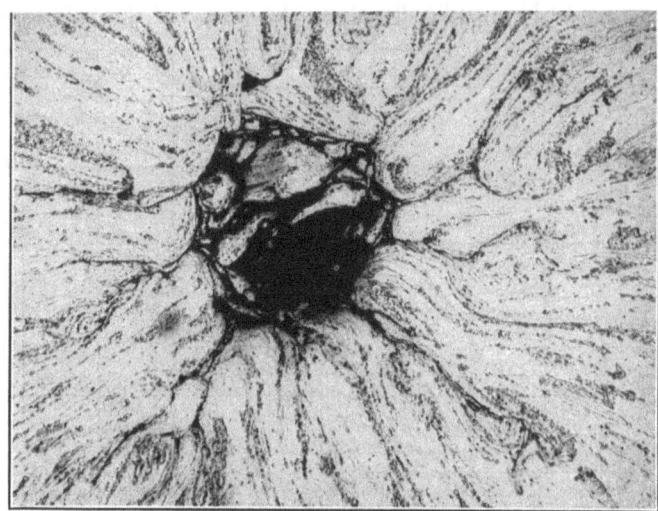

FIG. 77. Microstructure of the central part of cylinder of steel 14 after collapsing by a pressure of 355,000 psi.

FIG. 78. Microstructure of the central part of cylinder of steel 14 after collapsing by a pressure of 412,000 psi.

mentioned that there is probably grain growth under these conditions, the section of severely necked tension specimens eventually becoming square or angular, as if the entire section were occupied by a single grain. In steel such phenomena are certainly less prominent, but the possibility is not ruled out that even in steel there should be some grain growth at the inner surface of these collapsed cylinders. In any event, even if the grains remain of their original size, a point would presently be reached where the linear slip in the individual grains predominates, and the section is no longer circular. Plots such as Figs. 76 or 77, which assume a single internal radius and homogeneity, come to the end of their range of validity. Under such circumstances it is to be expected that the slip of the individual grains would be staggered with respect to each other, so that an open line along the axis no longer exists, and the hole would appear to an ordinary mechanical probing to be wiped out. If the specimen is cut open to find the hole under these circumstances, there is difficulty from burring of the edges and partial or complete closing of the hole under the action of the cutting tool or the polishing agent. It may even be that the hole is actually squeezed completely shut, as suggested by Fig. 78. But if this is the case it would seem not to be particularly pertinent for our present purposes, being a phenomenon connected with the individual crystal grains and not shown in homogeneous and isotropic material. It would appear that there is no necessary inconsistency with the early observations on copper, but the interpretation put on the early results must be altered.

The significance of the expression $P/\log_e (r_0/r_i)$ in the table and Fig. 76 is that this expression would be constant if the maximum stress difference criterion for plastic flow were valid, as may be shown by a simple analysis. The deviation of the expression from constancy therefore gives a partial indication of the degree of strain-hardening. The direction of variation is such as to indicate a positive hardening, rather than the softening suggested as a possibility in the introduction.

It would be desirable to get as much information as we can from these results about the precise dependence of strain-hardening on strain (or stress) under the conditions of these experiments. This requires a mathematical discussion of the stress-strain relations in the plastically strained cylinder.

Mathematical Analysis. The following treatment differs in several particulars from conventional ones. In the first place, it obviously is not adequate to treat strains and displacements as small. We shall use conventional cylindrical coordinates r, θ, and z, but these will refer to points in their finally displaced positions, not to their initial positions, as in the conventional treatments of elasticity theory. We shall denote by r_i the final internal radius of the cylinder, and by r_0 the final external radius. We shall denote the initial internal and external radius by a and b, respectively. Similarly the principal stresses, written in the conventional way as $\widehat{rr}$, $\widehat{\theta\theta}$, and $\widehat{zz}$, will be the stresses at the final points of the body. The stress equations of equilibrium under such circumstances, rigorously and without approximation, assuming an infinitely long cylinder without end effects, reduce to the single equation

$$r \frac{\partial \widehat{rr}}{\partial r} = \widehat{\theta\theta} - \widehat{rr} \tag{8-1}$$

Derivatives with respect to θ vanish, the stress $\widehat{rz}$ vanishes, $\widehat{zz}$ is constant along the axis for any fixed r, that is, $\partial \widehat{zz}/\partial z = 0$, and $\widehat{zz}$ is subject only to the condition

$$\pi r_0^2 P = - \int_{r_i}^{r_0} 2\pi r \widehat{zz}\, dr \tag{8-2}$$

where P is the external pressure. Equation (8-2) expresses the condition of balance between the total force exerted by the external pressure acting over the closed ends and the longitudinal stress in the walls of the cylinder.

The displacement has two components, radial and axial. The axial displacement must, as we have already seen, be such that there is no warping of the cross section, or in other words, the longitudinal strain must be a constant independent of r and z. The radial displacement ρ will be defined as that displacement which carries a point from its final position back to its initial position. Since the displacements and the strains are large, we shall use the so-called "natural strains" instead of the conventional strains, defining a natural strain e as $\log_e (L_{\text{final}}/L_{\text{initial}})$. Application of this definition gives at once

$$e_r = -\log\left(1 + \frac{d\rho}{dr}\right) \tag{8-3}$$

$$e_\theta = -\log\left(1 + \frac{\rho}{r}\right) \tag{8-4}$$

We assume that to a sufficient approximation the volume is constant, giving the condition

$$e_r + e_\theta + e_z = 0$$

or

$$\left(1 + \frac{d\rho}{dr}\right)\left(1 + \frac{\rho}{r}\right) = 1 + e'_z$$

Here e'_z is the conventional longitudinal strain, and $e_z = \log(1 + e'_z)$. e'_z is constant and the solution of the equation is

$$\rho = \sqrt{r^2(1 + e'_z) - c^2} - r \tag{8-5}$$

I am indebted to Professor Wendell Furry for the solution of this nonlinear differential equation.

We might now substitute into this solution the experimental result that $e'_z = 0$. It is, however, possible to prove that, subject to certain general assumptions about the equations of flow, e'_z (and e_z) must vanish, and thus to use the experimental fact that e_z does vanish to obtain certain presumptive information about the nature of the flow equations.

We assume the flow equations in the form

$$e_r = \alpha[\widehat{rr} - \tfrac{1}{2}(\widehat{\theta\theta} + \widehat{zz})] \tag{8-6}$$
$$e_\theta = \alpha[\widehat{\theta\theta} - \tfrac{1}{2}(\widehat{zz} + \widehat{rr})] \tag{8-7}$$
$$e_z = \alpha[\widehat{zz} - \tfrac{1}{2}(\widehat{rr} + \widehat{\theta\theta})] \tag{8-8}$$

In the general case these equations would have to be written in differential form, but under the specially simple conditions of this problem, in which the pressure increases monotonically, the integral form is permissible.

If flow takes place without strain-hardening, these three equations have to be supplemented by another equation, such as the von Mises plasticity condition or the condition of constant difference between the extreme principal stresses, and α becomes an unspecified function of the geometrical coordinates, determined by the special problem. Or, if there is strain-hardening, the auxiliary condition disappears and α becomes a specified function of stresses (or strains) and it may be stress history. The essential restrictions imposed by Eqs. (8-6), (8-7), and (8-8) are that the total volume change vanishes (the sum of the three equations vanishes identically) and that the strains are isotropic in the principal stresses. The latter is a restriction that is only approximately verified experimentally.

An experimental check on the assumption that there is no change of volume can be made by comparing the final external diameter, calculated under the assumption of no change of length from the original dimensions, and the final internal diameter, with the measured final diameter. This calculation was carried through for the 9-4 series. The discrepancy between calculated and measured diameter was, except in one case, consistently in the direction of a deficit of the calculated diameter. The maximum discrepancy was 0.0004 in., and the average 0.0003. The effect is so small that it may justifiably be neglected in the following calculations; however, the results were so consistent that they must be indicative of a real effect. The change of volume is in the direction of a decrease of density after plastic flow of about 0.2 per cent. It is well known that in other experiments a decrease of density is often produced by cold-work. These considerations apply to the permanent change of volume after release of stress. During the application of the stress there is also an elastic increase of density arising from the mean hydrostatic component of the total stress system. This has its maximum value at the external surface. In the extreme case above, the elastic part of the displacement of the external surface would be of the order of 7 per cent of the plastic part, and at the internal surface it would be vanishingly small. We take this to justify neglect of the elastic part of the volume change in this analysis, which can be only rough in view of the uncertainty of the fundamental assumptions.

Now solve Eq. (8-8) for $\hat{zz}$, obtaining $\hat{zz} = (1/\alpha)e_z + \frac{1}{2}(\hat{rr} + \hat{\theta\theta})$, and substitute back in (8-6), getting

$$\hat{\theta\theta} - \hat{rr} = -\frac{2}{3\alpha}(2e_r + e_z) \tag{8-9}$$

Substitute this back into the equation of equilibrium (8-1) and integrate:

$$\hat{rr} = -\int_{r_i}^{r} \frac{2}{3\alpha}(2e_r + e_z)\frac{dr}{r} \tag{8-10}$$

The constant of integration is so determined that $\hat{rr}$ vanishes at the internal surface, as it should. Since $\hat{rr} = -P$ at the external surface, we have

$$P = \int_{r_i}^{r_o} \frac{2}{3\alpha}(2e_r + e_z)\frac{dr}{r} \tag{8-11}$$

We may now substitute the value of $\hat{rr}$ from (8-10) back into (8-9) and obtain for $\hat{\theta\theta}$

$$\hat{\theta\theta} = -\frac{2}{3\alpha}(2e_r + e_z) - \int_{r_i}^{r} \frac{2}{3\alpha}(2e_r + e_z)\frac{dr}{r} \tag{8-12}$$

Either (8-6) or (8-8) now gives $\hat{zz}$

$$\hat{zz} = \frac{2}{3\alpha}(e_z - e_r) - \int_{r_i}^{r} \frac{2}{3\alpha}(2e_r + e_z)\frac{dr}{r} \tag{8-13}$$

Equation (8-2) now becomes

$$\pi r_o^2 P = \int_{r_i}^{r_o} \frac{4\pi}{3\alpha}(e_z - e_r)r\,dr + \int_{r_i}^{r_o} 2\pi r\,dr \int_{r_i}^{r} \frac{2}{3\alpha}(2e_r + e_z)\frac{dr}{r}$$

The double integral may be integrated by parts, giving

$$\int_{r_i}^{r_o} 2\pi r\,dr \int_{r_i}^{r} \frac{2}{3\alpha}(2e_r + e_z)\frac{dr}{r} = \pi r_o^2 \int_{r_i}^{r_o} \frac{2}{3\alpha}(2e_r + e_z)\frac{dr}{r}$$
$$- \int_{r_i}^{r_o} \frac{2\pi}{3\alpha}(2e_r + e_z)r\,dr$$
$$= \pi r_o^2 P - \int_{r_i}^{r_o} \frac{2\pi}{3\alpha}(2e_r + e_z)r\,dr$$

Substituting back leaves only the single term

$$-\int_{r_i}^{r_0} \frac{6\pi}{3\alpha} e_z r \, dr = 0$$

Since we already know from considerations of symmetry that e_z is constant across the section, and since r is always positive in the range of integration, the only solution is

$$e_z = 0 \tag{8-14}$$

We have thus proved that, if the equations of plasticity have the general form assumed, there can be no change of length. This agrees with experiment. However, we have not proved that if $e_z = 0$ the plasticity equations must have the form assumed. This probably is not true. A further investigation would be desirable to find under what still broader conditions it is a mathematical necessity that $e_z = 0$.

It may be remarked in passing that the same result, namely, $e_z = 0$, applies also to the case of a hollow cylinder with closed ends under internal pressure. The analysis above carries through with only the interchange of r_0 and r_i.

The same result, namely, that the plastic change of length vanishes when a cylinder closed with caps over the ends is exposed to either external or internal pressure, is implicitly contained in formula 19a of Beeuwkes and Laning,[1] derived on the hypothesis of small strains and a universal relation between the "octahedral" stresses and strains.

We may now substitute the result $e_z = 0$ back in the equations for the stresses, abbreviate by setting $I = \int_{r_i}^{r} \frac{4e_r}{3\alpha} \frac{dr}{r}$, and obtain

$$\widehat{rr} = -I \tag{8-15}$$

$$\widehat{\theta\theta} = -\frac{4e_r}{3\alpha} - I \tag{8-16}$$

$$\widehat{zz} = -\frac{2e_r}{3\alpha} - I \tag{8-17}$$

The relative magnitudes of the stresses are as indicated in Fig. 79. The principal stresses are always compressive; the numerically greatest principal stress is $\widehat{\theta\theta}$ at the external surface. In the elastic case, on the other hand, the greatest stress is $\widehat{\theta\theta}$ at the inner surface. At all points, in the plastic case, $\widehat{zz}$ is midway between $\widehat{rr}$ and $\widehat{\theta\theta}$, so that the maximum shearing stress is always given by the difference between $\widehat{rr}$ and $\widehat{\theta\theta}$.

[1] R. Beeuwkes, Jr., and J. H. Laning, Jr., Watertown Arsenal Report No. 660/16, p. 10.

It will be useful to have the explicit expressions for the displacement. Equation (8-5) gives, setting $e_z = 0$,

$$\rho = \sqrt{r^2 - c^2} - r$$

The constant of integration is determined by the conditions at the inner surface. When $r = r_i$, $\rho = a - r_i$. This gives

$$\rho = \sqrt{(a^2 - r_i^2) + r^2} - r \tag{8-18}$$

It follows at once that

$$e_r = \frac{1}{2} \log \left(1 + \frac{a^2 - r_i^2}{r^2}\right) \tag{8-19}$$

$$e_\theta = -\frac{1}{2} \log \left(1 + \frac{a^2 - r_i^2}{r^2}\right) \tag{8-20}$$

$$e_z = 0 \tag{8-21}$$

e_r is positive throughout; it has its greatest value at the inner surface where $e_{r_i} = \log(a/r_i)$, and its least value at the outer surface where $e_{r_0} = \log(b/r_0)$. Table XII contains the values of e_r at the inner and outer surfaces for the 9-4 series. In Fig. 80 is plotted e_r as a function of the radius of a cylinder that has collapsed from the given initial dimensions to a final internal radius of 0.007 in., corresponding very closely to the fifth specimen of the 9-4 group. The figure shows how greatly the high strains are concentrated toward the center. However, a more significant plot, as will appear presently because of the form of Eq. (8-22), is of e_r against $\log r$; this is shown in Fig. 81.

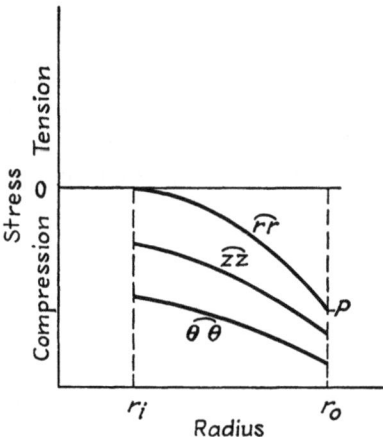

FIG. 79. The stress distribution in a collapsed cylinder.

We now have to discuss what information can be obtained about strain-hardening. The general method would be as follows. Suppose that we have determined α by independent experiment as a function of e_r. The condition that has to be satisfied is

$$P = \int_{r_i}^{r_0} \frac{4e_r}{3\alpha} \frac{dr}{r} = \int_{r_i}^{\sqrt{b^2 - a^2 + r_i^2}} \frac{4e_r}{3\alpha} \frac{dr}{r} \tag{8-22}$$

e_r is itself a known function of r_i, so that on substituting α as a known function of e_r the condition takes the form

$$P = \int_{r_i}^{\sqrt{b^2-a^2+r_i^2}} \phi(r,r_i) \, dr$$

where ϕ is a known function. The problem is solved by finding r_i so that this condition is satisfied. In general, graphical methods will have to be applied. The method is to assume a plausible value for r_i and with this value of r_i to find first e_r as a function of r and then $4e_r/3\alpha$ as a function of r. This function of r is then to be plotted against $\log r$; a plot will be obtained not much different in general shape from Fig. 81, since α is constant to a rough first approximation. The condition expressed by (8-22) is that the area under the curve between the lower limit r_i and upper limit $r = \sqrt{b^2 - a^2 + r_i^2}$ shall be equal to P. If the area turns out not to be equal to P, a different value for r_i is to be assumed and the solution found by trial and error.

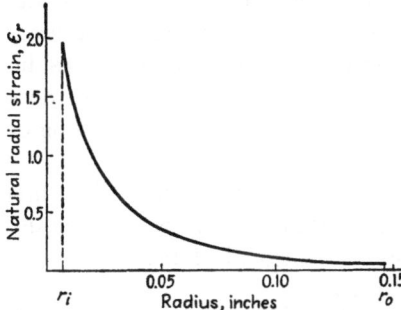

FIG. 80. Radial strain as a function of radius for the fifth specimen of 9-4 steel, collapsed to an internal radius of 0.007 in. by an external pressure of 412,000 psi.

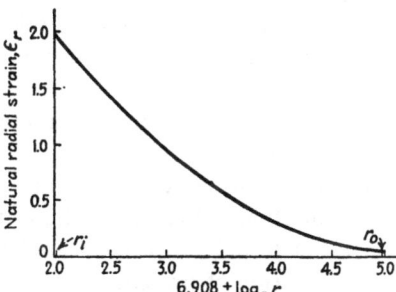

FIG. 81. Radial strain against $6.908 + \log_e r$ of the specimen of Fig. 80.

Let us first examine what happens if we assume no strain-hardening. Mathematically, this presents a degenerate case and is not to be solved by setting α equal to a constant in the formulation above. Under these conditions α becomes an undetermined function of the geometrical coordinates, and a new equation of condition makes its appearance. As this equation of condition we may assume either the von Mises condition or the maximum stress-difference condition. Which we assume is immaterial because under these conditions the two become identical, as the following analysis shows. The von Mises condition is

$$(\widehat{rr} - \widehat{\theta\theta})^2 + (\widehat{\theta\theta} - \widehat{zz})^2 + (\widehat{zz} - \widehat{\theta\theta})^2 = \text{const}$$

Substituting the special values of the stresses given by (8-15), (8-16), and (8-17), this becomes

$$\left(\frac{4e_r}{3\alpha}\right)^2 + \left(\frac{2e_r}{3\alpha}\right)^2 + \left(\frac{2e_r}{3\alpha}\right)^2 = \frac{3}{2}\left(\frac{4e_r}{3\alpha}\right)^2 = \frac{3}{2}(\widehat{rr} - \widehat{\theta\theta})^2 = \text{const}$$

But $\widehat{rr} - \widehat{\theta\theta}$ is the maximum stress difference, which completes the proof.

The solution for this degenerate case is perhaps most simply obtained by going back to the original equation of stress equilibrium. Put

$$\widehat{\theta\theta} - \widehat{rr} = K \qquad r\frac{d\widehat{rr}}{dr} = -K \qquad \widehat{rr} = -K\int_{r_i}^{r}\frac{dr}{r} = K\log\frac{r_i}{r}$$

with the boundary condition $-P = K \log (r_i/r_0)$. Hence

$$K = \frac{P}{\log (r_0/r_i)} \qquad (8\text{-}23)$$

Figure 76 shows that $P/\log (r_0/r_i)$ is not constant but increases with increasing pressure. We may take $P/\log (r_0/r_i)$ to give a sort of average stress difference. The conclusion is therefore that the average stress difference increases somewhat with increasing pressure, that is, with increasing flow toward the center. In other words, there is strain-hardening on the average.

The assumption of a constant value for K is, however, not a bad first approximation. Figure 75 shows the same thing. If K were constant, $\log (r_0/r_i)$ would be proportional to pressure instead of being merely a linear function of pressure as it is in Fig. 75.

We now inquire whether, assuming roughly $\widehat{\theta\theta} - \widehat{rr} = \text{const}$, the collapsing of cylinders yields flow stresses of the same order of magnitude as we would expect from other sorts of experiment. For example, are the flow stresses involved in the collapse of these cylinders the same as the flow stresses associated with simple tension? The stress-strain curve for simple tension has the general shape indicated in Fig. 82. At strains beyond the point where necking starts, which occurs at a natural strain of about 0.1, it is a matter of experiment that the stress-strain curve is straight. Below the necking point the curve falls below the straight line, as indicated, cutting the axis at the stress of first permanent set. If the cylinder is collapsed to a small fraction of its initial internal diameter, then the condition which sets the size is determined in large degree by the linear part of the stress-strain curve beyond 0.1, as is shown by Fig. 81, in which only a small part of the total area is contributed by strains under 0.1. For small collapsing pressures, however, e_r may be less than

COLLAPSE OF HOLLOW CYLINDERS

0.1 throughout the cylinder, and in this case the part of the stress-strain curve below 0.1 is determinative.

The limiting pressure at which plastic creep first starts can be obtained approximately from Fig. 75. The limiting value of log (r_0/r_i) is evidently log (b/a), which in these experiments was log $(0.3125/0.1000) = 1.139$. The intersection of the various straight lines of Fig. 75 with the horizontal drawn at the height 1.139 gives approximately, then, the minimum pressure of plastic flow. The value so obtained is slightly too low, because there is an intermediate stage between complete elasticity and complete plasticity in which the cylinder receives a permanent set after release of pressure, but in which some of the parts of the cylinder are still in the elastic range. In other words, plastic flow throughout the entire cylinder occurs at a somewhat higher value for log (r_0/r_i) than 1.139. This

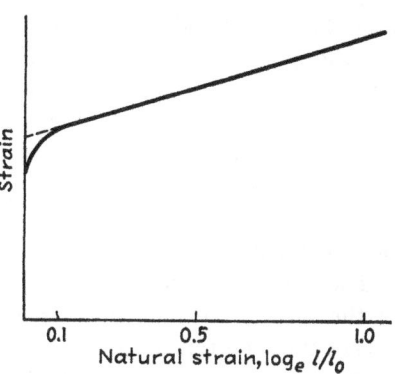

FIG. 82. Typical stress-strain curve for tension.

effect must be small, however, in view of the smallness of the strains in the elastic range. It is also shown by the lowest point of the line for steel 9-6 in Fig. 75, which falls very close to the point of intersection with the horizontal 1.139. The pressures of first plastic flow, obtained approximately in this way, are shown in Table XIII. In order to calculate from

TABLE XIII

Designation of steel	Pressure of first plastic flow, psi	Stress difference at first flow, psi	Tension for first plastic flow according to von Mises, psi	Observed tension at first plastic flow, psi	Observed technical tensile strength, psi
9-4	92,000	81,000	70,000	57,000	104,000
9-3	110,000	97,000	84,000	71,000	114,000
9-2	255,000	225,000	195,000	156,000	170,000
9-6	135,000	119,000	103,000		
14	58,000	51,000	44,000	24,000	57,000

these pressures the corresponding stress difference in the material of the cylinder, we assume the commonly accepted result that in the initial regions of small plastic flow the von Mises condition, or the condition of constant stress difference, expresses the relations with sufficient accuracy. This permits us to calculate the stress difference at flow by Eq. (8-23), setting log $(r_0/r_i) = 1.139$. The stress difference so calculated is given in the third column of the table. If the maximum stress-difference criterion is valid, this should also be the tension at which plastic flow begins in simple tension. If the von Mises condition applies, then a slightly different tension at first flow in tension is to be expected. This may be calculated as follows. Consider the two systems of stress: (1) the system at the inner surface of the collapsing cylinder, $\widehat{rr} = 0$, $\widehat{\theta\theta} = -K$, $\widehat{zz} = -\frac{1}{2}K$; (2) the system in simple tension $\widehat{rr} = \widehat{\theta\theta} = 0$, $\widehat{zz} = T$. Now at plastic flow $(\widehat{rr} - \widehat{\theta\theta})^2 + (\widehat{\theta\theta} - \widehat{zz})^2 + (\widehat{zz} - \widehat{rr})^2$ has the same value for the two stress systems. Hence

$$K^2 + \tfrac{1}{4}K^2 + \tfrac{1}{4}K^2 = 0 + T^2 + T^2$$

or $T = 0.866K$.

The tension for flow calculated in this way is given in the fourth column of the table. Tension measurements have been made on all the steels except 9-6; the tension of first flow is not very definitely marked, but it may be determined roughly and is given in the fifth column. Column 5 is of the same order of magnitude as column 4 and consistently less than it.

We have said that the tension at which flow first begins is not well marked; an upper limit can, however, be set, for it is surely less than the technical tensile strength, that is, the maximum load which the specimen will support divided by the original cross-sectional area. This figure is given in the sixth column of the table. We should expect that column 6 would be certainly greater than column 4. This is comfortably the case except for steel 9-2. Some special effect must be operative here that requires comment. 9-2 is the hardest of all these steels; it fractures with small elongation and reduction of area at high stresses. Furthermore, the initial flow point is close to the maximum. The softer steels show the phenomenon of a lower yield point in tension, which means that at low stresses strain-hardening is unimportant over a considerable range. With the harder steels like 9-2, on the other hand, hardening may rise sharply from the beginning, so that the assumptions of the analysis do not apply so well. This may give a partial explanation, but still it does not appear from these considerations why column 6 should ever be *less* than column 4. Whether the softer grades of steel show the "lower yield point" under the conditions of compressive stress prevailing in the cylinders might be made the subject of interesting experiment.

Consider next the relations at high pressures and at high strains toward the inner surface, in particular for the members of the 9-4 series compressed to the maximum pressure. The strain at the mid-point of the area of Fig. 81 (logarithmic plot) is about 1.25. The tensile stress at this strain in simple tension should be of the same order as the mean stress difference in the collapsing case as given by $P/\log(r_0/r_i)$. The former, according to the previous experiments, is 210,000 psi, and the latter, according to present experiments, as shown in Fig. 75, is 130,000. We have here a reversal of behavior at high strains. Whereas at low strains the flow stresses calculated from collapsing data are greater than those from tension data (columns 3 and 4 in Table XIII are greater than column 5), at high average strains the stresses calculated from collapsing data are lower than those calculated from tension, 130,000 against 210,000 in the example just considered.

This rather large reversal of behavior brings up the question of the reasonableness of our assumption that if we know when plastic flow occurs under one type of stress we can calculate when it will occur under other types of stress. Consider, for example, the von Mises condition $(\widehat{rr} - \widehat{\theta\theta})^2 + (\widehat{\theta\theta} - \widehat{zz})^2 + (\widehat{zz} - \widehat{rr})^2 = \text{const.}$ The signs of the stresses do not affect the result; according to this, flow occurs at the same numerical stress in simple tension as in simple compression. This may possibly be approximately the case at low strains, but at larger strains experiments to be described later show that it is definitely not true. Strain-hardening in compression is in general much slower than in tension; this is indicated by the figures 130,000 and 210,000 above. It would appear that at high strains the problem is much more complex than at low strains. It is probable that the strain-hardening function must be determined under conditions approximating as closely as possible the actual stress conditions. Thus the strain-hardening function α of Eq. (8-22) should be determined under conditions where the principal stresses are all compressive, and one of them midway between the other two and also under equivalent conditions with regard to strain, that is, with one component of strain vanishing and the other two numerically equal and with opposite signs. It is a matter for the experiment of the future to determine how much latitude is permissible in the simultaneous reproduction of all these conditions.

It would be feasible to carry the formal analysis above further. Thus, guided by analogy with the strain-hardening curve for tension, it might be assumed that α is a linear function of e_r, and then we might try to determine the two parameters of this linear relation in such a way as to reproduce the linear relation between collapsing pressure and $\log(r_0/r_i)$ shown in Fig. 75, a relation which hitherto has not been explained. In view of the approximate constancy of the stress difference one might anticipate considerable success in reproducing the linear relation with the help of two additional parameters. In fact, I have roughly carried such a calculation to the point where it appears that the linear relation can be approximately obtained with the help of two suitably chosen parameters, and also to the point where it appears that the parameter which expresses the slope of the strain-hardening curve is less by a factor of 2 or 3 than the slope of the strain-hardening curve in simple tension. This checks the conclusion already reached, that the hardening under the conditions of these experiments (and probably under compressive stresses in general) is less rapid than in tension. In order to connect these experiments with strain-hardening results obtained under other circumstances it would be necessary to make other experiments more nearly reproducing the conditions prevailing here, and in the absence of such experiments it is hardly worth while to elaborate the calculations.

Certain Early Results on the Collapse of Steel Cylinders under Pressure. In the experiments just described the permanent change of dimensions was determined after pressure had been applied and released. In the paper published in 1912, already referred to, in which it appeared that total collapse might take place at some finite pressure, the experimental arrangements were different from those in the later experiments in that the internal volume of the cylinder was measured as a continuous function of pressure during application of pressure. This made it possible to obtain different types of information than was possible in the later experiments, in particular the hysteresis between pressure and volume on applying cycles of pressure. For this reason these early experiments will be briefly alluded to here, although the pressure range was only about 90,000 psi.

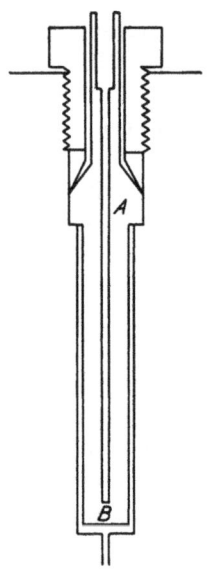

FIG. 83. Cylinder mounted so as to permit continuous measurement of the internal volume under changing external pressure.

The form of the cylinder which was subject to pressure is shown in Fig. 83. The cylinder was provided with a stem which emerged from the pressure vessel through suitable packing. A calibrated glass capillary was sealed to the emergent stem. The cylinder was filled with mercury, and changes of internal volume were indicated by the rise of mercury in the capillary. This is essentially the old Tate pressure gauge, carried into the region of plastic deformation. Experiments were made on seven cylinders. The part exposed to pressure was 6 in. long and 0.5 external diameter. Two were of soft "bessemer" steel, with internal diameters of 0.25 and 0.125 in., and five were of annealed drill rod (1.25 C) with internal diameters varying from 0.25 to 0.125 in. by steps of $\frac{1}{32}$. All seven were mounted in the same block simultaneously so that the pressure history of all seven was identical.

Figure 84 shows a representative set of results for the relation between pressure and decrease of internal volume for successive applications of pressure of progressively increasing magnitude. The rough qualitative behavior is what we have come to expect; that is, after every release and reapplication of stress plastic flow is not resumed until the previous maximum stress has been approximately reached. In finer detail, the diagram shows examples in which flow was resumed before the previous maximum had been quite reached and others in which the previous maximum was somewhat exceeded before resumption of flow. Which of these two finer patterns of behavior is followed depends on the time effects. If pressure is immediately reapplied after release, flow is resumed before the previous maximum stress is reached, whereas if there has been a more or less prolonged period of rest under no stress before reapplication, the previous maximum is exceeded by a greater or less amount. The original paper must be consulted for the details of the time schedule. There was a period of rest of 3 days between the second and third applications of pressure, when the overshooting of the previous maximum was most pronounced.

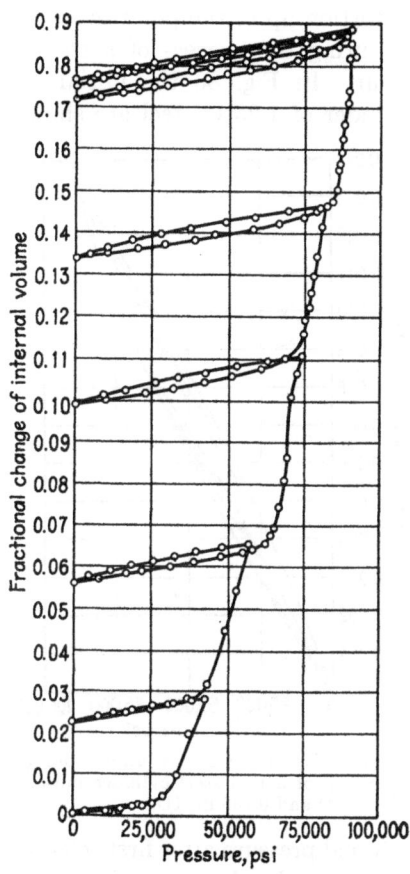

Fig. 84. Hysteresis loops obtained by varying the external pressure on a cylinder like that of Fig 83. This cylinder was initially 0.5 in. OD and 0.187 in. ID.

After the maximum stress has once been reached, the cylinder may be put into an approximately steady condition by subjecting it to a number of cyclic applications of the maximum stress. The final completely accommodated state is one in which hysteresis loops are described about a mean configuration very much like the well-known magnetic hysteresis loops. Figure 85 shows an example of such loops for the 1.25 C steel.

The amount of hysteresis varies greatly with circumstances. It increases rapidly with the pressure range. It is very much greater in the 1.25 C steel than in the "bessemer" and with a given steel is relatively greater in the cylinders with greater wall thickness. Cases have been found in which the breadth of the hysteresis loop is greater than the total elastic distortion up to the point of initial plastic yield.

During the process of accommodation, anomalous effects sometimes occur. In Fig. 86 the initial accommodation cycles for the heaviest cylinder of 1.25 C steel are shown. The initial effect of an increase of

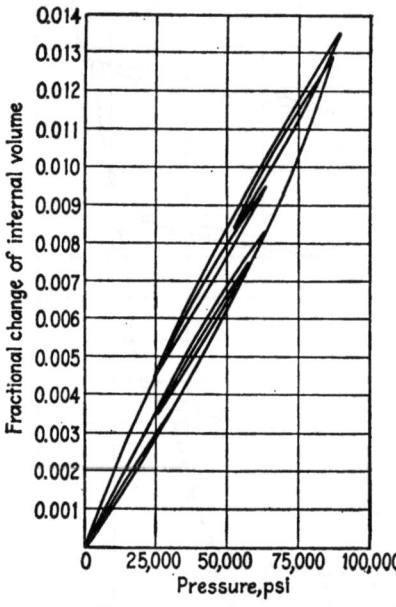 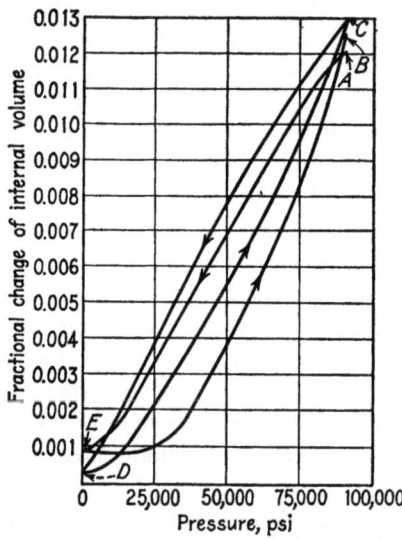

Fig. 85. Hysteresis cycles, after thorough seasoning, of a tool steel cylinder initially 0.5 in. OD and 0.156 in. ID.

Fig. 86. Abnormal hysteresis loops, before seasoning was complete, of tool-steel cylinder initially 0.5 in. OD and 0.125 in. ID.

external pressure after first release of the previous maximum is to *increase* the internal volume, as shown by the small negative slope at E. The effect gradually wears off with repetition of the cycle, but some memory of the effect always remains in an abnormally small initial slope, as at D. This points to a complicated internal distribution of stress and strain, much more complicated than is contemplated in the simple analysis attempted above. Such effects are greater in the higher carbon steel and in the cylinders with thicker walls.

The mean slope of the hysteresis loop, for instance, the slope of the line drawn from the initial to the final point of Fig. 85, is approximately what would be calculated by the ordinary equations of elasticity from the

COLLAPSE OF HOLLOW CYLINDERS 163

dimensions; that is, any change in the effective *elastic* constants produced by plastic flow is small.

The time rate of plastic yield undergoes modification as pressure increases. At low pressures, that is, in the initial stages of plastic flow when there has been little work-hardening, yield is at first rapid for a given increment of pressure, but the yield rapidly ceases. At higher pressures, that is, where the metal has flowed more and has experienced work-hardening, the initial rate of yield for a given pressure increment is less, but the yield persists through a longer time. Examples of the same sort of thing will be presented later in the discussion of two-dimensional compression.

The strain-hardening and accommodation under cycles of external pressure are an accommodation to that type of stress only. A cylinder collapsed in this way will burst under less than normal internal pressure.

CHAPTER 9[1]

THE EFFECT OF EXTERNAL PRESSURE ON CAVITIES IN BRITTLE MATERIALS

Introduction. If a material containing internal cavities is subjected to an indefinitely increasing pressure the presumption is that eventually the cavities must yield in one way or another. In the last chapter we have seen what happens if the material flows plastically. But if the material is incapable of plastic flow it is to be presumed that at sufficiently high pressure there will be fracture of some sort in the neighborhood of the cavities. It is of interest to find what the nature of this fracture is and what the stresses are that are required to produce it. In this chapter a number of experiments of this type on various materials ordinarily brittle will be reported. A number of these have not been previously described in print.

Elastic Solutions for Spherical and Cylindrical Cavities. The stresses and strains in a brittle substance must be given to a first approximation at least by the conventional elasticity analysis. The precise solution will of course depend on the geometrical details of the particular cavities, but an idea as to the orders of magnitude should be given by the solutions for two special cases, namely, a hollow sphere and a hollow cylinder closed at the ends subjected to external hydrostatic pressure. The case of the cylinder is the same geometrically as that treated in the last chapter.

The elastic solutions for these two cases are well known and are reproduced here for convenience of reference. The solutions as here given will be found in Love's "Elasticity."

1. *Hollow Sphere under External Pressure.* Denote the external radius by r_0, the internal radius by r_1, and the external pressure by p_0. The stresses are as follows:

$$\widehat{rr} = -p_0 \frac{r_0^3}{r^3} \frac{r^3 - r_1^3}{r_0^3 - r_1^3}$$
$$\widehat{\phi\phi} = \widehat{\theta\theta} = -\frac{1}{2} p_0 \frac{r_0^3}{r^3} \frac{2r^3 + r_1^3}{r_0^3 - r_1^3} \qquad (9\text{-}1)$$

The stress distribution is thus independent of the elastic constants. The formulas (9-1) show that the radial stress $\widehat{rr}$ varies from a numerical

[1] The material of this chapter, insofar as it has previously appeared in print, will be found chiefly in B2, B3, B4, and B5.

maximum of $-p_0$ at the external surface to zero at the internal surface. The circumferential stress $\widehat{\phi\phi}$ varies from a numerical minimum of $-\frac{1}{2}p_0[(2r_0^3 + r_1^3)/(r_0^3 - r_1^3)]$ at the external surface to a maximum of $-\frac{1}{2}p_0[3r_0^3/(r_0^3 - r_1^3)]$ at the internal surface. The limiting value of the maximum for infinitely thick walls is $-\frac{3}{2}p_0$. The stress difference varies from a minimum of p_0 at the external surface to a maximum of $\frac{1}{2}p_0[3r_0^3/(r_0^3 - r_1^3)]$ at the inner surface, the limit of the latter for an infinitesimal cavity being $\frac{3}{2}p_0$.

The radial displacement U has the value

$$U = -\frac{p_0 r_0^3}{r_0^3 - r_1^3}\left(\frac{r}{3\lambda + 2\mu} + \frac{r_1^3}{4\mu r^2}\right) \qquad (9\text{-}2)$$

From this the radial strain $e_{rr} = \partial U/\partial r$ may be at once obtained. For moderately thick cylinders e_{rr} changes sign from a compression at the external surface to an extension at the inner surface. Its maximum numerical value is $-p_0 \dfrac{r_0^3}{r_0^3 - r_1^3}\left(\dfrac{1}{3\lambda + 2\mu} - \dfrac{1}{2\mu}\dfrac{r_1^3}{r_0^3}\right)$ at the external surface. The circumferential strain is U/r, which is everywhere a compression. Its maximum value is $-p_0 \dfrac{r_0^3}{r_0^3 - r_1^3}\left(\dfrac{1}{3\lambda + 2\mu} + \dfrac{1}{4\mu}\right)$ at the inner surface. The shearing strain, or $e_{rr} - e_{\theta\theta}$, is $\frac{3}{4}(r_1^3/r^3)$. Its maximum value is $\frac{3}{4}\mu$ at the inner surface.

2. *Hollow Cylinder Closed with Caps over the Ends under External Hydrostatic Pressure.* Again denote the external and internal radii by r_0 and r_1 and the external pressure by p_0. The stresses are

$$\begin{aligned}
\widehat{rr} &= -p_0 \frac{r_0^2}{r_0^2 - r_1^2}\left(1 - \frac{r_1^2}{r^2}\right) \\
\widehat{\theta\theta} &= -p_0 \frac{r_0^2}{r_0^2 - r_1^2}\left(1 + \frac{r_1^2}{r^2}\right) \\
\widehat{zz} &= -p_0 \frac{r_0^2}{r_0^2 - r_1^2} = -\frac{\lambda}{\lambda + \mu} p_0 \frac{r_0^2}{r_0^2 - r_1^2} + e\frac{3\lambda + 2\mu}{\lambda + \mu}
\end{aligned} \qquad (9\text{-}3)$$

Again the stress distribution is independent of the elastic constants. In the above e is the longitudinal extension. Its value is

$$e = -p_0 \frac{r_0^2}{r_0^2 - r_1^2}\frac{1}{3\lambda + 2\mu}$$

It does not vanish, as it does in the plastic case.

The radial stress $\widehat{rr}$ has its maximum value $-p_0$ at the external surface and vanishes at the internal surface. The circumferential stress $\widehat{\theta\theta}$ has the value $-p_0[(r_0^2 + r_1^2)/(r_0^2 - r_1^2)]$ at the external surface, and $-p_0[2r_0^2/(r_0^2 - r_1^2)]$ at the internal surface, where it is a numerical maximum. The stress difference has its maximum value $2p_0[r_0^2/(r_0^2 - r_1^2)]$ at the inner surface.

The radial displacement U has the value

$$U = -p_0 \frac{r_0^2}{r_0^2 - r_1^2} \left(\frac{r}{3\lambda + 2\mu} + \frac{1}{2\mu} \frac{r_1^2}{r} \right) \tag{9-4}$$

From this the strains may be calculated. The radial strain $e_{rr} = \frac{\partial U}{\partial r}$ is

$-p_0 \frac{r_0^2}{r_0^2 - r_1^2} \left(\frac{1}{3\lambda + 2\mu} - \frac{1}{2\mu} \frac{r_1^2}{r^2} \right).$ For moderately thick cylinders it changes sign, being a compression at the external surface and an extension at the inner surface. Its value at the outer surface is

$$-p_0 \frac{r_0^2}{r_0^2 - r_1^2} \left(\frac{1}{3\lambda + 2\mu} - \frac{1}{2\mu} \frac{r_1^2}{r_0^2} \right)$$

and $-p_0 \frac{r_0^2}{r_0^2 - r_1^2} \left(\frac{1}{3\lambda + 2\mu} - \frac{1}{2\mu} \right)$ at the inner surface. The circumferential strain $e_{\theta\theta} = \frac{U}{r}$ is $-p_0 \frac{r_0^2}{r_0^2 - r_1^2} \left(\frac{1}{3\lambda + 2\mu} + \frac{1}{2\mu} \frac{r_1^2}{r^2} \right)$. It is a compression throughout and has the value

$$-p_0 \frac{r_0^2}{r_0^2 - r_1^2} \left(\frac{1}{3\lambda + 2\mu} + \frac{1}{2\mu} \frac{r_1^2}{r_0^2} \right)$$

at the outer surface and $-p_0 \frac{r_0^2}{r_0^2 - r_1^2} \left(\frac{1}{3\lambda + 2\mu} + \frac{1}{2\mu} \right)$ at the inner surface. The shearing strain, or $e_{rr} - e_{\theta\theta}$, is $p_0 \frac{r_0^2}{r_0^2 - r_1^2} \frac{1}{\mu} \frac{r_1^2}{r^2}$. Its maximum value is $p_0 \frac{r_0^2}{r_0^2 - r_1^2} \frac{1}{\mu}$ at the inner surface.

Experimental Results. *Glass.* 1. PRESSURE MEDIUM NOT WATER. The behavior of glass is sensitive to the liquid with which pressure is transmitted. Particularly if pressure is transmitted with water, which goes into solution to a certain extent in the glass under pressure, the results are markedly different from what they are when pressure is transmitted with a neutral liquid such as kerosene or petroleum ether or pentane. In the latter case it was early found (see B2 and B3) that external pressures as high as 400,000 psi may be supported without fracture by cylindrical capillaries with a ratio of external to internal radius of the general order of 3 to 1. The full details of these early experiments have been lost. It is probable that the glass was a soda-lime glass—pyrex was not available at that time. The specimens were of the general form indicated in Fig. 87, made with care by the glass blower so that the ends should be as nearly hemispherical as possible, and carefully annealed after blowing. If the ends were irregular geometrically or the annealing not thorough, cracks appeared in the neighborhood of the ends after exposure to pressure, in some cases the ends dropping off after removal from the pressure apparatus.

FIG. 87. General form of glass capillaries for collapsing experiments.

No permanent change in external dimensions could be detected after exposure to pressure by measurements to 0.0001 in. with a conventional micrometer. These early specimens were not preserved for subsequent examination as were later specimens.

The formulas above for the cylindrical case show that at the inner surface of these specimens there was a compressive stress and stress difference of the order of 800,000 psi, supported without fracture. The distortion may also be calculated if the elastic constants are known. For these assume mean values of $\mu = 3 \times 10^{11}$, $\lambda = \frac{3}{2} \times 10^{11}$ in absolute cgs units.[1] With these values it may be computed that at the inner surface there is a circumferential contraction not far from 10 per cent and a radial extension of about one-fifth as much. It is needless to say that these values of stress and strain are enormously greater than can be supported without fracture under normal conditions.

Although after exposure to pressure there is no change of dimensions measurable with the comparatively crude means employed, nevertheless some permanent distortion is produced as shown by the frequent spontaneous generation of internal cracks some time after release of pressure. In particular one experiment was made with a block of pyrex glass in which an ellipsoidally shaped bubble was blown, the long axis being approximately 50 per cent longer than the short axis. After careful annealing this was exposed to 380,000 psi external pressure in pentane. There was no general failure, but immediately after release of pressure four internal surfaces of separation appeared with a highly regular geometrical arrangement as shown in Fig. 88. These separation surfaces were doubtless generated on release of pressure as an accompaniment of the expansion at the internal surface, the internal surface having flowed slightly under pressure, so that on release the final strain at the inner surface was an extension with respect to the initial configuration. This

[1] See, for example, for typical constants Birch's "Handbook of Physical Constants," p. 84, Special Paper No. 36, Geological Society of America, 1942.

specimen was put by and observed from time to time. Two years later there was no detectable change in appearance, but after another year, 3 years after exposure, a new system of internal cleavage had developed, indicated by the dotted lines in the figure. This new cleavage is on a surface of revolution, and apparently has no connection with the originally developed cleavages. It is, however, evidence that between 2 and 3 years after exposure the process of internal readjustment was still going on and that there must therefore have been originally a microscopic amount of flow under the external pressure. There has been no further change in 3 more years.

2. PRESSURE MEDIUM WATER. If glass capillaries are exposed to pressure through the medium of water, other experience with the action of water suggests that results different from those above are to be expected. Of this other experience there is to be remembered in particular the extreme sensitiveness of glass rods in the "pinching-off" effect mentioned on page 108. Another effect of the same sort is the extreme sensitiveness to the action of water of glass windows for looking into the pressure apparatus. This effect first forced itself on my attention in connection with the great discrepancy between the results which Poulter was getting with his pressure windows and my results, the pressures which he could support being much higher than those which I could attain. This I finally traced to the effect of the pressure medium, water invariably breaking the windows at much lower pressures than could be supported by various other liquids such as oils, alcohols, or even glycerin. Work by Langmuir on the surface condition of glass had suggested that water goes into solution on the surface of

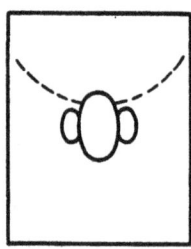

FIG. 88. The fracture in a bubble of pyrex glass after exposure to 380,000 psi. The dotted fracture developed between 2 and 3 years after exposure.

glass to a thickness of 200 molecules deep. It was to be expected that this effect would be magnified by pressure, with the possibility that water might be forced pretty deep into the glass. Consistent with this picture was the experimental fact that in some experiments the occurrence of fracture after exposure to pressure was sensitive to the rate of release of pressure. With slow release fracture would not occur, while it did occur with rapid release. It was as if the water entrapped by the pressure in the glass had to be given time to find its way out on pressure release; otherwise if it could not escape it would expand *in situ* on release of pressure and thus cause fracture.

It suggested itself that water might be actually forced through glass under proper conditions. To this end experiments were made with several capillaries of a 4 to 1 ratio of external to internal radius, OD 0.1 in., with hemispherically closed ends, of each of four varieties of glass: pyrex, soda-lime, Scotch combustion, and lead. These capillaries were exposed under water for a week to a pressure of 170,000 psi. The temperature was maintained at 50°C to hasten any possible action. Pressure was released slowly through 24 hours. The lead-glass capillaries were completely collapsed to a fine powder. There was no trace of moisture in the interior of the capillaries of the three other kinds of glass. It is evident therefore that any dissolved water that produces the fracture effects must be present in excessively minute quantities. This exposure to pressure produced no observable permanent change in the pyrex and soda-lime capillaries, but the Scotch-combustion glass showed a curious effect on standing after release of pressure. This presently separated itself into two concentric tubes, as indicated in Fig. 89. The outer tube subsequently developed longitudinal cracks and fell apart, while the inner tube remained intact. There has been no further change in the appearance of these tubes after 17 years.

FIG. 89. Shows how a tube of Scotch-combustion glass separated into two concentric layers after exposure to pressure in water.

A dummy experiment was made after these with water, the pressure medium being kerosene, and pressure being maintained at 170,000 psi for a week at 50°C. Quartz glass was added to the four other varieties. Pressure again was lowered through 24 hr. Again the lead-glass capillaries were found crushed, but there was no perceptible change in the appearance of any of the others or any change of dimensions measurable with a micrometer.

The consistent fracture of the lead glass in the two experiments invites speculation. It is known that lead glass has a low melting point. It may be that the prolonged exposure to pressure at the elevated temperature produced enough plastic flow to change the geometry sufficiently to permit fracture. The experimental arrangements were not such as to permit a decision as to whether fracture took place with increasing or decreasing pressure.

Materials Other Than Glass. The interior surfaces of the capillaries in the experiments just described were formed by fusion and therefore must have had a high degree of smoothness. In view of this the failure to produce fracture by stresses or strains as high as those reached in the experiments above acquires a certain plausibility. For if one inquires *where* at the internal surface fracture should start, one is hard put to find a unique place, the surface being homogeneous down to molecular dimensions. The fracture, therefore, having no place from which to start, does not start. But if the interior surface is less perfect, so that there are places of stress concentration, then it would be expected that fracture might occur. This is exactly what happens if the hole in the specimen is made mechanically, as by grinding. A number of experiments have been made on several varieties of single crystal and rock (quartz, tourmaline, calcite, feldspar, barite, porphyry, andesite, granite, and limestone) and are

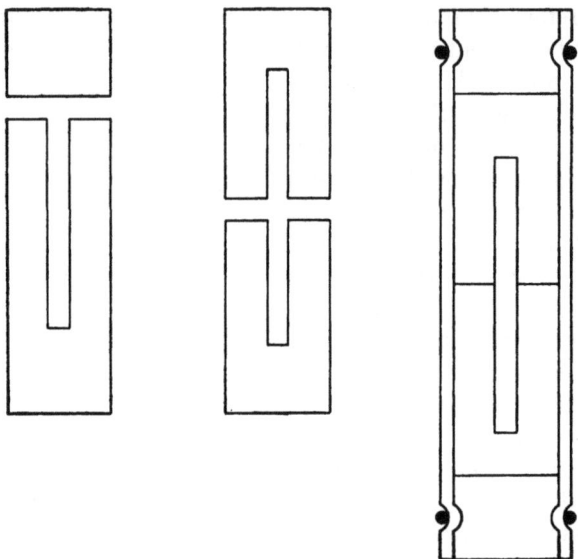

FIG. 90. The general shape of the specimens for collapsing tests on crystals and minerals, and the method of mounting for exposure to pressure. The abutting surfaces are optically flat.

described in B4 where the full details may be found. These specimens were exposed to external pressures up to 175,000 psi. The specimens were given the form shown in Fig. 90. Stress was transferred to the external surface from the pressure medium through rubber tubing as indicated, so that there was no complication from the surface action of the transmitting liquid.

Under these conditions failure always occurred, and in a new and characteristic way. Above a critical pressure, depending greatly on the material, small spicules were projected from the internal surface into the cavity. The rate at which these spicules were detached increased rapidly with increasing pressure. Eventually the cavity became full of the detached debris, which packed itself closely enough to exert an internal pressure which stopped the process. The cavity usually eroded itself to several times its original volume before the process stopped. The spicules were projected with considerable velocity; if a brass rod were placed on the axis of the cavity the spicules would embed themselves in it, giving the rod a frosted appearance like very fine sandpaper. One is reminded of the spalling which takes place in deep mines and which can be so dangerous. The mechanisms in the two cases may be similar.

A complication is introduced into this situation by the method of closing the cavity, and the erosion was usually greatest around the mouth of the borehole. This was in spite of the fact that the abutting surfaces at the closure were made optically flat. It was, however, possible to convince oneself by examination of the many specimens that, although the effects are exaggerated in the neighborhood of the mouth, they are characteristic of the cylindrical cavity and are not dependent on end effects.

In addition to the eroded cavity packed with debris, in many cases the outer walls of the specimens carried a few cracks. This was a different sort of phenomenon, however, and may have occurred on release of pressure. Specimens cracked in this way could be exposed again to external pressure with no apparent loss of strength for that type of stress.

A cylinder of optical glass was tried in addition to the materials mentioned above. This also fractured completely under an external pressure of 175,000 psi, although it withstood 70,000 psi without effect. This again would suggest the importance of the character of the internal surface.

The importance of the smoothness of the internal surface is confirmed by experiments on negative quartz crystals. It is well known to geologists that quartz crystals occur not infrequently in nature with internal cavities. These cavities may be partially filled with a liquid (CO_2) or may be empty. Many of these cavities have the geometrical shape of the crystal, with plane faces and sharp corners, and hence are called negative crystals. The faces of these negative crystal are as highly polished as the regular external surface of the normal crystal. A number of quartz crystals containing such negative crystals were exposed to a pressure of 250,000 psi in kerosene, with no obvious effect on the negative crystals. In some cases, however, the same crystals contained irregular-shaped cavities formed in some different way from the negative crystals. These cavities were sometimes, after exposure to pressure, filled with finely eroded sand as in the experiments on the drilled cylinders. Here we have, therefore, in a single experiment, no failure when the surface is perfect but failure by "flaking off" when the surface is less perfect.

The state of affairs with the negative quartz crystals described above was observed immediately after release of pressure. The crystals were then put away and were not looked at again until some 25 years later. It was now found that in a number of instances minute internal cracks radiated from the corners of some of the negative crystals. There must therefore have been, even in the case of crystal quartz, some flow in the regions of high stress concentration, a flow which slowly accommodated itself and eventually produced fracture.

The elastic distortion of the crystalline cylinders under external pressure differs from the solution written above for isotropic two-constant material. The rigorous solution for the crystalline cylinder seems never to have been obtained. It may be mentioned that I obtained an approximate solution by what is essentially a perturbation method, involving the variation of elastic constants from their isotropic values (see B4). Under external pressure a crystalline cylinder does not maintain its circular outline nor does the plane cross section remain plane; both section and the circular outline are distorted with a period depending on the crystal symmetry. Although the deviations from isotropic symmetry are small, they are nevertheless important, for unless the two halves of the specimen are fitted together in the orientation of the original crystal the fracture in the neighborhood of the closure is much more serious than otherwise.

Compacting of Powders. Closely related to the collapse of cavities in brittle substances under external pressure is the compacting of powders by pressure. The spectacular thing here is the resistance to complete compacting offered by any sort of powdered material. For instance, the cubic compressibility of finely powdered lead remains too high by the order of 50 per cent to pressures well above 400,000 psi. It becomes sensibly normal only in the neighborhood of 1,500,000 psi.

The resistance of powders to complete compacting involves two effects. There is in the first place internal friction. It is well known to powder metallurgists that if a powder is put in a long cylindrical mold it is practically impossible to compact it by pushing on it with a piston at the end. The reason is the friction on the walls by which the lower part of the charge is shielded from the pressure exerted by the piston at the top end. This effect is more or less adventitious and may be avoided by compressing a thin disk or by packing the powder into a sealed rubber or lead tube which is then subjected to external hydrostatic pressure. But under these conditions it is almost impossible to obtain a finally compacted product which does not fail by a few per cent of its theoretical density. An extreme example is powdered graphite, which can with the greatest difficulty be got up to 90 per cent of theoretical density. The volume compressibility of such compacted graphite remains abnormally high and the general response to pressure is pseduo-elastic.

All this means that in compressed powders there are cavities around which there must be very high values of stress, implying great strength. The example of glass shows that such great strength is to be expected if the surface conditions are right. In a compacted powder it would seem that such surface conditions might be produced automatically by the previous separation of fragments from the surfaces.

CHAPTER 10

WIRE DRAWING AND EXTRUSION UNDER PRESSURE[1]

The very high values of flow stress which can be imparted to steel at the large strains achievable under hydrostatic pressure cannot fail to be exciting to the cupidity of anyone capable of imagining the industrial possibilities of materials with such greatly enhanced properties. This vision is obscured by at least two clouds. There is in the first place the question as to whether the greatly enhanced properties imparted by pressure will be retained when pressure is released to atmospheric. This question will be discussed in great detail later when we have to consider the general question of the effect of various sorts of prestraining. For our present purposes the answer is essentially "yes," for it will prove that by far the greater part of the enhancement imparted by pressure is retained on release. The second cloud is the realization that this great enhancement of properties is highly localized in the neck where it can be exploited only with difficulty. If the region of enhanced properties could be extended into a wire, then we would have something worth while. Now in an ordinary wire we do have an enhancement of properties, and this is associated with the strain-hardening under the high strains produced by the wire-drawing process. But there is a limit to the hardening that can be imparted by ordinary wire drawing because the wire presently gets so hard that it breaks, and this has to be avoided by annealing, when the effect of the strain is lost and the hardening process has to begin over again. It suggests itself that if the entire drawing process could be conducted in a medium under high pressure we could take advantage of the ductility imparted by pressure to continue the drawing further before annealing and so realize greater degrees of strain-hardening and so greater strengths.

A first attempt to do this has been made and will be described here. The experimental details are here described for the first time; the results have been stated incidentally in the course of a discussion of a paper of Lynch, Ripling, and Sachs. The apparatus was more or less improvised out of pieces of apparatus already at hand and was restricted to operation at a pressure of 170,000 psi. The essential idea of the pressure apparatus

[1] Most of the material of this chapter has not been published. Some of the results have been referred to in *AIMME Tech. Pub.* 2449, p. 102, August, 1948, in connection with a discussion of a paper by J. J. Lynch, E. J. Ripling, and G. Sachs.

WIRE DRAWING AND EXTRUSION UNDER PRESSURE

is indicated in Fig. 91. The pressure vessel consists of two chambers with a connecting member. Pressure is produced by the advance of a piston into the first chamber. In the first chamber is the slug of metal to be drawn, so shaped that a short projection of the final diameter passes through the drawing die, the die also being in the first chamber. Attached

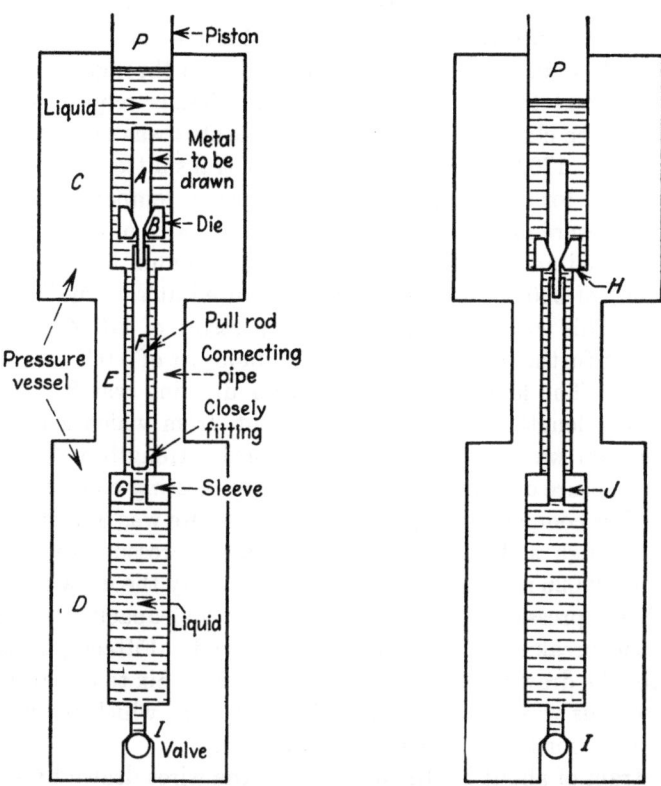

FIG. 91. Schematic diagram of apparatus for wire drawing under pressure, in the initial configuration.

FIG. 92. Schematic diagram of the apparatus for wire drawing under pressure, in the configuration immediately before the drawing, which is made to occur by releasing the valve I.

to the part projecting through the die is the "pull rod," which passes through the member connecting the two chambers and almost enters a closely fitting sleeve at the entrance to the second chamber. In operation, pressure is raised to the desired amount by advance of the piston. When the requisite pressure has been reached, the piston comes in contact with a shear pin, which is thereupon sheared through and releases a spring acting on sliding sleeves in a way not indicated in order not to increase the complication of the diagram. By the action of the spring the entire system consisting of die, slug of metal to be drawn, and attached pull rod is advanced against a stop at the bottom of the first chamber, the pull rod at the same time entering the sleeve at the entrance to the second chamber. The second stage, after release of the shear pin, is shown in Fig. 92. Pressure is now released in the second chamber by opening a valve at its lower end. The pressure in the first chamber thereupon drives the pull rod into the second chamber through the sleeve, carrying with it the end of the slug, which is thereby drawn through the die. It will be understood that in operation various details require attention which need not be elaborated here. The apparatus was used, as stated, at a pressure of 170,000 psi. Pressure was transmitted by kerosene, which is suitable for this range of pressure at room temperature.

Experiments were made on a commercial piano wire of diameter 0.076 in., which showed an initial tensile strength (maximum load divided by initial section) of 330,000 psi and which fractured at a "true" stress of 480,000 psi. This wire was drawn under a pressure of 170,000 psi in six passes, without further annealing, to a final diameter of 0.026 in. (natural strain of 1.9). The length of the wire after drawing was 8 in. From each drawing a length of about 1 in. was cut, from which a miniature tension specimen was fashioned, and the tensile strength and load and diameter at fracture determined in the miniature testing machine already described. For comparison, a piece of the same wire was drawn, without annealing, through a draw plate at atmospheric pressure to the same final diameter, 0.026 in. It was not possible to carry the drawing further at atmospheric pressure, the wire having lost all capacity for further extension, and breaking with no preliminary yield. Fifteen passes were necessary for the drawing at atmospheric pressure. Six tensile tests were also made on this wire, at the intermediate stages at which the diameter was the same as for the drawing under pressure.

Up to a strain of about 1.3 the behavior of the wires drawn by the two methods was indistinguishable within the rather wide scatter of the results. Tensile stress, true stress at fracture, and retained ductility were the same function of the drawing strain. Tensile strength and true stress at fracture were both increasing linear functions of strain. But beyond a drawing strain of 1.3 the wire drawn at atmospheric pressure began to deteriorate, the tensile strength and true fracture stress diminishing, until at a drawing strain of 1.9, the remanent ductility was zero, and the tensile strength and the true fracture stress coincided at 430,000 psi. The wire drawn under pressure showed no comparable deterioration; in fact, because of the scatter of the results it is not certain that there was any deterioration at all. At a drawing strain of 1.9 the remanent ductility was 0.29, the tensile strength 530,000 psi, and the true stress at fracture 720,000 psi. The maximum figures observed at any stage of the drawing process were 580,000 and 740,000, respectively. The maximum corresponding figures at any stage of the drawing process for the wire drawn at atmospheric pressure were 510,000 and 700,000, respectively, higher than any figures which I have seen quoted in the literature.

These results show the expected difference between the properties of wire drawn under pressure and at atmospheric pressure, and thus confirm the general point of view. The differences were, however, not striking enough to make further exploitation for commercial applications inviting

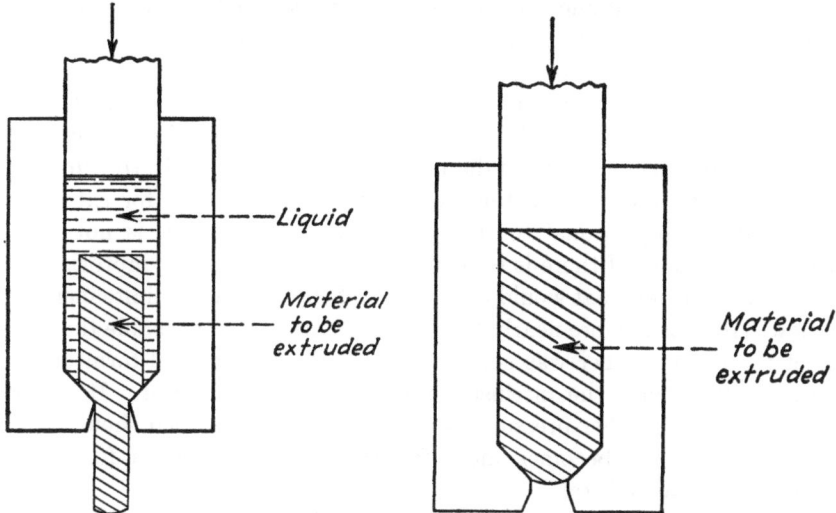

FIG. 93. Schematic drawing of apparatus for extrusion of material while under hydrostatic pressure.

FIG. 94. Extrusion apparatus without cooperation of pressure. The friction on the walls may be so great as to prevent extrusion completely.

at this time. If pressures twice as high could be handled, striking enhancement of properties might be anticipated. The reason that greater differences were not produced by a pressure of 170,000 psi is doubtless that in the throat of the die under normal conditions there are compressive stresses which have much the same effect as hydrostatic pressure. This is doubtless the reason that drawing is possible at all. In order that drawing under pressure should produce marked enhancement of properties it is to be expected that the hydrostatic pressure should be materially greater than the compressive stresses naturally occurring in the throat of the die.

Closely related to the production of wire by drawing in a liquid under hydrostatic pressure is the formation of wire by extrusion from a chamber under hydrostatic pressure. The general scheme is indicated in Fig. 93.

If one attempts to extrude the material without the aid of hydrostatic pressure, filling the chamber with the material and pushing directly on it with a piston, as indicated in Fig. 94, it will be found that pressure is not transmitted to the extruding orifice because of lateral friction on the sides of the chamber, so that pressures high enough to split the extruding chamber are reached before extrusion begins. No such difficulty is encountered, however, if the chamber is filled with liquid, and extrusion begins when the flow stress of the material is reached, which for the softer materials is much below the pressures which steel chambers can withstand. Of course, if one is contented with a small reduction of area at a single pass, steel may be extruded from steel chambers.

It is a property of this method of producing wire that much greater reductions of diameter are possible at a single pass than could possibly be reached by drawing. Thus I have extruded copper with a reduction of area of 16-fold (reduction of diameter to one-quarter). In the limits of my work, which was not extensive, the pressure required for extrusion was roughly proportional to the reduction of area. This is not what would be expected if there is for copper, for example, an asymptotic flow stress but is rather what would be expected if there is an indefinite increase of flow stress, as is more nearly the case for steel. It is probable that the effect found for copper was connected with the geometrical form of the die as well as with the intrinsic properties of the material. It is not unlikely that at higher pressures a more nearly right-angled shoulder on the extrusion dies becomes more favorable—there is room for much further experiment here.

In the limits of my work I found that another effect enters if one tries for too great reductions of area at too high pressures. This effect is a spitting out of the metal in gulps at the highest pressures instead of smooth emergence. Or, if the metal does not spit out, it may break along diagonal shear planes at the mouth of the die. Or there is still another limitation. At the higher pressures emergence may be so rapid as to result in large heating effects. If the wire is heated in the act of extrusion to its annealing temperature, benefit of the process is lost.

A quantitative study of the properties of extruded wires as compared with drawn wires was made by Mr. Howard R. Spendelow, Jr., of the Union Carbide and Carbon Corp. at my instigation and was reported on page 100 in the same discussion of the paper of Lynch, Ripling, and Sachs quoted above. These experiments were made on steel and copper. No difference was found in the properties of the wires made by the two methods, but the strain seemed to be the one significant parameter. It was verified that extrusion made possible greater reductions of area at a single pass. Except for this there seemed no difference which was inviting enough for industrial purposes to justify further investigation. I believe, however, that there is still work to be done here. The reductions of area by extrusion were not carried beyond those to which wire could be drawn and in this range were not carried to the point where the drawn wire begins to deteriorate. At greater reductions of area the extrusion method may offer possibilities. This will demand an elaborate study to find the reasons for the spitting out of the metal at the highest pressures and methods of preventing it. There may also be possibilities in the combination of drawing and extrusion under pressure.

Part II

OTHER TESTS INVOLVING LARGE DEFORMATIONS

CHAPTER 11

SIMPLE COMPRESSION[1]

Introduction. The production of large strains, in either extension or compression, encounters technical difficulties which require the use of special devices. In the case of extension, the tension specimen becomes unstable, with necking and fracture under normal conditions at natural strains seldom materially exceeding unity. By pulling the specimen under hydrostatic pressure we have seen that indefinitely large strains may be realized without fracture. In the case of simple compression the difficulty in reaching large strains is not instability or fracture, but change in the type of deformation due to barreling of the specimen and change in the type of stress due to end friction. The percentage effect of the failure of uniformity in either stress or strain becomes rapidly greater as the compression of the specimen proceeds with reduction of length and increase of diameter and consequent decrease in the ratio of length to diameter. This ratio may not be increased indefinitely before compression because of instability by buckling.

A method of avoiding the effect of progressive increase of nonuniformity of stress and strain as compression progresses is to perform the compression in stages, remachining the specimen back to the original proportions between stages with continual decrease in the absolute size. A disadvantage of this method is that stress and strain do not increase monotonically, but disturbances are introduced by the stepwise release and reapplication of stress. However, these disturbances are known to be comparatively unimportant, whereas the disturbances arising from nonuniformity, when a large compression is made in a single stage, are prohibitive. This method of compression in stages was apparently first applied by Taylor and Quinney[2] to a study of copper. They determined the strain-hardening curve in simple compressions up to reductions of length to one-fiftieth of the original length and natural strains of approximately 4 ($\log_e 50 = 3.91$) and found that the compressive stress reached

[1] This chapter is based on material contained in the fifth NDRC report, the eighth Watertown report, and B21.
[2] G. I. Taylor and H. Quinney, *Proc. Roy. Soc.* (London), **143**, 307–326, 1933–1934.

a limiting value asymptotically for the larger strains, quite unlike the behavior in extension. I have applied the method to an armor plate in a report to the NDRC (OSRD No. 3019) up to reductions in thickness to one-twentieth of the original, that is, to a natural strain of 3, with the result that the strain-hardening curve continued to rise with increasing strain, linearly to a first approximation, and to a second approximation at a slightly increasing rate. The behavior was essentially the same for specimens cut from three mutually perpendicular directions in the plate. I also confirmed in the NDRC report the result of Taylor and Quinney for copper. The behavior of steel is thus qualitatively different from that of copper. It is evidently important to extend this investigation to other grades of steel. This work was undertaken for the Watertown Arsenal in the eighth report, which is largely reproduced here.

Experimental Method and Details. These experiments are concerned with various aspects of the behavior of highly compressed steels of 10 different heat-treatments of specimens cut from the same original round bar 1.75 in. in diameter, and of the following composition, as determined by special analysis at the Arsenal: C 0.40, Mn 1.14, Si 0.48, S 0.019, P 0.016, Ni 0.26, Cr 0.38, Mo 0.13, V nil. Ten 6-in. lengths were cut from the bar. Five of these lengths were quenched and tempered back to 250, 350, 450, 550, and 650°C, respectively. The other five were austempered at 450, 500, 550, 600, and 650°C, respectively. From each of these 10 specimens two specimens for the compression tests were machined of 1.500 in. length and 1.500 in. diameter. Each stage of the compression was carried to a nominal reduction of length to two-thirds original, after which the specimen was machined back to the original 1 to 1 ratio of length to diameter, reducing the length during the machining by only the minimum amount necessary to true the ends. In detail, after the first stage of compression, the length was approximately 1 in. and the diameter 1.83 in.; the specimen was then machined to a length of 1 in. and diameter of 1 in. The second stage of compression reduced the length to 0.67 in., increasing the diameter to 1.22 in., after which it was machined to 0.67 in. length and 0.67 in. diameter, etc. One of the two compression specimens from each sample was carried through seven stages of compression to a final length of approximately 0.08 in. and total strain of approximately 2.8. The discrepancy between the final strain and the theoretical value 2.93 ($= \log_e 1.50/0.08$) arose from the slight amount of material removed on truing the ends. The second of the specimens from each sample was usually carried through three stages of compression to an approximate length of 0.42 in. and approximate strain of 1.25. It was then set aside and used in other experiments on the effect of prestraining which will be described later.

The successive stages of compression were performed in four hydraulic presses with piston diameters of 8, 3.5, 2.5, and 1.625 in., respectively. Specially designed packing reduced the friction of the pistons of the presses to negligible values. For the first stage of compression in the largest press platens of hardened steel were used, 2 in. thick. For all the later stages of compression platens of polished carboloy, reinforced with steel rings, were used. Carboloy is advantageous because of both the lower friction and the smaller deformability as compared with steel. The total compressive load on the specimen was calculated from the pressure actuating the press, given by Bourdon gauges calibrated against free piston gauges. Friction on the ends of the compression specimen was minimized by two thicknesses of 0.002-in. lead foil smeared with a lubricating paste of glycerin and colloidal graphite. At places where the lead remains in contact with the steel the frictional force cannot increase beyond the shearing strength of the lead. Previous measurements of shearing strength of lead under pressure show that at the highest pressures reached in this work (approximately 400,000 psi) the shearing strength of lead is less than 3,000 psi. However, the lead does not remain between specimen and platen over the entire compression face but cuts through at the edges, allowing steel to come directly in contact with steel or carboloy. This region of direct contact is manifested as a brightly burnished ring around the outer edge of the compression face; practically all the friction is contributed by this ring.

Fig. 95. Double bulge in simple compression specimens frequently obtained on the second compression after refiguring.

The width of the burnished ring varied with the conditions; as an average it was something of the order of 5 per cent of the diameter. The greater part of the compression face, where it was protected by the lead, showed characteristic patterns after the compression, reminiscent of the patterns obtained by deep etching. Flaws in the steel are interestingly brought out in this way, and a study of the patterns might well be worth while for its own sake. No systematic examination was made of this matter, however, in the present investigation.

After the first stage of compression the specimen was invariably distorted into a slight barrel shape, the maximum diameter at the center being of the order of 3 per cent greater than the minimum diameter at the ends for a shortening corresponding to an increase of initial diameter by 25 per cent. In the later stages of compression the distortion was not always to a barrel-shaped figure; distortion with a slight double bulge as shown in Fig. 95 was not uncommon. The explanation of this type of distortion is obviously the greater strain-hardening in the previous stage of compression of the equatorial regions due to greater distortion at the center of the bulge. This sort of distortion is self-correcting, and the difference between extreme diameters did not rise above a few per cent.

Simultaneous measurements were made of the pressure driving the hydraulic press and the distance between the platens of the press, the latter determined with a suitably mounted Ames dial gauge. For each stage of compression, with a shortening of approximately two-thirds, about 15 sets of readings were taken, spaced roughly uniformly over the entire range of distortion. The readings were made at intervals of 2 min, this interval being sufficient for the approximate cessation of creep. After completion of a stage of compression and removal from the press, the specimen was measured with a micrometer. The stress at the final point of a compression sequence was calculated from the final compressive load and the average final cross section. The strain corresponding to the final point was calculated as the natural logarithm of the ratio of initial to final length. The stress-strain values at intermediate points of each compression sequence were calculated from the running values of the load and the over-all shortening, the latter being corrected for distortion of parts of the press by suitable blank experiments. The running value for the cross-sectional distortion was distributed between initial and final values in proportion to the corrected values of the shortening. In some cases only the final points of the stages of compression were computed; the interest in computing all the points is greatest for the low strains during the initial stages where the strain-hardening curve is not straight. After each stage of compression the Rockwell C hardness was measured. For this purpose the two ends were trued in the surface grinder, flat and parallel, and the hardness determined on each end face at four points spaced at 90° around the periphery as near the edge as permissible. The average of the eight readings was taken as the hardness; the range of the extreme readings was seldom as much as two points. After the hardness measurements the diameter was reduced in the lathe to the value appropriate to the next stage of compression, without touching the ends again, except for stoning off the edges of the hardness pits. This procedure was applied to the first five stages of compression; by this time the diameter was so much reduced that peripheral readings were no longer permissible. The hardness for the last two stages of compression was determined from a reading at the center of a single face. The last two hardness readings may be expected therefore to have considerably more scatter and to be less significant than those for the first five stages.

Presentation of Results. The more important of the data, that is, the values for the final points of each stage of compression, are collected in Tables XIV and XV. In Figs. 96 and 97 the compressive stress of the final point of each stage of compression is shown as a function of the cor-

SIMPLE COMPRESSION

Table XIV

Treatment of steel	First specimen			Second specimen		
	Strain $\log_e \frac{l_0}{l}$	Compressive stress, psi	Rockwell C hardness	Strain $\log_e \frac{l_0}{l}$	Compressive stress, psi	Rockwell C hardness
Tempered at 650°C	0	....	22.0	0	94,000	21.5
	0.400	165,000	27.6	0.395	149,000	27.8
	0.802	179,000	31.8	0.801	178,000	31.6
	1.202	186,000	33.4	1.205	191,000	33.7
	1.597	195,000	35.0			
	2.013	211,000	37.1			
	2.362	227,000	38.4			
	2.790	236,000	40.1			
Tempered at 550°C	0	....	33.0	0	135,000	32.7
	0.402	193,000	32.8	0.325	196,000	33.1
	0.801	211,000	36.1	0.734	206,000	36.8
	1.216	222,000	38.2	1.139	250,000(?)	38.0
	1.586	232,000	39.8			
	1.989	256,000	40.7			
	2.390	280,000	42.3			
	2.838	296,000	44.1			
Tempered at 450°C	0	....	40.7	0	189,000	40.6
	0.220	211,000	38.7	0.223	219,000	39.4
	0.585	230,000	....	0.622	236,000	41.4
	0.668	249,000	42.5	1.023	262,000	42.5
	1.064	266,000	43.5			
	1.483	280,000	44.4			
	1.915	299,000	46.7			
	2.340	334,000	47.1			
	2.759	346,000	50.2			
Tempered at 350°C	0	233,000	48.3	0	213,000	47.7
	0.405	293,000	47.1	0.393	279,000	47.8
	0.804	307,000	48.5	0.702	290,000	48.6
	1.196	313,000	49.8			
	1.593	339,000	51.8			
	1.999	364,000	51.9			
	2.441*	384,000	53.2			
Tempered at 250°C	0	289,000	52.2	0	260,000	51.9
	0.392	357,000	51.7	0.406	346,000	52.9
	0.801	366,000	52.5	0.821	371,000	53.9
	1.200	(?)	53.8			
	1.594	394,000	55.4			
	Breaks					

*Breaks on next compression.

TABLE XV

Treament of steel	First specimen			Second specimen		
	Strain $\log_e \frac{l_0}{l}$	Compressive stress, psi	Rockwell C hardness	Strain $\log_e \frac{l_0}{l}$	Compressive stress, psi	Rockwell C hardness
Austempered at 650°C	0	57,000	14.5	0	56,000	13.8
	0.397	153,000	25.7	0.409	155,000	25.2
	0.808	169,000	29.2	0.829	171,000	29.7
	1.201	178,000	31.5	1.241	182,000	31.5
	1.609	191,000	32.5			
	2.004	202,000	34.7			
	2.386	226,000	36.3			
	2.810	249,000	38.2			
Austempered at 600°C	0	67,000	17.3	0	66,000	16.9
	0.400	156,000	26.1	0.400	159,000	26.0
	0.799	175,000	30.4	0.818	174,000	30.7
	1.214	183,000	32.1	1.227	185,000	32.4
	1.616	195,000	33.6			
	2.016	205,000	35.0			
	2.515	236,000	40.4			
	2.933	249,000	40.4			
Austempered at 550°C	0	84,000	24.4	0	87,000	23.8
	0.392	170,000	30.0	0.409	174,000	28.7
	0.789	183,000	32.9	0.826	188,000	32.5
	1.192	193,000	34.4	1.194	195,000	34.3
	1.597	208,000	36.0			
	1.980	216,000	37.0			
	2.474	252,000	40.7			
	2.876	294,000	41.0			
Austempered at 500°C	0	95,000	25.6	0	92,000	26.3
	0.381	173,000	30.7	0.391	176,000	30.5
	0.783	191,000	34.1	0.795	193,000	34.5
	1.175	206,000	35.7	1.203	211,000	36.0
	1.565	215,000	37.0			
	1.954	228,000	38.2			
	2.457	256,000	40.5			
	2.862	289,000	41.2			
Austempered at 450°C	0	88,000	26.8	0	90,000	26.3
	0.378	181,000	31.1	0.389	178,000	30.8
	0.777	196,000	35.0	0.798	196,000	34.6
	1.162	211,000	36.6	1.195	215,000	36.8
	1.559	222,000	37.8			
	1.960	229,000	38.5			
	2.451	252,000	43.6			
	2.860	287,000	43.0			

responding strain, in Fig. 96 for the tempered series and in Fig. 97 for the austempered series. The first three or four points plotted for each treatment are the average of the corresponding points for the two specimens; the later points are for the first specimen alone. Inspection of the tables shows close agreement between the two specimens in their common range, so that it is not necessary to try to distinguish between them.

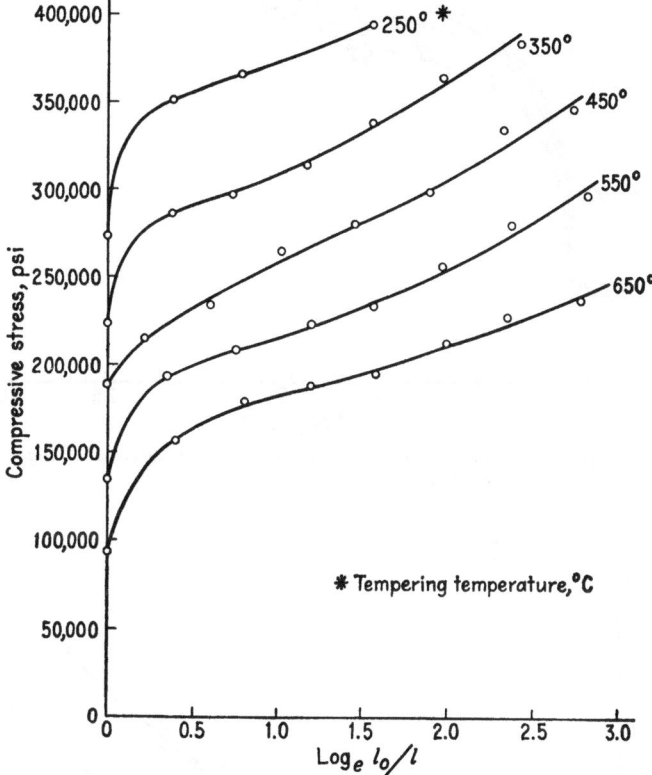

Fig. 96. Compressive stress against longitudinal strain for successive stages of simple compression of the tempered series.

All the hardness data of the tables are plotted in Fig. 98, the values for the tempered series by circles, those for the austempered series by crosses, with accents on the points pertaining to the virgin material, but with no other designation of the points according to specimen.

It would add unduly to the complexity to reproduce all the individual readings for each stage of compression. There are, however, certain transient phenomena connected with release and reapplication of pressure at each stage of compression which are of interest for themselves and which can be shown by a reproduction of all the readings. Two typical

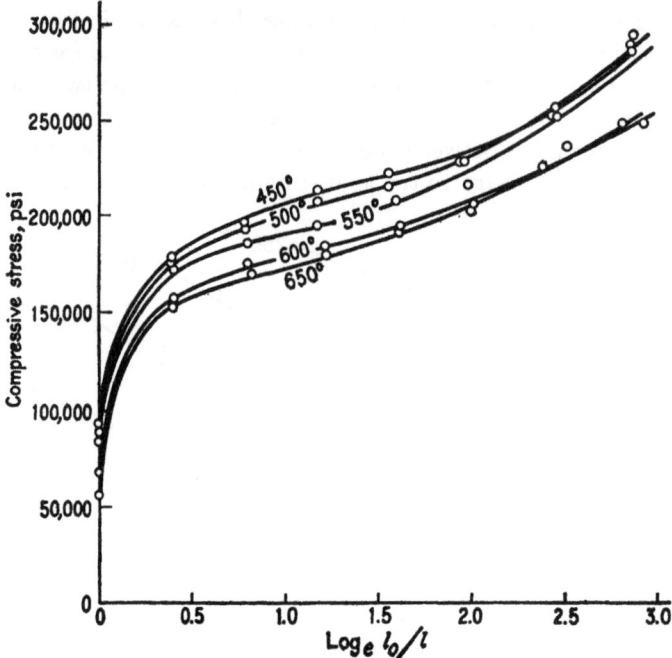

Fig. 97. Compressive stress against longitudinal strain for successive stages of simple compression of the austempered series.

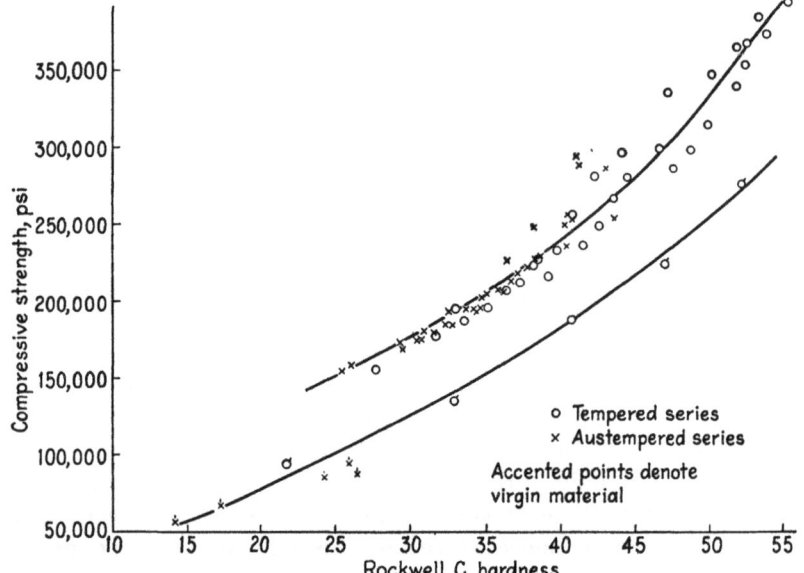

Fig. 98. Compressive strength against Rockwell C hardness for tests in simple compression with refiguring. The lower curve is for virgin material.

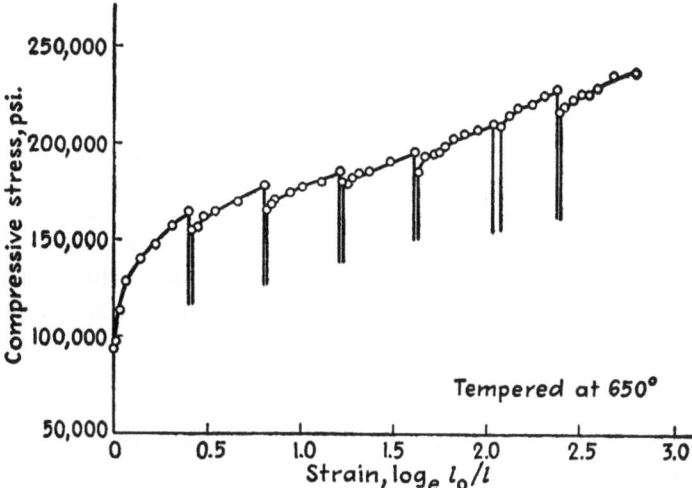

FIG. 99. Detailed results, showing all the experimental points for compressive strength against longitudinal strain for successive stages of simple compression with refiguring, of the specimen tempered at 650°C.

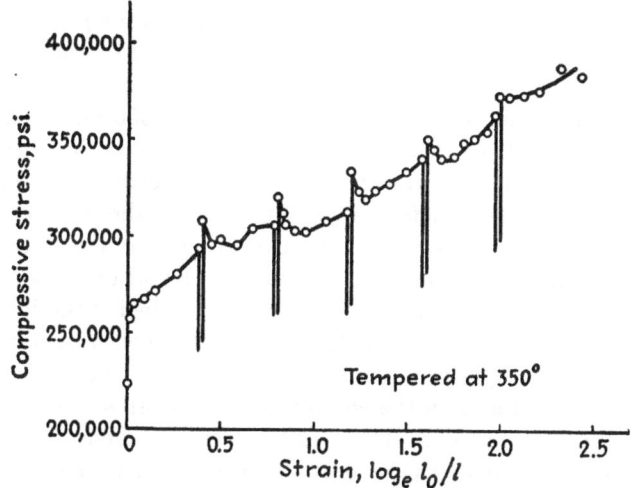

FIG. 100. Detailed results, showing all the experimental points for compressive strength against longitudinal strain for successive stages of simple compression with refiguring, of the specimen tempered at 350°C.

examples, illustrating the two qualitatively different types of behavior, are given in Figs. 99 and 100 for the specimens tempered at 650 and 350°C, respectively.

Discussion of Results. *The Strain-hardening Curve in Simple Compression.* Figures 96 and 97 show that the qualitative results found for armor plate in the report to NDRC apply to all 10 heat-treatments of the steel examined here, namely, the strain-hardening at large compressive strains (that is, strains greater than 0.5) rises with strain up to strains of at least magnitude 3 at a rate increasing slightly at the higher strains. This is to be contrasted with the behavior of copper, for which the strain-hardening approaches a limiting value. The average rate of strain-hardening in simple compression of these 10 heat-treatments is to a first approximation a constant, the same for all. To a second approximation the rate increases somewhat with the hardness. The range of stress values is much greater for the tempered series than for the austempered series. For the latter series a decrease in the tempering temperature from 650 to 450°C increases the flow stress by only one-half as much as is produced by the same decrease of temperature in the tempered series. The curves for the different temperatures in the austempered series lie so close together that the apparent crossing in places is probably within experimental error.

Temporary Effects on Release and Reloading. Figures 99 and 100, giving all the observations for the two series of specimens tempered at 650 and 350°C, show that the effect of releasing load and reloading either may be a depression of the yield point below that reached in the previous application of stress, as it is consistently for the softer steel tempered at 650°C, or it may be an elevation of yield point above the last previous, as it is consistently for the harder steel tempered at 350°C. The interval between release and reapplication of load varied for both series between 2 hr and 2 days with no apparent correlation between the interval of resting and the excess or defect of yield point. In the previous work for the NDRC on a single armor plate, however, there was a definite tendency for the previous yield point to be exceeded if the straining was resumed after a longer interval of time, and not to be reached if the interval was shorter. It will be recalled that this was also the behavior of the hollow cylinders collapsed by pressures up to 90,000 psi. The effect may well depend on the steel and on the complexity of the stress distribution.

Fracture in Compression. The two hardest specimens, namely, those tempered at 250 and 350°C, eventually fractured, as indicated in the tables and the diagrams. The absolute stress at the fracture point was about the same for both, 390,000 psi. The strain was greater for the softer steel, as is to be expected. In both cases failure was by slip along a shear plane at approximately 45° to the axis of compression.

Hardness as a Function of Compressive Strain. Figure 98 plots the compressive stress, that is, the compressive stress by which the material has been hardened by the previous deformation, against the Rockwell C hardness. Since the Rockwell hardness is itself to a large extent a measure of the flow brought about by compressive stresses, it is not surprising to find that the compressive stress is a single-valued function of the hardness, irrespective of the heat-treatment or the amount of strain-hardening. This holds except for the virgin specimens; the points for these lie on another curve, indicated by the symbols with accents, displaced by much more than any possible experimental error from the curve for previously strained material. The explanation of the displacement of this curve is clear in the light of the previous considerations. In the first place, the Rockwell hardness of truly virgin material can, by the very nature of the test, never be measured, since the hardness number is obtained from a deformation which involves flow and strain-hardening. To get back from the measured hardness to a rigorous characterization of the initial state a correction must be applied for the strain-hardening introduced by the act of measurement. If the slope of the strain-hardening curve is small, this correction will be small; if it is high, the correction will be high. Inspection of Figs. 96, 97, 99, and 100 shows that at low strains, that is, in the neighborhood of virgin metal, the strain-hardening curve rises very rapidly, but presently the slope decreases to a comparatively low and constant value. The correction is therefore large for the virgin material and small and constant for material prestrained to stresses corresponding to the straight part of the stress-strain curve, thus accounting for the different curves for virgin and precompressed material.

The explanation of another anomaly in the hardness is not so obvious, that is, the consistent decrease of hardness of the four hardest specimens, those tempered at 550, 450, 350, and 250°C, respectively, after the first stage of compression. It is probable that this effect is due to anisotropy of strain-hardening. In determining the hardness of the flat face the strain produced by the hardness test itself is predominantly compressive in the transverse direction, the direction of previous extensional strain. In the region of low strains the Bauschinger effect is operative, and this would result in a lowering of resistance to the force making the conical impression in the hardness test, and therefore in a lowering of the measured hardness, whereas the Bauschinger effect is not operative at higher strains.

Comparison of Stress-strain Curves for Simple Compression with Simple Tension. In connection with measurements to be described later on tension specimens cut from these specimens strained in simple compression, the tension curves were determined at atmospheric pressure on virgin specimens of all eight heat-treatments. These tension curves at atmospheric pressure give sufficient basis for a comparison of the curves for simple compression and tension. These tension curves are given later in Figs. 152 to 155; only certain of their properties will be referred to here. The tension curves have the well-established shape for such curves, sensibly straight beyond strains of 0.1 or 0.2, with a region of rapid rise at the low strains. The slope of the straight part is practically the same for all eight heat-treatments measured in the virgin condition, the flow stress rising at the rate of 63,000 psi for an increase of unity in the natural strain. This is to be compared with the rate of strain-hardening in compression, which may be obtained from Figs. 96 and 97. Comparison can be only rough, because the stress-strain curve in compression is not straight, but the slope of the line which best represents the results over the range of compressive strain examined is in general less than for the stress-strain line in tension, the rise of the line for compression varying from 30,000 to 50,000 psi for an increase of strain of unity. Furthermore, the absolute values of stress on the tension curves consistently stand several 10,000's psi higher than the stresses for the same strains on the compression curves. A further difference is that the initial curved part of the stress-strain curve extends to higher strains for compression than for tension. In both cases the extent of the initial curved portion decreases with increasing hardness. In summary, higher stresses are in general encountered in tension than at the same strains in compression.

CHAPTER 12

VOLUME CHANGES IN THE PLASTIC STAGES OF SIMPLE COMPRESSION[1]

Introduction. For most practical purposes it is a good enough approximation to assume that volume changes in the plastic range are unimportant. In the actual solution of specific problems in plastic flow this assumption about volume change can be treated in different ways. In the majority of cases it is sufficient to neglect the volume change altogether by imposing such a condition on the strains that the associated change of volume vanishes. For example, this was the assumption made in our analysis of the collapse of cylinders under external pressure.

A second stage of approximation, which neglects strain-hardening, treats the volume change as constant during plastic flow and equal to the elastic change under the yield-point stresses. The third stage of approximation, which I think represents the most advanced stage at present, takes account of strain-hardening and recognizes that the volume may vary during plastic flow because of the variation in stress associated with hardening. The simplest assumption here is that the volume change is proportional to the mean of the three principal stresses, with the same constant of proportionality which holds in the elastic range. This assumption would not be inconsistent with the assumption of isotropy in the strain-hardening range, but this assumption is known not to be a good approximation, except for small strains.

Although the factor of volume change is not important in the majority of cases, there are cases where it is important. An example would be the problem of finding the surface separating the plastic from the elastic regions in a massive block of material exposed to stresses above the yield point over parts only of its surface. Another conceivable application would be to the interpretation of seismic waves in the earth. If the effective volume compressibility were different during the process of plastic flow than during static conditions, or if there were anisotropies in the velocity of propagation associated with the direction of plastic flow, conclusions drawn from seismic records might be modified.

Very little has been done in the way of experimental attack on this problem. There is work at the Bureau of Standards by Stang, Green-

[1] This chapter is based exclusively on B27.

span, and Newman[1] on two aluminum alloys and two grades of steel strained in simple tensions up to elongations of 18 per cent. The principal strains were measured directly and the results expressed in terms of an effective Poisson's ratio. Most of the measurements were made on thin specimens cut from sheets, so that only two of the three principal strains were measured and the volume changes could not be obtained. The specimens of one grade of steel were circular in section, however, and the volume change was computed from the measured change of length and diameter, on the assumption of isotropy and unaltered Young's modulus on release. In these particular measurements the longitudinal strains were carried only to 1 per cent. A large part of the volume change in the plastic range was found to be permanent up to nearly 0.1 per cent decrease of density.

The enormous superiority of a direct measurement of the volume change as compared with its indirect determination from measurements of the change of longitudinal and lateral dimensions, as at the Bureau of Standards, is apparent from a simple example. Consider a cylinder of mild steel plastically shortened to 0.85 its initial length by an axial load of 100,000 psi. The change of radius under these conditions, assuming no change of volume, differs by only 0.07 per cent from that which would be calculated assuming the full elastic change of volume corresponding to this stress. This means that any significant deviations of the change of volume from that expected must be described in terms of measurements within a 0.07 per cent range. This is such a difficult problem as to be well-nigh hopeless, and the direct attack seems the only one worth considering.

Experimental Method. In the following a direct experimental attack was made on this problem by immersing the specimen undergoing plastic deformation in a dilatometer, which is filled with a liquid and provided with a capillary open to the atmosphere in which the liquid meniscus moves in response to changes of volume of the contents of the dilatometer. Simplicity of construction suggested simple linear compression as the type of plastic deformation that could be most easily handled. The apparatus is shown in Fig. 101. The specimen S, surrounded by a liquid in the chamber C, is compressed by the piston P_1 against the bed plate B. In order to compensate for the volume swept out by the piston as it descends compressing S, P_1 is coupled to another piston P_2 of exactly the same cross section, so that as P_1 enters the liquid-filled space P_2 is withdrawn by the same amount, and together there is no change of volume and no motion of the meniscus. The adequacy of the apparatus in this respect

[1] Stang, Greenspan, and Newman, *J. Research Natl. Bur. Standards*, **37**, 211–221, 1946.

can be simply checked by displacing the coupled pistons in the absence of a specimen. The motion of the liquid in the capillary is determined by the volume change of the specimen plus a contribution arising from the distortion of the end of the piston and of the chamber C arising from the action of the stresses in the supporting members. The effect of these distortions can be eliminated by using the differential displacement of the

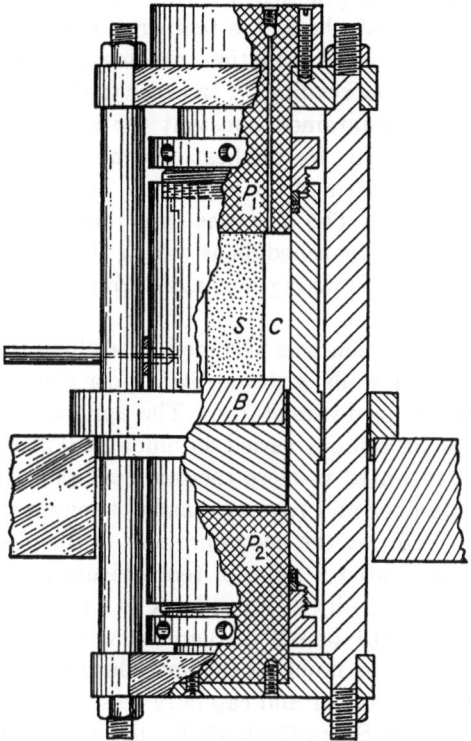

FIG. 101. Dilatometer apparatus for determining the change of volume of the specimen S during simple compression.

meniscus between that given by the plastically deformed specimen and that of a dummy specimen of hardened steel which undergoes only elastic deformation.

The compressive force was applied to the piston P_1 in one of my standard hydraulic pressures, with 3.5-in. piston, actuated by a hand pump having a maximum capacity of 15,000 psi. The pressure was altered in controlled steps by coupling the press to a free piston gauge, the weights of which were varied in appropriate steps. The specimens were almost always 2.5 long and from 1 to 1.5 in. in diameter, depending on the strength of the material. The load was increased in equal steps, usually about 15 in number, until a plastic shortening of the order of 16 per cent was reached. Pressure increments were applied on a uniform time schedule, usually one step every 30 sec. Every increment of load is of course accompanied by temperature effects. In the elastic range these temperature effects were beyond the sensitivity of the apparatus, but in the plastic range, where it is to be expected from the work of Taylor and Quinney[1] that practically all the work of permanent deformation will be converted into heat, the temperature effects in the last and largest stages of plastic deformation become so large as to disturb seriously the course of the readings. In order to eliminate this effect, the maximum load was maintained constant for a long enough interval, usually about 15 or 20 min., for the heat of compression to dissipate itself by conduction to other parts of the apparatus, as shown by the cessation of drift in the readings. The load was then released in the same equal steps and on the same time schedule as during the increasing part of the cycle. The possibility of making readings during unloading is an advantage of this method.

Temperature equalization was hastened by stuffing the space between specimen and chamber walls with narrow copper ribbon, such as is sold for scouring dishes in hardware stores. The apparatus was not thermostatted, but the room seldom varied as much as $0.2°$ during a run. A thermometer was mounted in close thermal contact with the outside of the dilatometer chamber and readings made of temperature drift before and after the run. During the maximum part of the plastic deformation this thermometer indicated a pulse of temperature rise of sometimes as much as $0.5°$, but this rapidly dropped back to a permanent temperature rise of not more than 0.1 or $0.2°$. Blank runs indicated that uniform temperature changes of as much as this introduced no appreciable error. The liquid filling the chamber and capillary was a mixture of water with a rust inhibitor sold by Sears Roebuck for use in automobile radiators. Water was chosen as the base because of its low thermal expansion.

In much of the preliminary work the capillary thread was mercury, separated by appropriate means from the solution in the chamber. The purpose of the mercury was to avoid error from the aqueous solution dragging out on the walls of the capillary, but the error was found to be inappreciable, and since considerable complication was introduced by the use of two liquids, particularly in getting rid of air, the use of mercury was abandoned in the later experiments. Complete absence of air bubbles in any part of the apparatus is essential. To this end the solution was freed from dissolved air by fresh evacuation before each filling, and the filling itself was conducted with the aid of a vacuum pump.

[1] G. I. Taylor and H. Quinney, *Proc. Roy. Soc.* (*London*), **A134**, 307, 1934.

The position of the meniscus in the capillary was adjusted after the apparatus was in place in the press by sucking out excess liquid with the air pump through a fine steel capillary thrust down inside the glass capillary. If there is air in any part of the apparatus the meniscus jumps on applying vacuum to the steel capillary; in this way satisfactory filling could be checked.

In order to function satisfactorily the apparatus must be well made; for this reason the chamber and its supporting rim were turned from the solid piece. The pistons were hardened and ground all over and polished on their cylindrical surfaces. The ends of the specimens were ground to parallelism in the surface grinder. The rubber washers where the pistons enter the chamber are a crucial matter; they must be thin to avoid excessive friction and close-fitting enough to avoid perceptible lead. Excessive friction at the packing reveals itself as an initial abnormal motion of the meniscus in the "wrong" direction on reversing the direction of motion of the piston. Part of this is due to elastic stretch in the tie rods coupling the two pistons, which can be minimized by making the rods large enough in diameter, and part is due to a frictional dragging of the rubber packing with the piston, which can be minimized by making the steel washers which confine the rubber a good fit for the piston. By proper attention to details this abnormal reverse motion of the meniscus was practically eliminated, but at first it proved troublesome.

In the elastic range elementary elasticity theory indicates that the volume change, as given by the motion of the meniscus, should be simply proportional to the load and the length of the specimen and independent of the cross section. For an isotropic material the volume change under a principal stress system which has only a single component Z_z along the axis of the cylinder is $\Delta V/V_0 = (\kappa/3)Z_z$, where κ is the ordinary coefficient of cubic compressibility. Substituting the relations $V_0 = \pi r^2 l$ and load $= \pi r^2 Z_z$ gives at once

$$\Delta V = \frac{\kappa}{3} l \times \text{load} \tag{12-1}$$

As already mentioned, the experiment has to be performed differentially in order to eliminate distortion in the apparatus. In the apparatus as constructed this distortion was comparatively large. When the dummy of hardened steel was used only about 25 per cent of the total motion of the meniscus was due to pure volume compression, the balance being due to distortion. This distortion arises almost entirely from change of length of P_1 under the compressive load. The capillary, which was calibrated for uniformity, had a cross section of 0.01 cm². This, with a specimen of hardened steel 2.5 in. long, gave a total motion of the meniscus of approximately 20 cm under the maximum load of 100,000 lb. Of this 20 cm, 5 cm arose from pure volume compression. The position of the meniscus could be read to 0.1 mm, thus making possible, as far as sensitivity of reading goes, a determination of the effective compressibility to a few tenths of a per cent. Actually, other irregularities made an accuracy as high as this illusory.

In addition to measurements of the volume from the position of the meniscus the change of length of the specimen was read with an Ames 0.0001-in. dial gauge attached to the piston of the press. With this gauge the beginning of plastic flow could be determined.

The fundamental question at issue was how the plot of ΔV against load continues beyond the plastic yield point; does it experience at once a change of character or does it suffer only slow change? One broad feature in the eventual behavior of the curve can be anticipated because it is known that many substances experience permanent alterations of density after exposure to plastic yield and release of the deforming stresses. The permanent alterations of density were given directly by the difference of position of the meniscus before and after the run. In a number of instances the permanent change of density so determined was checked by direct determination of the density before and after by weighings in air and water.

The simplest anticipation of what to expect is that on the first application of load the curve of ΔV versus load beyond the first yield point departs from the linear relation below the yield point by a gradually increasing amount which at the maximum is equal to the permanent change of volume found on release of stress, and that on release of stress the whole curve, except for a correction to be described presently, is displaced by an amount equal to the permanent change of volume, and that it is straight with the same slope as in the initial elastic range. On the second application and release of stress the simplest anticipation is that the curve, corrected as will be indicated, will be straight and have the same slope as in the elastic range. The correction indicated arises from the permanent change of length, since according to Eq. (12-1) the change of volume is proportional to load and length. After the first maximum load the length is shortened by about 16 per cent, so that on first release and second reapplication the changes of volume to be expected are 16 per cent less than at the same load in the initial elastic range. Correction was made for this by multiplying the observed volume changes, after subtracting off the correction for the distortion of the dilatometer obtained from the dummy run, by the ratio of initial to final length.

The simplest anticipation just outlined did not, as a matter of fact, turn out to be realized, but there were departures depending on the individual material. In all cases, and superposed on other complexities, there was hysteresis, and therefore failure of the complete isotropy assumed in deriving Eq. (12-1). This means, even for those parts of the curve which are approximately linear, that the constant of proportionality between load and ΔV is no longer simply connected with the coefficient of cubic compressibility, that is, with the volume change under hydrostatic pressure. This coefficient would, under the circumstances, have to be determined by direct experiment in which a hydrostatic pressure is actually applied to the specimen. However one may, if desired, retain the same equation and speak formally of an "effective compressibility" defined by the equation

$$\kappa_{\text{eff}} = \frac{3\Delta V}{l \times \text{load}} \qquad (12\text{-}2)$$

What the further significance of this effective compressibility is out of its immediate context would have to be found by other sorts of experiment, for obviously the present experiments cannot give a complete description of the behavior under all stress systems of material rendered anisotropic by simple compression.

The actual stress system to which the specimens were subjected was not the simple one component Z_z assumed so far in the discussion, but was complicated, after plastic yield had started, by the addition of frictional components on the ends, which manifested themselves as barreling of the specimen. The magnitude of the barreling was a function of the material; it was a maximum for copper, for which the plastic increase of diameter was 35 per cent less at the ends than at the center, and a minimum for iron, for which it was 21 per cent less. Since in the elastic range shearing stresses are accompanied by no change of volume, it is probably safe to assume that under these conditions the frictional stresses on the ends introduced no appreciable change of volume.

Some 40 experiments were made in all, of which 12 should be regarded as preliminary. The largest number of measurements were made on iron or steel from various sources, including cast iron, various low-carbon steels of commerce, Norway iron, a high-carbon steel, and an 18-8 stainless steel. The other metals were copper, brass, and duralumin. In addition, three rocks were tried: soapstone, marble, and diabase. Measurements were also made on two single quartz crystals. The results obtained with the rocks were unexpected and significant, opening up a new point of view with regard to what might be expected in general. Since the unexpected features shown by the rocks seem to be presented to a much less but still appreciable degree by some of the metals, it will probably conduce to clearness to describe the results for the rocks first.

200   OTHER TESTS INVOLVING LARGE DEFORMATIONS

I owe all the specimens of rocks to the kindness of Professor Francis Birch, who has had much experience in the preparation of cylindrical specimens of various materials for his geophysical experiments.

Experimental Results. *Soapstone.* Preliminary experiments with other specimens indicated the very narrow range within which this brittle material may be expected to support permanent deformation without fracture. The final specimen was exposed to two cycles of loading up to

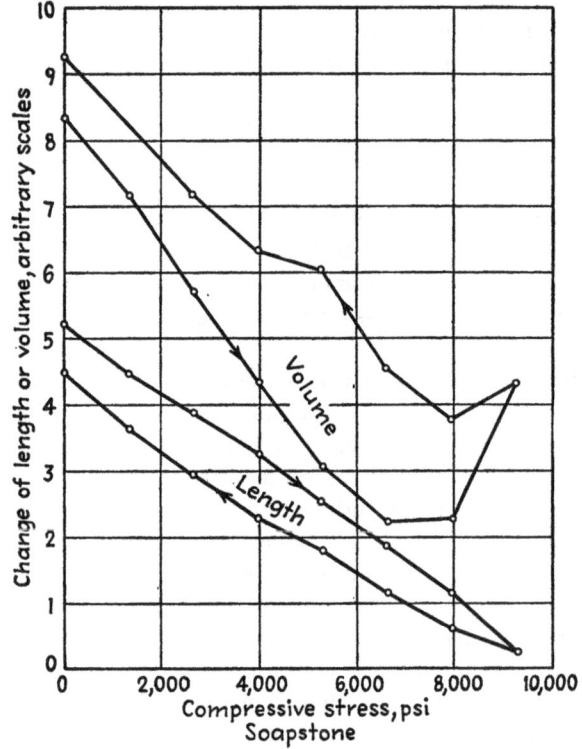

FIG. 102. The changes of length and of volume, both on an arbitrary scale, of soapstone as a function of simple compressive stress.

a maximum of 9,200 psi. The first application of load resulted in a permanent shortening of 0.07 per cent. On the second application there was further permanent shortening of 0.015 per cent, combined with a hysteresis of 0.07 per cent. After the first cycle of loading there was a permanent increase of volume of 0.0069 per cent, or one-tenth the fractional change of length, and after the second cycle an additional permanent volume increase of one-half as much. The specimen had received no externally visible permanent damage at the end of the two cycles. In Fig. 102 the volume change during the first cycle is plotted as a function of compressive stress; the plot for the second cycle is essentially similar. The volume change during increase of load appears as the sum of two effects. The first of these is the normal volume decrease, linear in the load, contributed by the volume compressibility as analyzed in Eq. (12-1). Superposed on this there is an effect in the opposite direction, that is, a volume increase, which becomes larger so rapidly at the higher loads that it dominates the situation and at the two highest loads the volume increases with increase of compressive load. This increase of volume can naturally be ascribed to an opening of interstices in the structure as a premonition of the fracture that would occur at a load only slightly beyond the maximum reached. However, the unexpected feature is that this opening of interstices in preparation for fracture has a very large recoverable component, so that in the initial stages of release of load the volume decreases instead of increasing as it would in the elastic range. This recovery proceeds further during release of load so that only a small fraction of the abnormal volume increase at the maximum load is permanently retained on total release of load.

This recoverable volume increase under compressive stress, probably associated with the opening of interstices, is the new feature disclosed by these measurements. Evidences of the same effect are to be looked for in other materials. I think it is natural to expect the effect to be largest in brittle crystalline materials.

Marble. Three specimens of superficially flawless marble from Darby, Vermont, were used. The recoverable volume effects are larger for marble than for soapstone, but on the other hand the balance between permanent plastic deformation and complete fracture is much more delicate, and three specimens were used before satisfactory readings were obtained in the region of recoverable volume increase. The relation between volume and load for the third specimen is shown in Fig. 103. The maximum load reached was 7,200 psi. Application of this load was immediately followed by such rapid creep of the volume readings that the load was at once decreased as rapidly as possible to the next lower step, 7,100 psi, in order to avoid fracture. Although good readings could not be made in the region of rapid motion, the meniscus was momentarily observed several centimeters beyond the maximum recorded in Fig. 103, corresponding to a recoverable volume increase of perhaps 0.1 per cent. An even more extreme case of recoverable volume change was observed in one of the preliminary specimens in which a volume recovery of 20 cm of the capillary, or 0.3 per cent of the total volume, was observed during the rapid manipulations incident to saving the specimen from catastrophic fracture. Returning now to the third specimen, Fig. 103 shows a permanent volume increase after release of load of 0.04 per cent.

The specimen was so badly flawed and disintegrated in parts that a second application of load was not attempted. Marble differs from soapstone in that the recoverable volume change during release of load is not spread so uniformly over the entire range of load but is confined practically entirely to the first 20 per cent of release from the maximum. Over the mid 60 per cent on releasing load the curve is parallel to the same part of the curve for increasing load, unlike soapstone. The volume compressibility, to be calculated according to Eq. (12-1), does not agree with the volume compressibility given directly by measurements with hydrostatic pressure. The compressibility so calculated is abnormally high in the first 10 per cent of the range, being about 38×10^{-7} (kg/cm² unit), and in the mid-range, where the relation is approximately linear, is abnormally low, being 7.6×10^{-7}, whereas the measurements of Adams and myself[1] suggest a value in the neighborhood of 13.5×10^{-7}. The change of length as a function of load is also shown in Fig. 103. The permanent fractional decrease of length was 0.25 per cent, 6 times as great as the permanent fractional increase of volume.

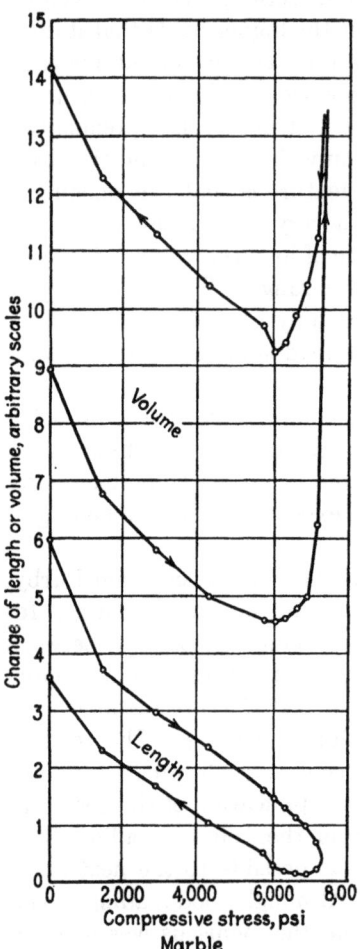

FIG. 103. The changes of length and of volume, both on an arbitrary scale, of marble as a function of simple compressive stress.

Diabase. This proved to be an extremely brittle material, much more brittle than soapstone or marble. Measurements were made on two specimens. The first specimen fractured at a maximum load of 35,400 psi, after practically no warning of impending catastrophe from the length measurements, the last length measurement before fracture differing from a Hooke's law linear relation by only 0.01 per cent of the length. The volume measurements were much more sensitive, however, perceptible devi-

[1] L. H. Adams and E. D. Williamson, *J. Franklin Inst.*, **195**, 493, 1923; P. W. Bridgman, *Am. J. Sci.*, **7**, 96, 1924.

ation from Hooke's law in the "abnormal" direction having already begun at one-half the final load, and at the last reading having progressed so far as to result in a reversal in direction of motion of the meniscus. In the initial linear part of the curve the cubic compressibility calculated according to Eq. (12-1) was 16×10^{-7} against a value in the general neighborhood of 11×10^{-7} to be expected from direct measurements by Birch.[1] The second specimen was handled more gingerly than the first. Increase of load in the smallest increments, corresponding to 540 psi, was initiated at a somewhat smaller load than before, and load was immediately reversed on the first reversal of motion of the meniscus.

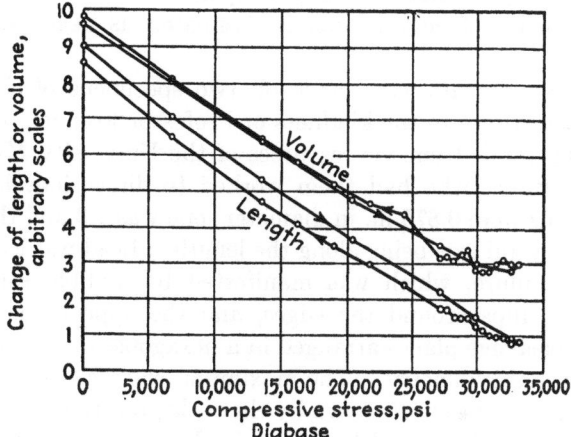

FIG. 104. The changes of length and of volume, both on an arbitrary scale, of diabase as a function of simple compressive stress.

The maximum load reached on the first cycle was 32,600 psi. A second cycle was then applied to a maximum of 29,600 psi. There was no perceptible permanent change of volume after either of these cycles. A third loading was then applied to the previous maximum of 29,600 psi, where the specimen broke with great brittleness after holding the load for a few seconds. The volume changes and the change of length during the first cycle are shown in Fig. 104. During the first loading the "abnormal" component of volume increase is plainly evident in the increase of curvature culminating in final reversal of direction at the last point. This abnormal component of volume increase throws the material into a highly complex state, as shown by the oscillations in the curve on release of load. These oscillations in the volume curve are reflected in less extreme fluctuations in the curve for change of length. The effect of the abnormal volume component gradually disappears on releasing load; after

[1] Francis Birch and Richard B. Dow, *Bull. Geol. Soc. Am.*, **47**, 1235, 1936.

one-half load it is no longer evident, and the specimen returns to zero load with no certain permanent effect on the volume. Loading in the second cycle was not carried far enough to give reversal of the volume curve, but the same reverse curvature was found beyond one-half loading on both ascending and descending branches. A new feature on the second cycle was the appearance of hysteresis on the volume curve, amounting at the maximum to 0.00004 of the total volume. This hysteresis was distributed smoothly over the entire extent of the curve, vanishing only with complete release of load.

In all these experiments on the three rocks the distortions beyond the elastic limit were so small that any temperature effects, even assuming that all the work of compression is dissipated as heat, were entirely negligible.

Quartz. Experiments were made on two specimens of single-crystal quartz. These I owe to the kindness of Professor Frondel, who selected two flawless crystals from the resources of the University Museum, and to Professor Birch, who had them worked to dimensions in his shop. Both specimens were 0.875 in. in diameter; one was 2.0 and the other 2.5 in. long, the crystal axis being along the length. Load was carried to the beginning of failure, which was manifested by audible snapping, the flaking off of chips around the edges, and the appearance of internal longitudinal cleavage planes arranged in a hexagonal pattern corresponding to the crystal symmetry. The maximum load was 118,000 psi. No trace was found of the effect shown by the rocks, but the relation between both length and volume and load remained linear up to the maximum load, and on release there was no appreciable hysteresis and no permanent change of either length or volume. This agrees with what might be expected from other experiments with quartz crystals, for permanent plastic deformation in quartz crystals at room temperature has seldom if ever been realized.

We now turn to a description of the experiments with metals. In assessing the meaning of the results, the suggestions arising from the experiments with rocks are to be kept in mind, namely, the entrance of a new volume effect at the upper end of the range of applicable stress, consisting of a component of increasing volume under increasing compressive load, which is recoverable in whole or in part on release of load. This increase of volume is most naturally thought of as due to the opening of interstices of one sort or another under the action of the stress system. These interstices may perhaps be fairly large, as between the grains or around mechanical impurities, or they may be on the much smaller scale of "dislocations" in the lattice. The precise shapes of these interstices and their orientation with respect to the stress system may be expected to vary from material to material, resulting in the appearance of different kinds of anisotropy. In addition to these interstices which increase in number under the action of a stress, it is to be anticipated that there may also be interstices initially present which tend to close and disappear as the compressive stress increases. Altogether, there are possibilities here for many types of behavior, of great complexity.

Mild Steel. Measurements were made on eight specimens in all. This included four specimens of a 1035 steel which were machined to 1 in. in diameter from the residue of a 2-in. bar supplied by the Watertown Arsenal during the war for measurements of plastic distortion in connection with a war contract. This bar was specially selected by Dr. J. H. Hollomon for soundness and isotropy, a condition which was checked by the previous failure of simple compressive tests on 1-in. cubes in different orientations to show any detectible anisotropy. The other four specimens of mild steel included two specimens cut from a 1-in. bar of "screw stock" from the stock of the laboratory machine shop, and two specimens from a 1-in. bar of an ordinary commercial "cold-rolled" steel.

The cold-rolled steel contained gross imperfections, evidenced at the termination of the run by longitudinal slip lines and extruded seams on the surface. Load was carried to a maximum of 104,000 psi with shortening to 0.86 of the initial length. Measurable plastic yield in length began at 67,000 psi. At a load considerably less than this, at 44,000 psi, there occurred an abrupt discontinuity in the tangent of the volume curve in the direction of an abnormally high compressibility. The easy explanation of this is that it was due to the closing of imperfections initially present. At the upper end of the range any volume changes in the opposite direction, such as might be suggested by the experiments with rocks, were mostly masked by the temperature changes, but there was a small outstanding residue in this direction. An increased curvature in the initial stages of release was evidence of the same effect. There was no measurable permanent change of volume on conclusion; perhaps the two volume effects canceled.

The "screw stock," judging by the appearance of the external surface of the yielded specimens, was essentially homogeneous and entirely different in character from the cold-rolled. One of the specimens of screw stock was used for the express purpose of getting as much information as possible about the temperature effects during plastic yield. For this purpose the load was increased in a few seconds, immediately after the first appearance of yield, to its maximum value and maintained there for 23 min until creep in the volume readings had ceased, indicating the attainment of temperature equilibrium. First yield occurred at 74,000 psi; the maximum was 102,000 with shortening to 0.85 of the initial length.

The temperature rise to be expected was computed on the basis of direct measurements of the constants of the apparatus. This involved finding the motion of the meniscus when the temperature of the entire apparatus was varied by varying the temperature of a water bath in which it was immersed, and by approximate calculations, involving the weights of the various parts, as to the probable initial distribution of temperature throughout the apparatus. Agreement within 1 per cent was found between the measured temperature effect and that computed on the assumption that all the mechanical work of plastic deformation was dissipated as heat. This agreement is, however, fortuitous, the approximations made in the calculations being crude. Taylor and Quinney have found, by experiments designed especially for the purpose, that 85 per cent of the work of torsional deformation of a steel bar is dissipated as heat. It cannot be claimed that the present measurements contradict this.

The conventional series of measurements was made on the other specimen of screw stock. Yield point, maximum load, and permanent plastic yield were essentially the same as before. Two cycles of loading were applied; there was a permanent increase of proportional volume after the runs of 0.00014. On the initial application of load there was a break in the tangent of the volume curve at a load 10 per cent higher than that of first perceptible yield, in the same abnormal direction as shown by the experiments on rocks, that is, in the direction of a component of increasing volume. At the maximum, this increase was about 50 per cent greater than the increase of volume permanently retained on release of load. This 50 per cent excess is the "recoverable" volume effect and is distributed more or less uniformly over the entire range, resulting in a too small apparent compressibility on the first release and second cycle of loading. The volume curves after the first maximum were not entirely smooth but indicated the presence of minor anisotropies. An effect shown by all three curves is a volume compressibility in the upper range of load nearly equal to normal, with a cusp-like decrease at a stress of 36,000 psi.

Measurements on the four specimens of 1035 steel gave results agreeing with each other in broad outline but differing in finer detail. The yield point, the maximum load, and the permanent shortening were not essentially different from the screw stock. There was a permanent proportional decrease of density after the runs of approximately 0.0001, given consistently both by the capillary and the weighings in air and water. On the first application of load all the specimens showed, in the general neighborhood of the yield point in length, but sometimes somewhat above and sometimes somewhat below the yield point, a change in direction of the volume curve, indicating the entrance of a component of increasing volume under compressive load.

Figure 105 shows in detail the results for that one of the four specimens for which the results were most complete. Unlike Figs. 102 to 104, it is the differential $\Delta V/V_0$ which is shown, obtained by calculation from the meniscus readings from which the readings obtained during the dummy

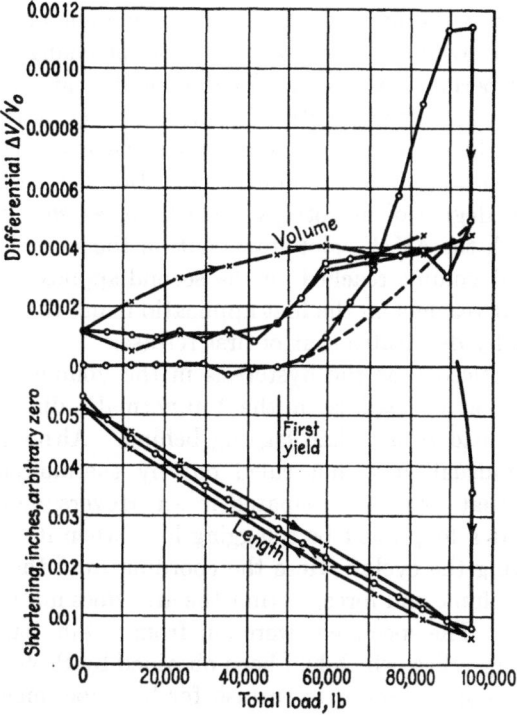

FIG. 105. Results for a 1035 steel. In the upper part of the diagram the differential proportional change of volume is shown as a function of total load, and in the lower part the shortening in inches on an initial 2.5-in. length, taken from an arbitrary zero, is also shown as a function of total load.

run with hardened steel are subtracted off. A differential $\Delta V/V_0$ constant with load means a cubic compressibility, or rather an "effective cubic compressibility," the same as that of hardened steel. The latter is known to be about 1 per cent greater than that of pure iron. After the first maximum of load the differential $\Delta V/V_0$'s of Fig. 105 have been corrected by a factor for the permanent change of length, as already explained. The figure shows that on the first application of load the differential $\Delta V/V_0$ remained approximately zero up to the yield point, indicating a cubic compressibility approximately the same as that of pure iron. On passing the yield point, the entrance of a component of increasing volume is rapidly manifest. The true extent of this is at first masked by temperature effects, but on attaining temperature equilibrium a component of permanent volume increase of approximately 0.00005 under the maximum load is indicated. On release of load this component of volume increase is rapidly relinquished until the load of the first yield point is reached; beyond this the component remains practically constant at 0.0001, which is the permanent increase of volume on total release of stress. The "recoverable" component of volume change is here 4 times as large as the permanently retained component, and much larger than for the screw stock. The "effective" compressibility on release is, below the yield point, the same as initially and the same as for pure iron. This does not mean, however, that the initial condition has been regained, for there is a high degree of anisotropy locked up, as shown by the large slope with which the $\Delta V/V_0$ plot starts out on the second application of load. The final volume reached on the second application is essentially the same as that reached on the first application, and the curve of second release practically duplicates that of first release.

It is to be noticed that the hysteresis in the volume relation on the second application of load is in the "abnormal" direction, being in advance of the load rather than lagging behind. Although this sort of hysteresis is unusual, it is not ruled out by thermodynamics, which demands only that work be dissipated in an irreversible cyclic process. This means that a loop must be a lagging loop when its area represents work done during the cycle. Here the coordinates of the loop have the dimensions of volume and force, so that the area does not represent work.

Norway Iron. The specimens were cut from a 5-in. bar in the collection of laboratory relics, which had been obtained by Professor E. H. Hall at least 50 years ago in order to furnish for his experiments a sufficient and reproducible stock of the purest iron then available. It was completely permeated by large-scale mechanical imperfections as shown by the seams and slip lines which appeared on the external surface after plastic yield. These imperfections result in a gross behavior unlike that of any other substance. Two specimens were used, both 2.5 in. long. The first was 1 in. in diameter; this was too small for the range of the apparatus and only a single cycle of stress was applied to it. The initial yield point was 22,300 psi and the maximum load 70,000, where the shortening was to 0.83 the initial length. The abnormal component of volume increase manifested itself almost at once on passing the yield point and increased so rapidly that the volume change was retrograde over nearly one-half the total stress range. Only a small part of the retrograde motion was due to temperature effect. After release of load there was a permanent proportional decrease of density of 0.0025. On release of load the curvature of the volume curve was in the abnormal direction, indicating a higher effective compressibility at the higher stresses. This means that the interstices opened to a certain extent on release of compressive stress, and can be understood if the interstices were flat on the average and oriented across the direction of stress.

The second specimen of Norway iron was 1.25 instead of 1 in. in diameter and could be carried over the normal load range of the apparatus. The first yield occurred at 23,900 psi; load was increased to a maximum of 65,000 psi, where the length was reduced to 0.83 initial. It is to be noticed that the strain-hardening characteristics of this iron are distinctly different from the steels, strain-hardening extending over a considerably wider range of stress. On release of load a second loading cycle was applied with no further permanent change of volume and very little hysteresis. The proportional permanent change of density was 0.002, consistently by both methods and somewhat less than for the first specimen. This volume increment functioned in the same way as for the first specimen, resulting, after the initial application, in too high an apparent compressibility in the upper stress range. The plot of ΔV versus stress on the second cycle consisted of two stretches, each approximately linear, with a knee and change of direction at approximately the stress of initial yield. This would indicate that the anomalous flattening of the interstices begins with increasing stress in the neighborhood of 24,000 psi, and plays itself out on release at the same stress.

High-carbon Steel. A single experiment was made on a high-carbon steel, annealed drill rod from the laboratory machine shop stock of approximately 1.25 C. The behavior of this was quite different from the low-carbon steels. The onset of plastic yield was much more gradual; it began to be perceptible at about 45,000 psi. Stress was carried to a maximum of 98,000 psi where the length was reduced to 0.86 initial. On release of stress there was a permanent proportional *increase* of density of 0.00015. Corresponding to this unusual increase of density the character of the curve of volume change versus stress was different. At 42,000 psi there was a slight change in direction corresponding to the expected opening of interstices with increasing stress, but this tendency persisted only to 57,000, where there was an abrupt change of direction corresponding to the entry of the opposite effect, or closing of cavities with increasing stress. This dominated the situation above 57,000 and was retained on release of load in the permanent increase of density. The volume curve of release of load exhibited several consistent episodes, but these were quite minor in character, and the curve was linear nearly within experimental error.

Cast Iron. Measurements were made on a single specimen of gray cast iron from the stock of the laboratory machine shop. The specimen was 2.5 in. long and 1.25 in. in diameter, turned from a 1.5-in. bar free from obvious imperfections. There was no proper yield point, but deviations from a linear relation between stress and shortening were detectable from the lowest stresses. Stress was pushed to a maximum of 45,000 psi, where the permanent shortening was 1.5 per cent. After release of load there was a permanent increase of proportional volume of 0.0019. The entry of a component of volume increase was evident on the volume curve as low as 7,000 psi; from here it rapidly increased, resulting in retrograde motion above 31,000 psi. At the higher stresses this component of volume increase continued to increase with time under constant load, eventually reaching an asymptotic value. This creep is in the opposite direction from that arising from temperature effects, which were probably negligible because of the comparatively small amount of deformation. The maximum creep observed was equivalent to about one-tenth the permanent volume change. On release of load the curve of volume versus stress was smooth, but with very marked curvature, corresponding to an effective compressibility rapidly increasing at the low stresses. The effect is in the direction to be explained by the open spaces acting like a more compressible substance coupled to the iron.

Stainless Steel. Measurements were made on a single specimen of "Type 303" from a bar in the laboratory machine shop stock. Type 303 is described as an 18-8 steel, with 0.20 maximum carbon. Plastic yield in length was first perceptible at 51,000 psi; loading was increased to 103,000 psi, where the length was decreased to 0.865 initial. On release of load there was a permanent *increase* of proportional density of 0.00012. At no part of the loading or unloading process was there any evidence of the anomalous component of volume increase exhibited by other materials. On the first loading there was a break in the course of the curve of ΔV versus load immediately beyond the first yield point in the direction of increasing effective cubic compressibility, indicating a closing of interstices rather than an opening. At the maximum load this additional compression amounted, as well as could be judged, to 2 or 3 times the compression permanently retained on release of load. On release, practically all the difference between the maximum and the permanently retained excess compression was relinquished in the first 20 per cent of the unloading, the curve being approximately linear in the last 80 per cent of its course.

Copper. Measurements were made on four specimens, two of them preliminary before the apparatus was functioning satisfactorily. The specimens were all 2.5 in. long and 1.5 in. in diameter, taken from a 1.5-in bar of hard-drawn copper in the stock of the laboratory machine shop. One of the preliminary specimens was annealed; the other three were left in the hard-drawn condition. The plastic shortening of these four specimens varied from 0.90 to 0.84. All four showed a permanent *increase* of proportional density varying from 0.00021 to 0.00049, with a rough correlation between the amount of plastic shortening and increase of density. In this correlation the annealed specimen did not fall out of line with the others.

The results in detail for the fourth specimen (hard-drawn), for which the apparatus functioned most satisfactorily, were as follows. First plastic yield in length was perceptible at a load of 26,400 psi; loading was continued to a maximum of 43,300 psi, where the shortening was to 0.870 initial. On release of load the proportional density was increased by 0.00049 according to measurements by capillary and by 0.00047 according to weighing in air and water. On first application of load the curve of volume versus load passed smoothly through the yield point without alteration in direction. Practically all the disturbance of this curve occurred in the last step of loading, where 0.8 of the total plastic shortening occurred. The increase of density, once acquired, was permanently retained at a constant value independent of the load during the subsequent unloading and second cycle. This means that the effective cubic compressibility given by Eq. (12-1) was constant, except during the last step of the first loading, over the entire process of two loadings and two unloadings. The numerical value for this was 7.2×10^{-7} (kg/cm^2 unit), which checks exactly with my previous value[1] for copper directly determined by applying hydrostatic pressure.

Brass. Three sets of measurements were made, two of them with the preliminary apparatus. The specimens were 1.25 in. in diameter, from a bar of the same diameter of commercial brass from the stock of the laboratory machine shop. These were used "as is," and were presumably hard-drawn. All three specimens gave permanent *increases* of density after plastic yield. The two preliminary specimens gave results of little value except for the permanent increase of density. The third specimen yielded first under a load of 42,000 psi; it was carried to a maximum of 70,000 psi, where the shortening was to 0.89 initial. The permanent decrease of proportional density was 0.0002 by the capillary and one-half as much by weighing. The results for this material were more complicated than for most of the others, doubtless a consequence of the high internal strains which every machinist knows are likely to be locked up in a bar of drawn brass. No disturbance in the ΔV versus load curve was evident on the first loading on passing the yield point, and the final equi-

[1] P. W. Bridgman, *Proc. Am. Acad. Arts Sci.*, **77**, 206, 1949.

librium point lay on a smooth extrapolation from those points at loads low enough not to be disturbed by temperature effects. The curve on first release had a very pronounced S shape with minimum and maximum, and this S shape was retained, but less accentuated, on the second loading and unloading. The simplest description of the behavior is that it is as if there were two components in the volume change arising from the operation of the interstices. One of these is a closing of the pores under the application of load which has a permanently retained part and also another part of approximately equal magnitude which is relinquished in the first quarter of the unloading process. The second component is a recoverable and reversible opening of pores under load, uniformly spread over the entire stress range, which produces too low an effective compressibility as calculated by Eq. (12-1). The mean effective compressibility, calculated from the second loading cycle, disregarding the upper quarter of the range where the other component of volume change is dominant, is 7.1×10^{-7} against 9.2×10^{-7} (kg/cm^2 unit) expected from direct measurements under hydrostatic pressure.

Duralumin. Two preliminary measurements were made and one final one. The final specimen was 2.5 in. long and 1.25 in. diameter, cut from a 1.5-in. bar from the laboratory machine-shop stock. First yield occurred at 47,500 psi. Load was increased to a maximum of 65,000 psi, where the length was 0.903 initial. The yield curve was of unusual character; it started with unusual abruptness and then continued with a nearly linear relation between length and load to the maximum. Any permanent change of density was too small to detect certainly. No change could be established from the capillary, and the weighings gave 2.8080 for the density at room temperature of the virgin specimen, and 2.8079 after deformation. There was an abrupt and large change of slope of the ΔV versus load curve on passing the first yield point in the direction to be expected from the anomalous component of increasing volume under increasing compression. This increasing component at the maximum load, estimated by an extrapolation, had risen to the equivalent of a decrease of proportional density of 0.00046. None of this was apparently permanently retained on release of load, but the precise way in which it was relinquished on release of load was not apparent, since the curve consisted of several episodes with at least two points of inflection and one abrupt break in direction near the yield point. The effective cubic compressibility of duralumin calculated by Eq. (12-1) from the mean of the rising and falling curves was 11.7×10^{-7} (kg/cm^2 unit). There is apparently no directly measured value for comparison. The direct value for pure aluminum, however, is 13.4×10^{-7}, suggesting that 11.7 would be smaller than what would be found by direct measurement.

Discussion and Summary. It has been found that complicated changes of volume accompany the longitudinal plastic yielding of cylindrical specimens under simple compressive stress. It is simpler to describe these volume changes in their own terms than to introduce an effective Poisson's ratio, since even in the ideally simplest case the curve of Poisson's ratio is not simple in the region of large plastic yield. It is found that, superposed on the volume decrease under compressive stress that would be calculated by extending into the plastic range the simple linear relation which holds in the elastic and isotropic range, there are other components of volume change which may vary in a complicated way over the range of stress. These may be components either of volume increase or of decrease—the former would indicate the opening of the interstices and the latter their closing. Both these effects may be manifested as permanent alterations of density on release of stress.

Of the materials examined in this paper, permanent increases of density were shown by annealed high-carbon steel, an 18-8 stainless steel, copper, and hard-drawn brass. The opposite effect, a permanent decrease of density, was shown by the three rocks (soapstone, marble, and diabase), by two kinds of mild steel, by Norway iron, and by gray cast iron. Vanishing volume change was found for quartz crystal, for a cold-rolled steel, and for duralumin. The volume increases of Norway iron and cast iron were comparatively large and doubtless the result of large-scale imperfections of structure, and therefore not of particular significance. The component of volume increase shown by mild steel and the rocks probably involves a mechanism which is significant for theories of plastic flow and fracture. This component of volume increase is itself a strong function of stress. For the three rocks it increases so rapidly in the region just below fracture as to dominate the other factors so that the total volume change becomes retrograde. Furthermore, this component of volume increase is largely recoverable and reversible on release of stress. The same effect is also shown by the 1035 steel, although not to so large an extent as to result in retrograde change of volume. Four-fifths of the component of volume increase of this steel is relinquished on release of load and only one-fifth retained as permanent increase of volume. In duralumin there is a component of volume increase at maximum load amounting to 0.00046, all of which is relinquished on release of load in the region above the initial yield point. In many cases the behavior of the volume on release of load and a subsequent cycle of reloading retains the memory of the initial plastic yield in an elbow on the volume curve at the stress of the initial yield point.

The new picture presented by these experiments is that fracture is prepared for, in at least some cases, by the *reversible* creation by the stress itself of alterations in the structure; when these alterations have proceeded to a critical degree the structure becomes unstable and fracture ensues. Perhaps the simplest way to think of this is as an increase in the number of "dislocations" which are normally present in the lattice. If the lattice under normal conditions tends to assume a certain equilibrium concentration of dislocations, it is natural to suppose that this equilibrium number may be increased with stress. It is probable that the same mechanism operates in simple tension as in simple compression, the energy differences which enter into the conditions of equilibrium being proportional to squares of the stress. It would, however, be more difficult experimentally to detect the effect for tension than for compression. Beside this new effect of a component of increasing volume under increasing compressive stress, some substances, notably copper, show the more naturally to be expected closing of imperfections under compressive stress. This may be retained as a whole or in part as a permanent increase of density on release of stress. There appears to be no reason why the two mechanisms should not operate simultaneously, and this does indeed seem to be the case for a material like brass with complicated behavior.

CHAPTER 13

TWO-DIMENSIONAL COMPRESSION[1]

Introduction. The stress conditions under which plastic flow has been studied up to the present have been usually restricted because of the experimental inconvenience of applying stresses which are arbitrarily variable in different directions. The most common method of realizing more than one arbitrary component of stress has been by the simultaneous application of longitudinal tension and internal hydrostatic pressure to thin tubes. By this method it is possible, within certain limits, to apply two arbitrary components of tensile stress. The plastic flow which can be produced in this way is limited to comparatively small values by the incidence of tensile fracture. Much larger strains could be realized without fracture under compressive stresses, but few, if any, such experiments appear to have been made, although it would seem that it would not be especially difficult to expose a tube to external pressure with simultaneous longitudinal compression or tension. It would, however, probably be necessary to use tubes of appreciable thickness in order to avoid buckling under the external pressure, and in this case the strains would not be homogeneous and the interpretation of the results would be complicated.

In the following, experiments are described in which rectangular blocks are exposed to arbitrarily variable compressive forces on two of the three pairs of faces, the remaining pair of faces being free from force. The stresses and the strains are homogeneous, and the mathematical interpretation of the results is especially simple, the experimental test block playing the role of the mathematical element of volume of the fundamental equations. The compressive forces are applied by hard platens pressing against the faces. Obvious geometrical limitations arising from mutual interference of the platens impose rather definite restrictions on the strains. There are doubtless various ways of meeting the geometrical conditions. In the following the strain as well as the stress was kept two-dimensional, on the gross average. Since two-dimensional stress in general produces three-dimensional flow, restriction of the strain approximately to two dimensions might appear to limit seriously the part of the two-dimensional stress domain open to exploration. However, small deviations from a rigid two-dimensionality in the strain were instru-

[1] This chapter is practically a rearrangement of B21.

mentally feasible, which gave sufficient flexibility to permit a very considerable variation in the stresses, so that a very appreciable part of the total region of plastic flow under two arbitrary compressive stress components could be explored in spite of the geometrical restrictions.

The Apparatus and Method. The direction of larger compressive stress will be denoted by z, that of minor compressive stress by x, and the direction of no stress by y. The corresponding stresses will be written as Z_z, X_x, and $Y_y(=0)$. The Z_z stress is applied by a pair of opposing rectangular blocks, of indefinite length in the y direction and with a width in the x direction approximately equal to the x dimension of the specimen. The X_x stress is applied by two hard blocks of indefinite extent in both the z and the y direction; the distance between these two blocks in the x direction must be kept not far from the initial width of the test block. The arrangement is indicated in Fig. 106. If the distance between the x blocks becomes less than the initial x dimension of the z blocks, these will be pinched and the measured stresses will be in error. On the other hand, if the distance becomes too great, the test block will flow plastically into the narrow channel between z and x blocks, destroying the homogeneity of both stress and strain. In the following experiments the initial x dimension of the test block was made 0.502 in. and of the z block 0.500 in. During the experiment the distance of separation of the x blocks was kept between 0.501 and 0.504 in.; within these limits there was neither pinching nor appreciable plastic flow into the crevasse at the edge, which was only 0.002 in. wide at the most. The x component of force must be maintained in such adjustment as to keep the x dimension in these limits; a rather wide latitude proved possible in meeting these conditions.

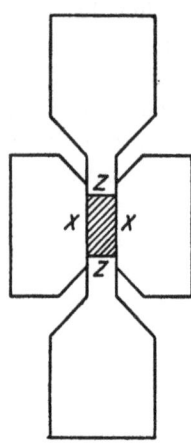

FIG. 106. Section through the platens for simultaneous compression in two directions at right angles. The specimen is shown shaded. The z blocks and the x blocks are pushed together by two independent hydraulic presses, which are so manipulated that the x blocks do not pinch on the z blocks.

The z blocks and the x blocks were pushed together by two independent hydraulic presses. The press driving the z blocks was a standard press that has been used in much of my other work; it has a 3.5-in. plunger, a free space of 10 in. between the rods, and is driven by a hand pump with capacity of 15,000 psi. The x blocks were pushed together with a specially constructed press with 2.5-in. plunger. This press was machined from a single piece of steel 4 in. in diameter and was compact enough to be mounted transversely in the 10-in. free space between the tie rods of the 3.5-in. press. Pressure was furnished by a second hand pump. This press was freely suspended from a point in the vertical line with its center of gravity, so that the whole press floated and could follow the motion of the specimen as it was plastically deformed, without exerting any extraneous force on it. The maximum displacement of the x press was 0.002 in. in the x direction and 0.125 in. in the z direction. Pressure was led into it through copper tubing of 0.25 in. OD and 0.062 in. ID; the tube was made long enough to exert no appreciable constraint.

Two different methods of applying the loads were employed. In the first, the pistons were advanced by pumping until approximately the desired increment of strain was reached. Pumping was then stopped and time allowed to elapse until a steady state was approximately attained. This steady state was reached by a simultaneous further slight automatic change of both strain and pressure. Since the flow in the z direction was, in general, many-fold greater than in the x direction, most of the manipulations were connected with the z press. A fraction of a stroke of the pump of the x press at the beginning of the step usually was sufficient, the pressure of the x press then automatically taking care of itself. The steady-state pressures were read on calibrated Bourdon gauges in the lines of the two presses. In the second method of applying load, measurements were made at controlled intervals of total load. This was accomplished by interposing in the lines leading to the presses two dead-weight free piston gauges. Pressure was adjusted by setting the appropriate weight on the pan of the dead-weight gauge, and pumping until the piston floated. Since the motion in the x direction was so slight, automatic rise or fall of the free piston of the x gauge was sufficient to maintain the load constant within the limits of x deformation. The motion in the z direction was sufficiently large, however, to demand frequent pumping to keep the z free piston floating. For the more rapid rates of flow in the z direction the sluggishness of response of the free piston gauge was great enough so that the load would have fluctuated within undesirably wide limits even if the piston were kept always floating by rapid pumping. Especial measures had to be taken to meet this condition. In the first place, the piston of the free piston gauge was made much looser than is convenient for normal operation. The most important modification, however, was to provide volume flexibility in the line to the z press by attaching a reservoir of 750 cm^3 capacity containing compressed nitrogen, obtained initially from a storage flask carrying 1,500 psi. Up to pressures on the 3.5-in. piston of 1,500 psi, pressure was obtained from the compressed storage flask by admitting the requisite amount of gas through a valve. Above 1,500 psi, the storage flask was cut off and pressure obtained by the hand pump. The maximum pressure on the 3.5-in. plunger was of the order of 7,000 psi. A simple calculation showed that even under the most rapid plastic flow of these experiments inertia effects in the moving parts of the presses are inappreciable.

The running dimensions of the test block in the z and x directions during the course of the experiment were measured by measuring the distance of separation of the z and the x blocks. This was done with Ames dial gauges, mounted on yokes attached to slender rods straddling the specimen and passing through appropriate holes drilled in the blocks. Since the motion in the z direction was large, reaching a maximum of 0.25 in., it was sufficient to measure the z displacement with an ordinary gauge, graduated to 0.001 in., with 1-in. total stroke. The x displacements, being much smaller, were measured with a jeweled gauge graduated to 0.0001 in. with total stroke of 0.2 in. Any lagging of the gauge readings due to friction was eliminated by introducing into the whole system a very slight amount of mechanical vibration by pressing an electric razor against an appropriate member of the apparatus. Consistent differential readings of the x dimensions could be made to 0.00001 in. or even better if the pointer of the dial happened to be exactly on a division.

Under stress, the dimensions of the test block differ from the dimensions indicated by the Ames gauges by the elastic distortion of the z and the x blocks. A special study was made of the elastic distortion by stressing hardened test blocks of various dimensions. Any correction of this sort was found to be unimportant. The method finally adopted for dealing with it was to obtain the elastic distortion at the maximum stress by comparing the Ames gauge readings with the micrometer dimensions of the test block after removing from the apparatus, and to distribute the maximum correction over the intermediate stresses in proportion to the load.

The x blocks were made of tool steel, hardened to Rockwell C 67. In general the stress which the x blocks have to carry is one-half that of the z blocks. The z blocks were also at first made of the same tool steel as the x blocks. Their performance was later somewhat improved by setting in carboloy inserts. Carboloy can be highly polished so as to minimize the friction; furthermore, carboloy, being only one-third as deformable as steel, does not permit the test specimen to embed itself to the same extent, so that the effective friction is reduced as the test specimen extends itself plastically along the platen, pushing its advancing edge up the incline of the depression in the platen.

One of the objects of the experiments was to study the rates of flow for various combinations of stress. The "flow" understood here is the "primary" flow which occurs immediately after application of an increment of load and does not include such long-range phenomena as elastic aftereffects and creep. Of course, there is no sharp dividing line between these various effects; the time intervals involved in these measurements were not greater than the order of 15 min. At first, observations were made with manual recording and an ordinary watch, but these proved inadequate, and the final study of the rates of flow was made with photographic recording. The two Ames dial gauges were mounted side by side in the field of an Eastman Cine Special, for 16-mm film, so arranged that single frames could be exposed, either by hand operation for moderate rates of flow, or for more rapid rates by motor control, permitting up to a maximum of four exposures per second. The full opening, $f/1.9$, was used, as well as maximum film speed, corresponding to 64 frames per second for normal operation, and one-quarter shutter opening. Illumination was by two suitably mounted 500-watt floodlights. It was necessary to use a film of high contrast, in which the exposure latitude was accordingly small; the film used was Eastman positive. I am much indebted to Mr. Paul Donaldson for skillful development of the film and advice with regard to photographic details. The time was provided by a synchronous Telechron motor connected so as to give one revolution per second, the hand sweeping over a dial graduated to hundredths. The total number of revolutions (that is, seconds) was counted by a conventional Veeder counter mounted so that it was also in the field of the camera. The current weights on the free piston gauges were also recorded photographically, suitable labels being placed manually in the field when the weights were changed. An additional check on the z load was obtained by also recording the stretch of the tie rods of the z press; this stretch gives the actual load on the specimen without the component due to friction of the ram. The stretch of the tie rods was measured with an Ames 0.0001-in. jeweled gauge, connected to the tie rods through a lever giving 5-fold multiplication, so that changes of length of 0.000002 in. could be established. This corresponds to a Z_z stress on the test specimen of about 30 psi. With the tie rod gauge it was possible to establish that there was perceptible sluggishness in the response of the stress to changes of load on the free piston gauge, in spite of the amelioration by the use of the gas reservoir. This sluggishness, however, was confined to the initial stages immediately after change of load when the rate of flow was high and was therefore not a matter of importance. On the conclusion of a run, the specimen was removed and several photographs taken of it on the same film, as a matter of record. The film was then developed, transferred to a couple of spools which permitted it to be conveniently examined under a microscope, and the desired data read off from the film and recorded. The photographic record was sharp enough so that times could be read to a fraction of a hundredth of a second, and the Ames gauges to a tenth of their smallest divisions.

Theoretical Background. The experimental procedure was guided by the accepted equations of plasticity; investigation of the validity of these equations was one of the principal objects of this work. We shall write these equations in the form for flow:

$$\frac{\dot{\epsilon}_x}{X_x - \frac{1}{2}(Y_y + Z_z)} = \frac{\dot{\epsilon}_y}{Y_y - \frac{1}{2}(Z_z + X_x)} = \frac{\dot{\epsilon}_z}{Z_z - \frac{1}{2}(X_x + Y_y)}$$

In these equations the x, y, and z axes are assumed to be the principal axes both of stress and of strain. This demands that the axes do not rotate in space during the straining process; this condition is obviously satisfied in the present experiments. The strains are the "natural" strains, that is, $\epsilon_x = \log_e (l/l_0)_x$. The usual sign convention will be followed; a tensile stress will be taken as positive and a strain of extension positive. Under the conditions of our experiments, Y_y vanishes identically, so that the equations can be specialized as follows:

$$\frac{\dot{\epsilon}_x}{X_x - \frac{1}{2}Z_z} = \frac{\dot{\epsilon}_z}{Z_z - \frac{1}{2}X_x}$$

Furthermore, Z_z is always larger numerically than X_x. If we put $X_x = \alpha Z_z$, α will vary between zero and unity. Under these conditions $\dot{\epsilon}_z$ always has the sign of Z_z. The flow along the x axis vanishes for $\alpha = \frac{1}{2}$, has the same sign as $\dot{\epsilon}_z$ for α greater than 0.5, and the opposite sign if α is less than 0.5. The ratio of the two velocities expressed in terms of α is $\dot{\epsilon}_x/\dot{\epsilon}_z = (2\alpha - 1)/(2 - \alpha)$.

In the following experiments the conditions are investigated under which there is no flow in the x direction, and also, when there is such flow, whether the velocity of flow in the x direction bears the theoretical ratio to velocity in the z direction in terms of α. These questions are to be investigated at various points on the strain-hardening curve; it is conceivable that one relation might hold in the initial stages of flow and another in the later stages when there has been appreciable strain-hardening. The basic physical postulate back of the above expressions for flow is the postulate of isotropy. One of the objects of the following experiments may therefore be phrased: to find to what extent the basic assumption of isotropy is satisfied.

It is obvious that if α varies arbitrarily during the straining process the flow equations cannot be simply integrated, and we may *not* write the conventional expression

$$\frac{\epsilon_x}{\epsilon_z} = \frac{X_x - \frac{1}{2}Z_z}{Z_z - \frac{1}{2}X_x}$$

In these experiments α did vary during the straining process, so that the simple integrated form is not admissible. In general, it is obvious that no integrated form of the equations, in which the history of the variation of α or something equivalent to it does not appear, can be better than an approximation. The mere existence of an integrated form of the flow equations obviously imposes restrictions. It is customary to suppose that not only are the restrictions met, so that the history of α does not enter the final equations for strains, but that there is such a simple connection between strains and stresses of different types that if the relation between strain and stress is known for one type of stress it may be inferred for other types of stress. Various suggestions have been made as to the nature of the generalized relation. One of the most recent, and also one of apparently wide applicability, is that the "significant" stress $\bar{S}$ is a universal function of "significant" strain $\bar{\epsilon}$ for all types of plastic flow. The significant stress and strain are defined as follows:

$$\bar{S} = \{\tfrac{1}{2}[(X_x - Y_y)^2 + (Y_y - Z_z)^2 + (Z_z - X_x)^2]\}^{1/2} = Z_z(1 - \alpha + \alpha^2)^{1/2}$$
$$(\text{if } Y_y = 0, X_x = \alpha Z_z)$$
$$\bar{\epsilon} = [\tfrac{2}{3}(\epsilon_x^2 + \epsilon_y^2 + \epsilon_z^2)]^{1/2} = \epsilon_z(\tfrac{4}{3})^{1/2} \quad (\text{if } \epsilon_x = 0, \epsilon_y = -\epsilon_z)$$

The curve which gives $\bar{S}$ as a function of $\bar{\epsilon}$ may be described as the generalized strain-hardening curve. One of the objects of the following experiments was to determine the generalized strain-hardening curve under the present conditions (that is, strain approximately two-dimensional, with two-dimensional stress), and to find whether significant stress is indeed a universal function of significant strain by comparing with the plastic flow obtained under other conditions of stress, in particular under simple compressive stress ($X_x = Y_y = 0$).

Returning now to the rate of flow, we have so far considered only relative rates of flow in different directions. There is also the more fundamental matter of absolute rate of flow. To consider this question the flow equation must be expanded into the form

$$\frac{\dot{\epsilon}_x}{X_x - \tfrac{1}{2}Z_z} = \frac{\dot{\epsilon}_z}{Z_z - \tfrac{1}{2}X_x} = \phi$$

A complete investigation of the phenomena demands that we find of what arguments ϕ is a function, and what is the value of the function for various values of its arguments. It is obvious physically that for points in the stress-strain plane below the strain-hardening curve ϕ must vanish identically, there being no plastic flow here, and that ϕ must be everywhere positive in a region above the stress-strain curve. The mere existence of a strain-hardening curve would seem to suggest that ϕ gets larger very rapidly as the distance from the strain-hardening curve increases. Too great displacement from the strain-hardening curve would lead into the domain of fracture. ϕ must then, at the very least, be a function of stress and strain. It is such a function that for every value of stress, within limits, there is some strain which makes ϕ vanish; in general, the larger the stress, the larger the corresponding strain. In addition to being a function of stress and strain, ϕ may also well be a function of history. If it is not a function of history, as perhaps we may assume provisionally as a first approximation, then $\phi = 0$ is the equation of the strain-hardening curve. Considerations of symmetry demand that under these conditions the stresses and strains enter through their invari-

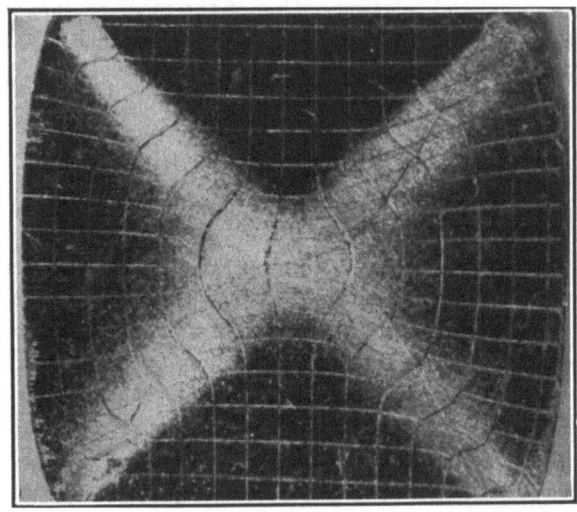

Fig. 107. The typical shearing cross which develops in two-dimensional compression under insufficient lubrication. The rolling of the originally free lateral face onto the compressed face is to be noticed.

ants, or ϕ have the form $\phi[\text{stress invariant} - f(\text{strain invariant})]$. Comparison with the definitions for significant stress and strain shows that this is equivalent to $\phi[\tilde{S} - f(\tilde{\epsilon})]$.

It was one of the purposes of the following experiments to get some indications as to the form of ϕ and to find to what extent the above simple assumptions about its nature are justified.

In actually carrying out the measurements the conditions are by no means as simple as in the ideal mathematical discussion above. The chief disturbing factor is terminal friction on the z faces. It is well known that in a simple compression test, with free lateral expansion, the specimen has a tendency to barrel out because of the frictional constraint on the ends. If friction is reduced by proper lubrication, barreling disappears and the strain becomes homogeneous. Under our conditions, compression with one component of lateral flow prevented, the effects of terminal friction proved to be much larger and were also much more difficult to eliminate. The nature of the distortion can be made visible by scratching on the X surfaces of the virgin specimen a rectangular grid, together with a couple of diagonals to make the distortion at the corners more manifest. Figure 107 is a photograph of the x face of a compressed specimen in which the lubrication of the z face was not adequate. Part of the original y face has rolled up onto the z face, the edge still being, however, a more or less sharp right angle. This would suggest that there must be some sort of mathematical singularity at the edge. In fact, a moment's consideration shows that in a compressed specimen flowing over the platen with friction there must be a mathematical singularity at the edge. For in the z face, because of friction, there is at the edge a nonvanishing tangential stress Y_z, whereas in the y face at the edge the tangential stress component Z_y vanishes identically merely because the surface is free. Here is a contradiction, because the conditions of mathematical equilibrium demand that everywhere $Y_z = Z_y$. An exact mathematical characterization of the details of the singularity demanded to avoid this contradiction would depend in any particular case on the elastic deformation of the platen and would doubtless be difficult to obtain. The general nature of the singularity, however, is probably the

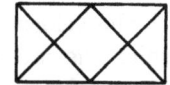

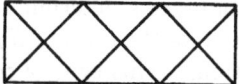

Fig. 108. Multiple shearing crosses developed when there is friction on the compressed face and an integral ratio of breadth to height.

same whatever the details and may be obtained from any particular solution. There is one particular solution in the literature due to Nadai.[1] Another particular solution, involving only very simple mathematical functions, has been found by Westergaard but has not been published. Both these solutions agree in that the singularity at the edge involves a shearing stress which is a maximum on the plane approximately bisecting the edge angle. In the present work there was frequently observed a region of maximum shearing distortion in a 45° zone; it is plainly evident in Fig. 107. Slip in this zone gets started with great ease and propagates itself so as to amount to an essential instability. This instability may be favored by the geometry of the specimen. This sort of slip starts most easily if the x face is a square (z and y dimension equal). If the initial dimensions of the block are in the proportion of 3 to 2, 45° slip is not inclined to start at first, but as the distortion proceeds and the ratio of the sides passes through 1 to 1, diagonal slip is very likely to start. In fact, if the ratio of the y dimension to the z dimension is integral, a

[1] A. Nadai, *Z. Physik.*, **30**, 115, 1924.

shearing pattern on 45° lines is very apt to develop; Fig. 108 suggests such patterns that have been observed for the 2 to 1 and the 3 to 1 ratio.

These regions of concentrated shear may be suppressed by suitable lubrication; the requirements are much more difficult to meet than they are for ordinary simple compression. This seems natural, for we are here concerned with slip on a single set of planes, whereas in simple compression, cross slip on two interfering systems of planes is involved, with greater destruction of the microscopic structure. In the description of the detailed experiments the details of the lubrication will be described, and also the investigations to find what error is introduced by the inhomogeneities that actually occur.

The Measurements. Three types of measurements were made: those in which an approximately steady state was reached, those in which the velocity of flow was measured under constant load, and those in which the rate of flow was maintained constant and the stress determined as a function of strain.

1. *Approximately Steady-state Measurements.* These measurements embrace a determination of the strain-hardening curve, and a determination of the ratio of X_x to Z_z (that is, of α) for vanishing flow in the x direction. This type of measurement was made first and involved considerable preliminary work, the nature of which will be indicated.

The original program called for a series of measurements corresponding to those already made in simple compression; in these experiments the height of the block was reduced to two-thirds its initial height in a single stage of compression, the block was then remachined back to its original proportions, and the process repeated seven or eight times until the total reduction was some 20-fold. It was soon obvious that there would be great difficulties in carrying through an analogous program for two-dimensional compression because of the much greater tendency to distortion during compression. Even a single stage of compression with a reduction of height to two-thirds demanded much more perfect lubrication. A number of experiments were made to find the best lubrication. The method used successfully for simple compression was the same as that used for the lubrication of the external conical surfaces of my high-pressure containers, namely, two thicknesses of 0.002-in. lead foil smeared on all sides with a thin paste of colloidal graphite, glycerin, and water. The distortion with this lubrication proved undesirably large under present conditions. Other simple lubricants were tried: Scotch tape, stearic acid, lead oleate, gold leaf, either alone or in various combinations with the lead foil, but were not satisfactory. A thin layer of soft solder applied with a soldering iron was not bad but was not used for fear of annealing effects from the heat of the iron. Experience with the coat of solder was useful, however, in indicating that a plastic film tightly attached to the steel is desirable. If the film has not some sort of attachment it will be entirely squeezed out; this was the trouble with the lead. Various schemes were tried on the idea of a guard ring, that is, a separate layer of the same material in contact with the platens, the idea being that the deformation due to friction would be confined to the guard layer, leaving the central portion homogeneously strained. Any arrangement analogous to a guard ring proved, however, extraordinarily sensitive; it would either flow completely away from the central block or else embed itself in it, and the scheme was abandoned. The method finally adopted was to copperplate the z faces with a film of copper about 0.001 in. thick, and then put on top of the copper the sandwich of lead foil and graphite paste described above. Some difficulty was found in making the copperplating sufficiently adherent; the final technique was to start the plating

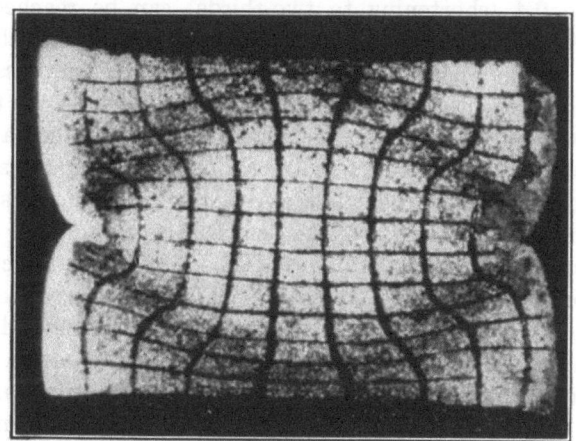

Fig. 109. A type of instability that often develops on second compression after refiguring.

in a cyanide bath and finish in an acid sulphate bath. The x faces are not so sensitive to friction effects as are the z faces; for the x faces the simple lead foil sandwich was adequate.

After the first compression, the equator was always bulged slightly, even with the best lubrication. Under the best conditions the strain at the equator might be 2 or 3 per cent greater than at the ends. If now the compressed specimen is machined back to the original proportions and subjected to a second compression, the equator at the beginning of the second compression will be more strain-hardened, so that the ends will start to flow first. It might be thought that this condition would be self-correcting, since an excess flow at the ends would harden them and the flow would be thrown back to the equator. Actually, however, initial excess flow at the ends is followed by the development of shearing instability and the block folds into a crease based on the equator, as shown by Fig. 109. It might appear that this sort of thing could be avoided by making the end friction somewhat greater on the second application, but this proved too sensitive to manage, the second compression being almost certain to develop an instability like that of either Fig. 107 or Fig. 109, if the initial stage of compression was pushed as far as a shortening to two-thirds. It therefore proved infeasible to carry through the original program of pushing the strains to large values by the device of multiple compressions with refigurings. If, however, the strain of the individual stages is kept to something of the order of 10 per cent, multiple compressions with refiguring can be carried out, and runs were made on two different steels with six stages of compression with a total shortening to one-half.

The extra labor of making six stages in order to reach a strain of 0.7 (shortening to one-half) is not justified, however, because in a single stage strains up to 0.4 (shortening to two-thirds) can be reached with no appreciable error from lack of homogeneity of flow, although the inhomogeneity may be too great to permit a second stage of compression. The question of the character of the inhomogeneities and the error that might be introduced by them was examined with some care. The nature of the plastic flow was studied with the help of a network of squares scratched on the x faces, as already indicated. After compression the squares are deformed, and a measurement of the deformed squares under a micrometer microscope gives the local deformations. If the end friction is too great, the sort of plastic flow indicated in Fig. 107 is produced. The y face rolls up onto the z face at the edge, the angle at the new edge remaining comparatively sharp. The edges are, therefore, a region of infinite strain. From the edges a shearing cross is developed diagonally across the x faces; this shearing cross is marked by surface roughening of the steel, plainly visible in the photograph, and by great angular distortion of the squares. Based on the z faces are two triangular prismatic wedges which are pushed almost bodily into the remaining material. In this wedge the squares preserve nearly their original dimensions. At the center of the x face, where the shearing diagonals cross, the square is deformed to a rectangle, without angular rotation.

If the deformation were homogeneous, the initial height being h_0 and the final height h, it is easy to see that a square would be deformed to a rectangle, the ratio of whose sides is $(h/h_0)^2$ (for the height of the square is reduced in the ratio h/h_0, and the width is increased in the ratio h_0/h). The deviation of the ratio of the sides of the deformed square from $(h/h_0)^2$ may be taken as a measure of the failure of the deformation to be homogeneous. With good lubrication this ratio was found to have its theoretical value within experimental error. Thus, a block of soft Cr-V steel compressed to approximately two-thirds its initial height gave a ratio for the sides of the deformed square at the center of the x face of 2.23 against an $(h/h_0)^2$ of 2.28, and another block of the same steel under approximately the same compression gave 2.32 and 2.33, respectively. On the other hand, a specimen in which there was a well-developed shearing cross and marked rolling at the edges gave for the ratio at the center of the x face 5.01 and at the base of the prismatic wedge 1.70, against an $(h_0/h)^2$ of 2.19.

Another method of measuring the failure of homogeneity, simpler and not so detailed, but adequate for most purposes, is merely to determine the ratio of the y dimension at the equator of the x face after compression to the average of the y dimension at the two ends of the x face. For good specimens this differs from unity by only a few per cent.

In determining the strain-hardening curves it is the average stresses and the average strains which are determined, and it might be suspected that these averages would be relatively insensitive to local deviations from homogeneity. It turned out, as a matter of fact, that the averages were so insensitive that one could practically disregard any inhomogeneities within the limits of compression mentioned. Thus, in one experiment with a soft Cr-V steel, the lubrication on the z face was intentionally omitted so as to develop the maximum inhomogeneity. The compressed specimen had a strongly developed shearing cross, and the rolling at the edges was so great that 16 per cent of the final z face had originally been part of the y face. The strain-hardening curve for this specimen lay only 5 per cent below the average of the curves for several other specimens, the deformation of which was essentially homogeneous. It was a general result that, if anything, the average stress for nonhomogeneous distortion is less than for homogeneous distortion.

The general method of determining the strain-hardening curve with approximately zero velocity of flow in the x direction was as follows. The hand pumps were direct connected to the presses, without the intermediary of a free piston gauge, and the pressures were read from the corresponding Bourdon gauges. The dead spaces in the presses and connections were minimized so that a maximum pressure increment would be produced by a given increment of strain. A preliminary run was made to determine the approximate values of the x pressure required for zero x flow corresponding to various z pressures and strains. Two observers took part in the final run; the function of one was to observe the x pressure gauge and of the other to perform the pumping and observe the z gauge. Both presses were first brought to pressures slightly below the initial plastic yield point. The x press was then raised somewhat, and the z pump then operated slowly. On passing the yield point the x pressure started to drop slowly, since it had been intentionally set too high, at the same time the x dimension decreasing. As the z pressure continued to rise with continued flow in the z direction, the drop of x pressure slowed down, ceased, and then reversed. Constant x pressure means zero velocity of x flow. The instant the x observer detected a reversal, the z pumping was stopped and the various readings recorded, thus obtaining the data necessary for a calculation of a point on the strain-hardening curve and for the simultaneous values of X_z and Z_z which make $\dot{\epsilon}_x$ vanish. The x pressure was then raised so as to be slightly "too high," and the process repeated. After familiarity had been acquired, it was possible by cautious procedure to obtain from 5 to 10 points on the strain-hardening curve in a total range of z strain of 0.4. This "familiarity" involved so choosing the increments of x pressure that the total flow in the x direction never passed the limits of 0.002 in. No rigid time schedule was adhered to; the pumping was "slow," which meant that from 5 to 10 min was used in giving an increment of strain of 0.05. The x Bourdon pressure gauge was graduated to 10 kg/cm^2; it was constructed by the Société Genvoise in the early 1900's, and had a very good multiplying mechanism, which gave readings with no perceptible backlash. It was possible to establish the point of reversal within 1 kg/cm^2 or better. From the known dimensions of the dead space in the x press and the compressbility of the transmitting oil it is possible to calculate that a pressure increment of 1 kg/cm^2 should correspond to a distortion of the specimen of approximately 0.000003 in.; that is, the pressure gauge should provide a sensitivity in the detection of changes of the x dimension several times as great as that of the Ames 0.0001-in. gauge; this indeed proved to be the case.

The effects of friction were very small. The pistons of the presses were packed with the minimum amount of packing, and special tests showed that friction was not more than 2 or 3 per cent under unfavorable conditions. Under the conditions of the experiment, the stress Z_z is smaller than that calculated with no allowance for friction because the z piston is advancing, overcoming both friction and the stress Z_z, whereas the stress X_x is greater than that calculated with no allowance for friction, because the stress X_x is advancing the x piston against friction and against the pressure of the x gauge. The calculations in the following were made with no allowance for friction. The corrected Z_z and X_x would be brought nearer together by a correction for friction; the maximum such effect is somewhere of the order of 5 per cent.

The strain-hardening curves under conditions of zero x flow, with the simultaneous values of X_x, were determined for the following materials: annealed "Solar" tool steel (typical composition: C 0.50, Mn 0.40, Si 1.00, Mo 0.50), eight different blocks cut in various orientations from a bar of 2-in. round stock; annealed Cr-V steel (typical composition: C 0.40, Mn 0.25, Si 0.20, V 0.22), one series on six blocks of various orientations and with various methods of lubrication, one series of six compressings with refiguring on the same block, and another series on another block

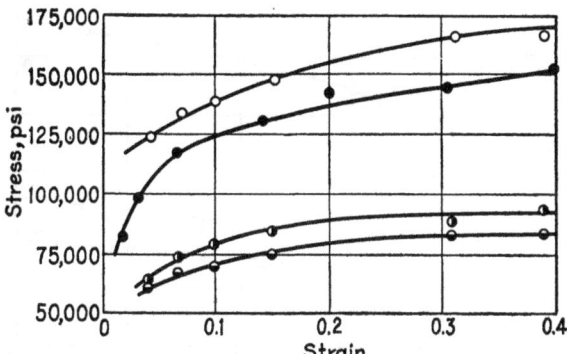

FIG. 110. Stress-strain relations for 1045 steel, austempered at 650°C. Reading from the top down, the first, third, and fourth curves relate to the block compressed, by suitable manipulation of X_x, so that flow in the x direction vanishes. The first curve plots Z_z, the third X_z, and the fourth $\tfrac{1}{2}Z_z$, which by elementary theory should be equal to X_z. The second curve is for another block with no lateral support.

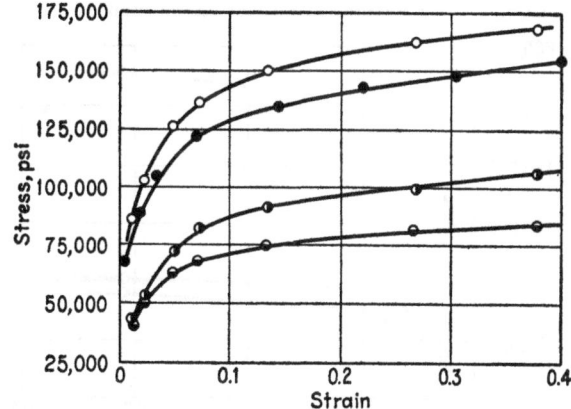

FIG. 111. Stress-strain relation for soft 1045 steel, austempered at 600°C. Reading from the top down, the first, third, and fourth curves relate to the block compressed, by suitable manipulation of X_x, so that flow in the x direction vanishes. The first curve plots Z_z, the third X_z, and the fourth $\tfrac{1}{2}Z_z$, which by elementary theory should be equal to X_z. The second curve is for another block with no lateral support.

for four compressings with refigurings; a stainless steel of unknown composition, six compressions with refiguring on a single block; a 1045 steel provided by the Watertown Arsenal in the following heat-treatments: austempered at 650, 600, 550, 500, and 450°C, quenched and tempered at 650°, 550°, and 450°C. The latter were the same as the steels used in simple compression.

Runs were made on one specimen of each of these heat-treatments, or in a couple of cases on two or three. The initial dimensions of the speci-

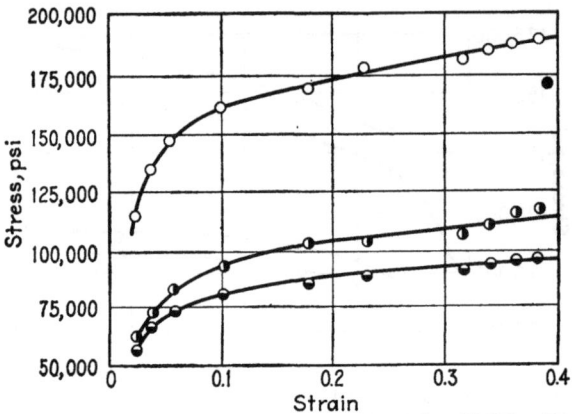

FIG. 112. Stress-strain relations for 1045 steel austempered at 550°C. Reading from the top down, the first curve shows Z_z, the second X_x, and the third $\frac{1}{2}Z_z$ for a block compressed in such a way that flow in the x direction vanishes. This block was compressed in two stages; the points for strains greater than 0.2 are for the second stage, after the stress had been released, the block refigured, and again exposed to stress. The solid black disk shows a single point for another block compressed without lateral support.

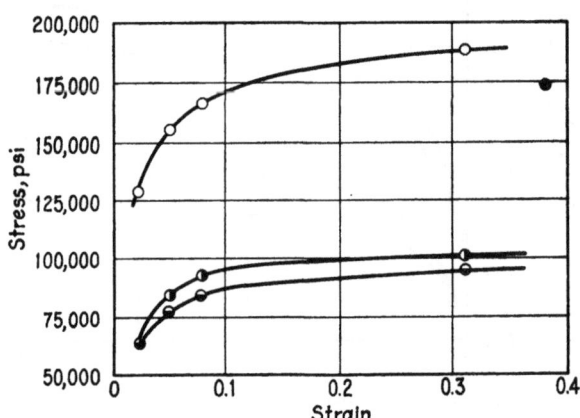

FIG. 113. Stress-strain relations for 1045 steel, austempered at 500°C. Reading from the top down, the first curve shows Z_z, the second X_x, and the third $\frac{1}{2}Z_z$, for a block compressed in such a way that flow in the x direction vanishes. The solid black disk shows a single reading for another block compressed without lateral support.

mens were: z, 0.750 in.; y, 0.500 in.; x, 0.502 in. The compressions, except for the multiple compressions with refiguring, were pushed to a reduction of height of two-thirds.

In addition to the data for two-dimensional flow, the strain-hardening curves in simple compression ($X_x = Y_y = 0$, $\epsilon_x = \epsilon_y = -\frac{1}{2}\epsilon_z$) were also determined, either by special experiment or else taken from previous data. The results are shown graphically in Figs. 110 to 118.

An examination of the curves shows that universally X_x is greater by up to 20 per cent than its theoretical value, $\frac{1}{2}Z_z$. The discrepancy is

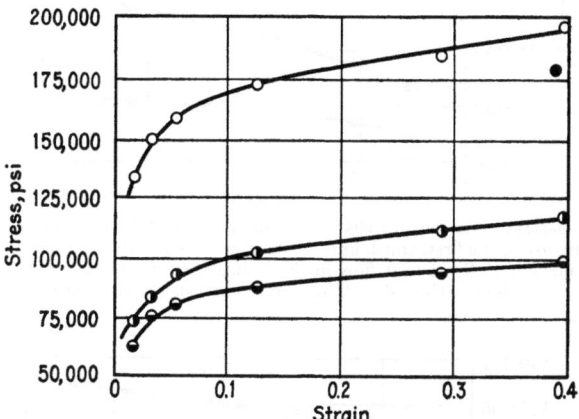

FIG. 114. Stress-strain relations for 1045 steel, austempered at 450°C. Reading from the top down, the first curve shows Z_z, the second X_x, and the third $\frac{1}{2}Z_z$ for a block compressed in such a way that flow in the x direction vanishes. The solid black disk shows a single reading for another block compressed without lateral support.

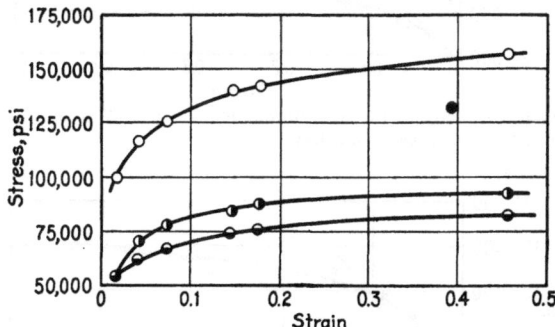

FIG. 115. Stress-strain relations for 1045 steel, quenched and drawn back to 650°C. Reading from the top down, the first curve shows Z_z, the second X_x, and the third $\frac{1}{2}Z_z$ for a block compressed in such a way that flow in the x direction vanishes. The solid black disk shows a single reading for another block compressed without lateral support.

greatest for the specimens austempered to the highest temperatures. Correction for friction would accentuate the discrepancy. In the initial stages of plastic flow X_x tends to be equal to $\frac{1}{2}Z_z$; the discrepancy reaches its maximum in the neighborhood of a strain of 0.1, and beyond this there does not seem to be usually any marked tendency for the discrepancy to become accentuated.

232   OTHER TESTS INVOLVING LARGE DEFORMATIONS

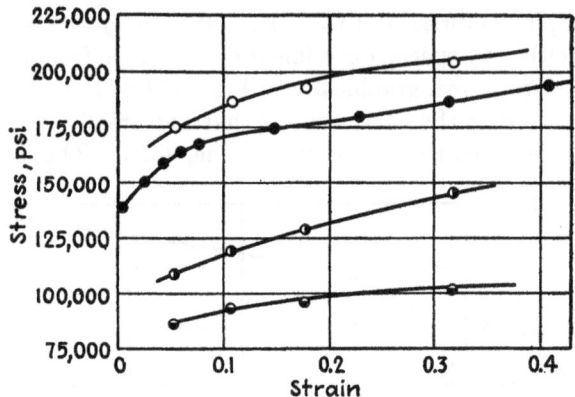

FIG. 116. Stress-strain relations for 1045 steel, quenched and drawn back to 550°C. Reading from the top down, the first, third, and fourth curves relate to the block compressed with suitable manipulation of X_x so that flow in the x direction vanishes. The first curve plots Z_z, the third X_x, and the fourth $\tfrac{1}{2}Z_z$. The second curve is for another block with no lateral support.

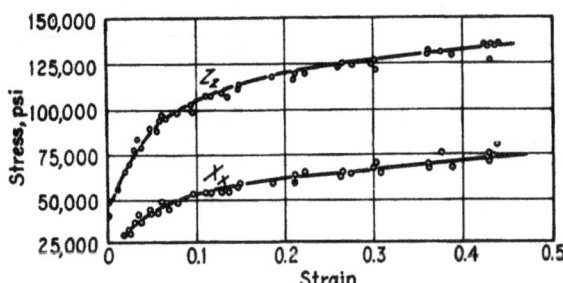

FIG. 117. Stress-strain relations for six different blocks of Cr-V steel, X_x being so manipulated that flow in the x direction vanishes.

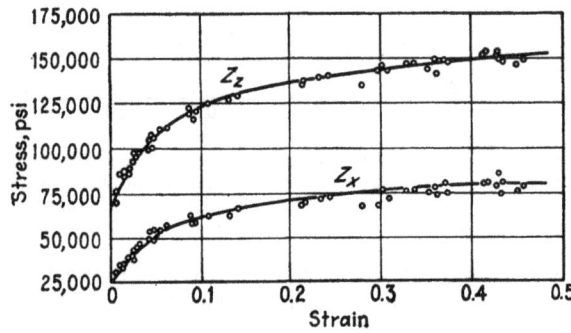

FIG. 118. Stress-strain relations for eight different blocks of Solar steel, X_x being so manipulated that flow in the x direction vanishes.

The result signifies, among other things, that the material is no longer isotropic after plastic flow in compression has taken place.

In addition to the experiments with steel, a single run was made with a block of commercial annealed copper. In the initial stages of plastic flow simultaneous values of Z_z and X_x were 39,500 and 21,500 psi, respectively; at the end of the test when ϵ_z had reached a value of 0.404, the corresponding values were 55,000 and 30,000. Again, as with steel, X_x is somewhat more than one-half Z_z, indicating a failure of the perfect isotropy assumed in the equations. A marked difference between the behavior of steel and copper was that the steady state was reached more quickly with copper after pumping stopped.

2. *The Velocity of Flow.* The use of photographic recording in measuring the velocity of flow has already been described. The velocity measurements were made at constant load; this demanded the use of the free piston gauges, as already indicated. With constant pressure on the x press, the stress X_x is rigorously constant, because the area of the x face does not change during flow. The z face does, however, increase in area as flow progresses, so that constant z load does not ensure constant Z_z. The dependence of Z_z on ϵ_z at constant z load may be calculated as follows. We have

$$Z_z = \frac{\text{load}}{\text{section}} \qquad \text{Section} \times z = \text{vol}$$

Hence

$$Z_z = \frac{\text{load}}{\text{vol}} z$$

$$\left(\frac{dZ_z}{dz}\right)_{\text{load}} = \frac{\text{load}}{\text{vol}} = \frac{Z_z}{z}$$

$$\epsilon_z = \log \frac{z}{z_0}$$

$$dz = z \, d\epsilon_z$$

and

$$\left(\frac{dZ_z}{d\epsilon_z}\right)_{\text{load}} = Z_z$$

The usual procedure was to keep the x load constant at some selected value and determine the flow for successive increments in the z load. Each increment of z load was suddenly applied, by suddenly opening a valve to the gas reservoir in which the requisite pressure had been previously established. The initial stages of flow were comparatively rapid, and the camera exposures were made rapidly, at intervals of a second or less. The first exposure was made a fraction of a second after the application of load, as soon as it was safe to assume that inertia effects had vanished. As the flow slowed down, exposures were made at roughly equal increments of strain. Exposures were continued until the rate of flow in the z direction had dropped to something of the order of 0.0005 in./min. This procedure gives the velocity of flow at different points on the inclined parts of the sawtooth pattern indicated in Fig. 119. The slopes of the inclined part are given by $dZ_z/d\epsilon_z = -Z_z$ (ϵ_z drawn positive in the diagram). Successive increments of z load were made in this way at constant x load until the total deformation in the x direction had reached its permissible limit; the x load was now increased sufficiently to reverse the direction of x flow, and the stepwise increase of the z load was resumed. This was continued until the total shortening was about 33 per cent. The total number of z steps varied from 10 to 25, average somewhere around 15, with a total number of photographic exposures of several hundred. Not all the exposures were usually used in taking the record off the film.

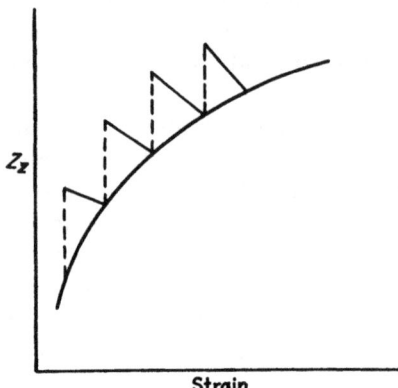

FIG. 119. Shows the general connection between stress and strain in determining the rate of flow off the strain-hardening curve.

The flow phenomena of the following materials were studied: an annealed 1035 steel, five specimens in two-dimensional compression and one in simple compression; annealed Omega tool steel, typical composition C 0.55, Mn 0.80, Si 2.30, V 0.25, Mo 0.50, P 0.015, S. 0.025, four setups with two-dimensional compression and one with simple compression; stainless steel, one specimen in two-dimensional and one in simple compression; five heat-treatments of the 1045 steel from the Watertown Arsenal, one specimen each in two-dimensional and in simple compression.

The photographic record gives the strains and the times. The rates of flow were calculated from the record by dividing increments of strain by increments of time. These rates were then plotted in one way or another depending on the point at issue. The photographic record contains the material for determining the strain-hardening curve from the limiting readings when the steady state was approximately reached. The strain-hardening curves were calculated from the photographic data and compared with those obtained by the first method of procedure. The agreement by the two methods was within experimental error; it is not worth while reproducing the details.

Although principal interest in the measurements was in the z rates of flow, the photographic record also contains the material for a calculation of the x rate of flow, but with less accuracy than for the z rate. This permits a calculation of $\dot{\epsilon}_x/\dot{\epsilon}_z$. Since the corresponding values of X_x and Z_z are known, we have the material for a comparison of the actual value of $\dot{\epsilon}_x/\dot{\epsilon}_z$ with its theoretical value in terms of the stresses, or in terms of α. In Fig. 120, $\dot{\epsilon}_x/\dot{\epsilon}_z$ is plotted as a function of $R[\equiv (X_x - \frac{1}{2}Z_z)/(Z_z - \frac{1}{2}X_x)]$ for all the samples of steel for which $\dot{\epsilon}_x$ could be determined with any accuracy. The wide scatter of the points is understandable in view of the very small x strains. The conventional equations of plasticity demand that these points all lie on the 45° line. In spite of the scatter of the points, it is obvious that the 45° line definitely does not represent the results. So far as they can be represented by a line, the line has a considerably smaller slope and furthermore passes somewhat below the

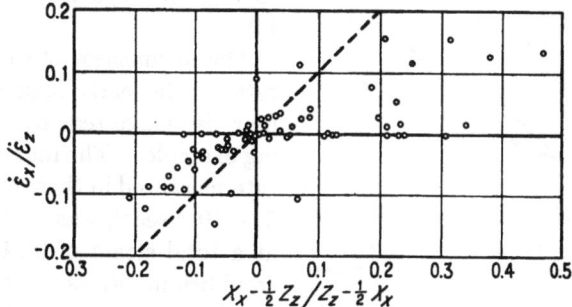

FIG. 120. Plot of all the observed ratios of rate of flow in the x direction to that in the z direction against the ratio of the "effective driving stresses" in these two directions. According to the elementary theory the points should lie on the dotted 45° line.

origin instead of through it. This agrees with the results of the first part, namely, $\dot{\epsilon}_x$ vanishes for a value of α somewhat greater than 0.5. No correlation could be found between the deviations of the points from a line and any other obvious single factor, such as grade of steel, location on the strain-hardening curve, or absolute value of the flow velocity. Only one point was found for which the *sign* of $\dot{\epsilon}_x/\dot{\epsilon}_z$ was definitely "wrong" in terms of the stresses. This particular point is hard to explain by experimental error, the increments of both ϵ_x and ϵ_z being too large to allow any chance for an error of sign. It may well be that the points actually should be scattered, and that the ratio $\dot{\epsilon}_x/\dot{\epsilon}_z$ depends on some combination of parameters more complicated than any that were obvious.

Much numerical material was collected for the velocity of primary flow $\dot{\epsilon}_z$, and many curves were drawn. The maximum observed rate was 5×10^{-3} sec^{-1}; the minimum was 0.00×10^{-5}. The material is too extensive and complicated to attempt to reproduce in detail. There are, however, certain outstanding approximate generalizations. As seemed natural, the velocity of flow was found to increase rapidly with increasing distance from the limiting strain-hardening curve. To a rough first approximation the increase is exponential, that is, at points not too near the origin the logarithm of the rate is approximately linear in terms of the displacement from the strain-hardening curve. This displacement may be taken in terms either of stress or of strain. The rate at which the rate increases on moving away from the strain-hardening curve is itself a strong function of location on the strain-hardening curve, being much greater in the early stages of deformation, that is, for small strains. The qualitative nature of the effect is suggested in Fig. 121, the successive curves being drawn for the successive steps indicated in Fig. 119.

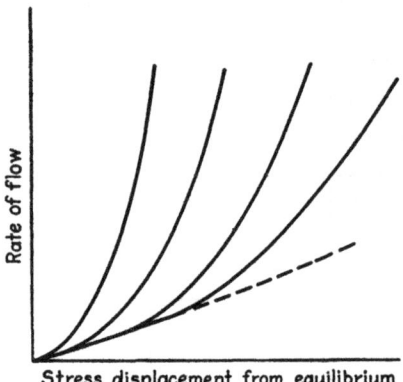

Fig. 121. Shows the progressive change in the character of the flow curves for increasing degrees of strain-hardening. The steeper curves are for smaller strain-hardening.

The abruptness of the rise of the rate in the early stages of plastic flow is suggested by the following example. The maximum strain rate measured in these experiments, 5×10^{-3} sec^{-1}, was for Omega steel at a total strain of 0.04, where the equilibrium stress is 122,000 psi. This rate was produced by a stress displacement from the equilibrium curve of 1,850 psi or 1.5 per cent. The rate just mentioned was the average over a time interval of 0.4 sec; the initial rate must have been much higher.

The curves of Fig. 121 have a linear envelope, which approaches the origin with a finite slope. If the displacement from the strain-hardening curve is measured in terms of normal distance, instead of along the stress axis, the drawing apart of the curves becomes even more marked than in the figure. The general order of magnitude of the drawing apart of the rate curves may be suggested by a single example. For the 1045 steel, austempered at 650°C, the rate of flow was 1.0×10^{-3} sec^{-1} for a displacement from the strain-hardening curve of 185 psi in the early stages where the equilibrium stress was 82,000 psi, whereas in the later stages, where the equilibrium stress was 156,000 psi, a displacement from the strain-hardening curve of 6,700 psi, or 36 times as much, was necessary to reach the same rate of flow. The slope of the linear envelope shown in Fig. 121 depends markedly on the grade of steel, being less for the harder steels. As an example, this slope was 2.7×10^{-6} cm^2 sec^{-1} kg^{-1}

for the stainless steel, and 7×10^{-7} cm^2 sec^{-1} kg^{-1} for the 1045 steel tempered at 450°C.

No material difference could be found between the character of the rate curves for two-dimensional and simple compression. The very rapid increase in the rate of flow for slight displacements from the strain-hardening curve in the early stages of flow explains the possibility of approximately calculating the velocity of propagation of the plastic front from the statically determined stress-strain curve. The loss in the abruptness of the flow phenomena at higher strains, shown by the drawing apart of the curves in Fig. 121, would suggest that in a wire which had

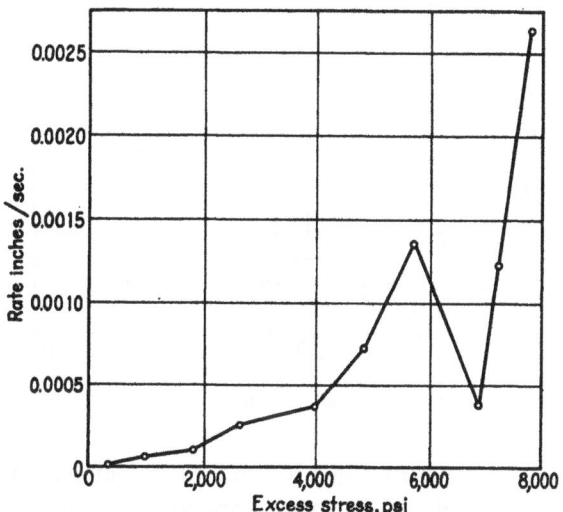

FIG. 122. Shows, for soft Omega steel, regeneration of rate of flow after a drop from a high initial value.

been previously severely strained the velocity of propagation could not be calculated from the stress-strain curve with as close a degree of approximation as for the virgin wire.

The actual rate curves, as a general rule, have the smooth idealized configuration shown in Fig. 121 over their entire extent only for the early stages of the strain-hardening curves, that is, only for small strains. For larger strains the smooth-rate curves are often interrupted by episodes with an abnormally high rate. This holds for both simple and two-dimensional compression. Figure 122 shows an example. The location of the episodes may be anywhere on the curve and would seem to be a probability affair. One is practically certain to find several of these episodes, if one waits long enough, in the region close to the strain-hardening curve where the rate on the average is approaching zero. It was not at all uncommon for the strain to hang at a steady value, constant within 0.00001 in. for 1 or 2 min, and then within 2 or 3 sec increase by 0.001 in.

This flow in jumps is not contained in the conventional equations of plasticity but is obviously a matter of considerable importance in its suggestions as to the mechanism of plastic flow. There would seem to be little doubt that it is a real phenomenon and not an artifact. One might be inclined, at first, to think it an effect of friction. But it cannot be friction of the x face, because the jumps are shown equally in simple compression where the x face is free. Neither can it be friction on the piston of the press, because the effect occurs only when the *specimen* has received comparatively large strains. Similar effects are found in other circumstances. One is reminded of deep-seated earthquakes. The continued

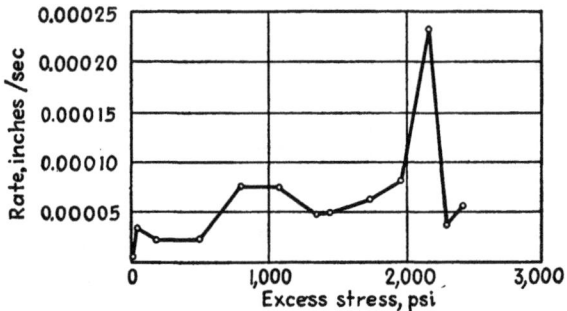

FIG. 123. Shows, for soft Omega steel, attainment of a relatively high rate of flow after a slow beginning.

internal fractures with self-healing which I have found when shearing is combined with high hydrostatic pressure are doubtless somewhat similar.[1]

It would be interesting and important to establish whether a jump in the z flow is accompanied by a jump in the x flow. The answer to this question could not be certainly found from the data, the absolute value of the x flow usually being too small. In two or three instances, however, there were slight jumps in the x flow accompanying those in the z flow. The only conclusion that is justified is that there is no present evidence against supposing that jumps in the two flows occur together, or in the theoretical ratio.

In addition to the flow in jumps there is another phenomenon sometimes found at the larger strains. On increasing the load suddenly at constant strain, flow is very likely not to begin at once, but the flow hangs and requires time to build up to its maximum rate. This is not an inertia effect, for it is found only in the later stages of flow, and the time intervals

[1] P. W. Bridgman, *J. Applied Phys.*, **8**, 328, 1937; *Proc. Am. Acad. Arts Sci.*, **71**, 387, 1937.

are too long. This effect was found only in the 1035 and Omega steels. An example is shown in Fig. 123. I have previously found a similar effect in tension. Tension specimens, pulled under hydrostatic pressure to an elongation far beyond the elongation at fracture under normal conditions, and then repulled at atmospheric pressure, may show a similar initial hanging of the flow. The mere existence of the effect shows that strictly the velocity of flow cannot be a function of stress and strain only but must involve the history to some extent. The same conclusion, of course, could have been drawn from the existence of jumps in the flow.

For the more rapid rates of flow, the rate is not sensitive to X_x; under the conditions of these experiments the regular sequence of curves indicated in Fig. 121 was usually not affected by a discontinuous change in

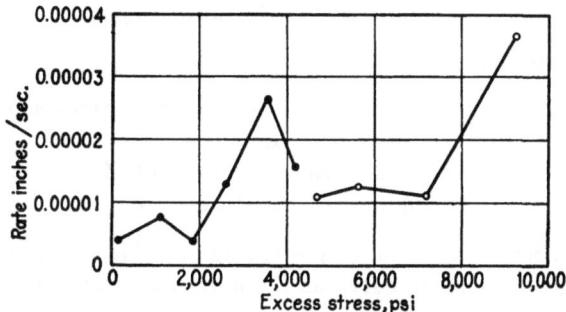

FIG. 124. Shows, for 1035 steel, an increase flow velocity along z brought about by an increase in X_x. At excess stresses greater than 4,200 psi, the value of X_x was 63,000 psi. Below 4,200, X_x was increased to 80,000 psi. Z_z remained approximately constant at 115,-000 psi.

X_x. At the smaller rates, where the effect to be expected from a change of X_x is more nearly of the same magnitude as the rate itself, the effect of changing X_x is more easily exhibited, and in particular the paradoxical effect already suggested, namely, that an increase of X_x at constant Z_z may increase the rate of z flow. Figure 124 shows for the 1035 steel the effect of an increase of X_x in increasing $\dot{\epsilon}_z$. This increase of rate is contained in the equations of plasticity, as a consequence of an increase of the function ϕ due to an increase in the stress invariant part of the argument. Not all increases of X_x bring about an increase of ϕ, but obviously any increase of X_x from a value greater than $\frac{1}{2}Z_z$ to a value still greater than $\frac{1}{2}Z_z$ will result in such an increase; these are the conditions prevailing in Fig. 124.

The Generalized Strain-hardening Curve. The problem of deducing the strain-hardening curve for one type of deformation from measurements on another type of deformation has been much discussed and is one on which the present measurements have some bearing. The two types of deformation with which we are concerned are simple compression and two-dimensional compression. Three possibilities in the way of generalizing the strain-hardening curve will be considered.

1. The maximum stress is a universal function of the maximum strain. The maximum stress is Z_z and the maximum strain ϵ_z for both simple and two-dimensional compression. If this function is the correct one, then Z_z plotted against ϵ_z should give coincident curves for simple and two-dimensional compression.

2. The maximum shearing stress is a universal function of the maximum shearing strain. For simple compression,

$$X_x = Y_y = 0 \quad \epsilon_x = \epsilon_y = -\tfrac{1}{2}\epsilon_z$$
$$\text{Maximum shearing stress} = \tfrac{1}{2}Z_z$$
$$\text{Maximum shearing strain} = \tfrac{3}{4}\epsilon_z$$

For two-dimensional compression,

$$Y_y = 0 \quad |X_x| < |Z_z| \quad X_x \text{ and } Z_z \text{ of same sign}$$
$$\epsilon_x = 0 \quad \epsilon_y = -\epsilon_z$$
$$\text{Maximum shearing stress} = \tfrac{1}{2}Z_z$$
$$\text{Maximum shearing strain} = \epsilon_z$$

If this is the correct function, Z_z plotted against $\tfrac{3}{4}\epsilon_z$ for simple compression should give the same curve as Z_z plotted against ϵ_z for two-dimensional compression.

3. The "significant stress" $\bar{S}$ is a universal function of the "significant strain" $\bar{\epsilon}$. The definitions are as follows:

$$\bar{S} = \{\tfrac{1}{2}[(X_x - Y_y)^2 + (Y_y - Z_z)^2 + (Z_z - X_x)^2]\}^{\frac{1}{2}}$$
$$\bar{\epsilon} = [\tfrac{2}{3}(\epsilon_x^2 + \epsilon_y^2 + \epsilon_z^2)]$$

For simple compression,

$$\bar{S} = Z_z \quad \bar{\epsilon} = \epsilon_z$$

For two-dimensional compression,

$$\bar{S} = Z_z(1 - \alpha + \alpha^2)^{\frac{1}{2}}$$

where $X_x = \alpha Z_z$

$$\bar{\epsilon} = \epsilon_z(\tfrac{4}{3})^{\frac{1}{2}} = 1.155\epsilon_z$$

If this is the correct function, Z_z plotted against ϵ_z for simple compression should give the same curve as $Z_z(1 - \alpha + \alpha^2)^{\frac{1}{2}}$ plotted against $1.155\epsilon_z$ for two-dimensional compression.

Under present experimental conditions there is one qualitative difference between the various criteria. If either of the first two is correct, Z_z plotted against ϵ_z for two-dimensional compression should give a curve passing smoothly without jump over any points where X_z may be changed discontinuously. If, however, the third criterion is correct, there should be a jump in the Z_z versus ϵ_z curves provided that there is a jump in $\bar{S}$ on changing X_z, for $\bar{S}$ versus $\bar{\epsilon}$ is smooth. The larger the jump in $\bar{S}$, the more critical should be the test. Now $\bar{S}$ as a function of X_z or of α has a flat minimum in the neighborhood of $\alpha = \frac{1}{2}$, so that any jumps of X_z which keep it in the general neighborhood of $\frac{1}{2}$ will not afford a sensitive test. It is for this reason that all the quasi-static measurements described under the heading on page 224 are not well adapted to bring out the distinction that we are seeking. The curves specially determined for rate of flow by photographic recording in some cases embraced changes of X_z in a more favorable domain, and it is the limiting curves given by these measurements that were used in comparing the three criteria. It has already been mentioned that the limiting strain-hardening curves determined in this way were the same as those obtained by the first method.

As it turned out, the cases in which there were favorably situated jumps in X_z did not permit a decision between the various criteria. In one instance, that of Omega tool steel, at a point where there was a jump in α from 0.45 to 0.72, there was a greater jump in the curve of $\bar{S}$ versus $\bar{\epsilon}$ than in the curve of Z_z versus ϵ_z, the former jumping up by 5 per cent and the latter jumping in the same direction by only 2 per cent. The 1045 steel, tempered at 650°C, gave two opportunities for comparison. There was one jump of α from 0.38 to 0.68; at this point there was no discontinuity in the $\bar{S}$ curve, whereas Z_z jumped down by 3 per cent. At a somewhat greater strain for the same material there was a jump down in α from 0.64 to 0.50. The $\bar{S}$ curve now jumped down by 2 per cent, whereas the Z_z curve was sensibly continuous. Various points of discontinuity in α for other steels did not permit a decision between a jump or an abrupt change in direction of the curves, the experimental points not being sufficiently closely spaced. On the whole it appeared that the tendency to a break of some sort was greater on the $\bar{S}$ curves than on the Z_z curves.

The data would appear, therefore, not to permit a clean-cut decision on the basis of the breaks in the curves, and the decision must be sought from a comparison of the whole general course of the curves.

In making this comparison it is to be kept in mind that $\bar{S}$ and $\bar{\epsilon}$ are identically equal to Z_z and ϵ_z, respectively, for simple compression. In anticipation, neither of the criteria is met within the limits of error of the measurements, and it becomes a question of choosing the one which is the best approximation. As regards the criterion that Z_z is a universal function of ϵ_z, the Z_z curve for two-dimensional compression consistently lies above that for simple compression for all the steels except the 1035 steel. For strains above 0.1 the order of magnitude of the excess is 10 per cent. For the 1035 steel the curves cross at a strain of 0.07, and at a strain of 0.15 the curve for simple compression has risen to 10 per cent above the curve for two-dimensional compression. For all steels the curve of $\bar{S}$ for two-dimensional compression lies below that for simple compression. The discrepancy between the curves may be either greater or less than for the Z_z curves but on the whole is inclined to be rather less, so that on the whole the $\bar{S}$ criterion is to be preferred to the Z_z criterion. With regard to the maximum shearing-stress criterion, the curve for maximum shearing stress for two-dimensional compression consistently lies above that for simple compression, except for the 1035 steel. The discrepancies are again of the general order of 10 per cent, but detailed comparison would show that for the present selection of steels the maximum shearing-stress criterion is on the whole the best by a narrow margin. For the stainless steel, either the significant-stress or the maximum-shearing-stress criterion holds with much less error than for the other steels. In fact, the general behavior of the stainless steel is different from that of the others; its strain-hardening curve, for example, is sensibly linear up to a strain of 0.5.

All the above remarks apply in the region of strains of moderate magnitude, say, above 0.02. In the initial stages of flow, with the single exception of the 1035 steel, the stress for two-dimensional flow is less than for the three-dimensional flow of simple compression. It is natural to associate this with an initial slip on a single set of slip planes in two-dimensional flow as contrasted with cross slip on systems of interfering planes in simple compression.

CHAPTER 14

MIXED COMPRESSION[1]

During the war experiments were made on the effect of cold-working armor plate by a hybrid compression process, which cannot be properly described as either one- or two-dimensional, for which I have coined the description "mixed compression." The work itself was done under a rather hybrid combination of auspices. The original idea was mine, namely, that if an armor plate could be cold-worked by the same type of deformation as that imparted by the penetrating projectile its ballistic performance might be improved. One of the principal effects of the penetrating projectile is to push the material of the plate from its path, thus thickening the plate. This is evidenced as a rim around the projectile hole on both front and rear surfaces of the plate. It was indicated, therefore, that the plate should be cold-worked by thickening it. This is a rather unusual sort of cold-working and would obviously demand special means to produce it. Experiments were started in my laboratory and presently a method worked out for thickening plates large enough to permit the performance of ballistic tests on them. I had no facilities for making the ballistic tests and secured the cooperation of the Princeton Ballistics Laboratory, which had just begun operation under the auspices of the NDRC. This promise of cooperation was an entirely informal affair, arranged on the basis of my personal acquaintance with the men at Princeton. It also appeared that the effect of heat-treatment on the plates should be studied, and in this the Watertown Arsenal promised cooperation, again on a completely informal basis and with no connection with any contract. The work was presently completed, with, as it will appear, no results that could possibly affect the progress of the war, and the question of the publication of the results presented itself. It appeared that the Princeton Laboratory would have been willing to publish the results of their ballistic tests as part of their regular reports to the NDRC, but they were also willing that I should publish them, since the initiation was from me. There was some correspondence about it, copies of which were forwarded by them as a matter of routine to Washington, with the result that the NDRC office in Washington was presently suggesting to me that I publish an NDRC report on the experiments as one of my regular series under my contract with them. It was wartime, and I did not

[1] This chapter is based on the third NDRC report.

care to bring up the point that the NDRC really had no connection with it, certainly no contractual connection, so that I took the easiest way, which I am sure was completely improper and uncanonical, of publishing it as one of my NDRC reports. This report was at first classified as confidential, a classification which has since been completely lifted.

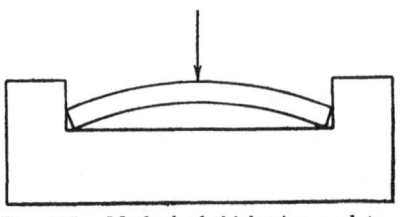

Fig. 125. Method of thickening a plate by flattening after bending.

The present interest of this work does not justify reproducing most of the experimental details, but because of the novelty (as far as I know) of the fundamental method of attack, certain of the results will be described here which perhaps may have a more general interest.

The first problem was how to thicken the plates by cold-working. The difficulties were obvious enough if a straightforward attack was made by subjecting the plate to compressive forces in the plane of the plate. Enormous forces would be required, and buckling would have to be prevented. At first the thickening was attempted by more indirect means. Inquiry had elicited the information that adequate ballistic tests could be made on square pieces, 2.5 in. on a side and 0.25 in. thick. These dimensions were small enough to make it possible to make the first attempt at thickening the plate by first bending it and then pressing it flat while held in a clamp such that the length of the chord along the bent arc is held constant. The method is suggested in Fig. 125. The method was successful in thickening plates which were soft enough to withstand the requisite bending, but a difficulty was that most of the plates of a hardness great enough to make them of interest ballistically were too hard to tolerate the requisite bending without fracture. But, quite apart from this, it appeared that probably there would be no advantage from this procedure, because of the entirely unexpected result that the hardness of plates cold-worked in this way was decreased rather than increased. Five plates were treated in this way, reduced from squares 3 in. on a side to squares 2.5 in. on a side by first compressing along one direction and then along the other. The average Rockwell C hardness of these plates was diminished by this treatment from 23.9 to 20.8. These plates were later subjected to ballistic tests with results to be expected from the diminished hardness.

A more direct method of thickening the plates was therefore demanded. This was finally accomplished by upsetting the plates by compressive forces along the edges while the faces were held in a massive clamp to prevent buckling. The arrangement is shown in Figs. 126 and 127. The requisite lateral motion of the faces during deformation was provided by steel ferrules under the bolts of the clamps which plastically shortened as the plate expanded laterally. Suitable dimensions for the ferrules were found by trial. The lateral compression was performed in a hydraulic press; a total load of about 250,000 lb was required. The final square figure was obtained by first compressing along one dimension and then along the other. In one set of experiments the final figure was obtained in two steps, from 3 by 3 to 3 by 2.5 to 2.5 by 2.5 in. In another series four steps were made: 3 by 3 to 3 by 2.75 to 2.75 by 2.75 to 2.75 by 2.5 to 2.5 by 2.5 in. No difference was apparent in the final result. The thickening accompanying this amount of upsetting is 44 per cent $[3^2/(2.5)^2 = 1.44]$.

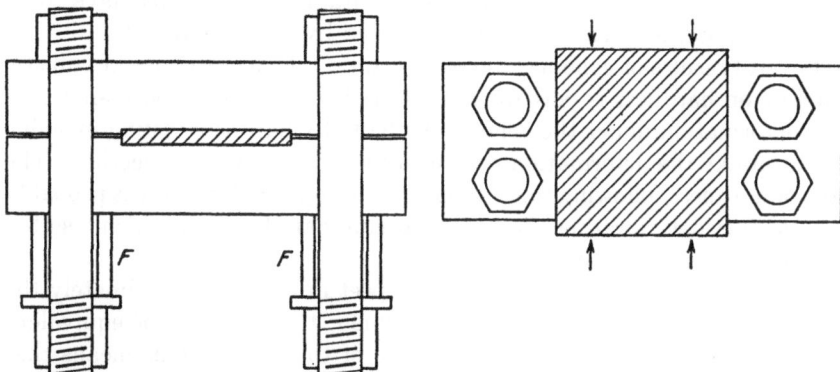

FIG. 126. Clamp in which the plate is held during upsetting. The ferrules yield plastically and so permit lateral expansion of the plate.
FIG. 127. Elevation of Fig. 126. The arrows indicate the direction of action of the platens of the hydraulic press.

Deformation in this manner resulted in the expected increase of hardness. The first series of plates treated in this way comprised 23 samples of the same 0.45 C steel in the lower hardness ranges from a Rockwell C of 20 to 27. Hardness increases were found of as much, in one instance, as from 20.6 to 27.9.

The Watertown Arsenal now entered with the preparation of an elaborate series of plates of a range of hardness. Some 50 plates were prepared altogether, all from the same 0.45 C steel, all quenched into water from 1525°F, and then drawn back in groups to temperatures of 800, 900, 950, 1000, 1100, 1150, 1200, 1250, and 1300°F, respectively. Half of these were upset as described to a thickness ratio of 1.44 and the other half were retained as blanks. All were machined to the same final dimensions, 2.5 by 2.5 by 0.25 in., and sent to Princeton for the ballistic tests. Increases of hardness were produced in the various hardness groups as follows: from 28.0 to 34.6, from 32.3 to 36.2, from 38.6 to 40.5, from 39.8 to 42.5, from 40.6 to 41.7, and from 44.0 to 46.2.

The results of the ballistic tests can be simply described. For all the plates of the same composition ballistic behavior depends on a single parameter, the hardness, independent of whether the hardness has been acquired by heat-treatment or cold-work. Furthermore, the hardness limit at which the ballistic behavior of the plate begins to fall off is the same whichever way the hardness was acquired. Hence, although the anticipated improvement of ballistic behavior of the plates was thus demonstrated to be brought about by suitably directed cold-work, there is no military advantage, because obviously it is cheaper to produce a given hardness by heat-treatment than by cold-work. It is not ruled out, however, that cold-working might be profitable if ever plates have to be made of materials that cannot be hardened by heat-treatment.

Several results of general interest remain. The dependence of ballistic quality on hardness is perhaps not surprising when it is considered that the deformation under the cone or the ball of a hardness testing machine is not unlike the deformation produced by a penetrating projectile. The loss of hardness following thickening by bending and flattening is probably a secondary deteriorating effect of the preliminary bending, which resulted in the fracture of the harder plates.

Geometrically, the strain in a plate upset in this way is ultimately the same as in the neck of a tension specimen, the direction of equivalent tension being at right angles to the face of the plate. This means that the hardness of the neck of a tension specimen, tested on a surface cut across the neck perpendicular to the axis, is increased. I do not know whether such measurements have been made; they would be difficult for small specimens. An increase of hardness would appear probable in view of the generalization, presently to appear from the experiments on prestrained specimens, that prestraining of any sort, if carried to a sufficiently great degree, results in hardening for any other type of subsequent straining. At low prestrains this of course is not the universal effect, because we have the Bauschinger effect, in the opposite direction.

The natural strain associated with a thickening of the plates in the ratio 1.44 is 0.36. In the data already given for the strain-hardening process by simple tension examples may be found in which the flow stress F is increased by 10 per cent or more for a strain of 0.36. This is of the same order as the increases of hardness found for these upset plates. The two effects are thus not out of line with each other.

Much more information could have been obtained from these "mixed compression" tests if the upsetting force could have been measured as a function of strain. The constraint exerted by the clamps was, however, too irregular to permit this.

CHAPTER 15

TORSION COMBINED WITH SIMPLE COMPRESSION[1]

Introduction. Although no feasible method has been found for conducting torsion experiments in a medium under hydrostatic pressure, something not dissimilar is possible by combining torsion with simple axial compression along the axis of twist. Qualitatively, the same increase of ductility or of twist before fracture that would be anticipated if the torsion could be applied under hydrostatic pressure may be expected here; because, if fracture tends to initiate itself by the slip of one plane past another, the longitudinal compression will tend to push the planes back into contact, there will be self-healing, and twist may be carried further without large-scale fracture than without the longitudinal load. In addition to the qualitative increase of ductility under compression, such experiments should make possible a quantitative comparison of the rate of strain-hardening brought about by simple shearing distortion as against the strain-hardening produced by strains of extension which we have already considered. There has been much speculation and study of the possible equivalence of various sorts of plastic deformation with regard to strain-hardening, but most of this has been restricted to comparatively small distortions.

Search failed to disclose that experiments of this kind had been made before, although there have been studies of the torsion of long wires under tension. Under such conditions the torsional ductility is decreased, as might be expected.

Geometrically, because of stability considerations, the experimental arrangements for realizing torsion combined with simple compression differ materially from the simpler arrangements which are suitable for combining torsion with tension. The use of long specimens is ruled out because of buckling. A way of applying the compressive force without end friction must be employed. If possible, the strains should be homogeneous throughout the specimen in order to avoid the complications met in torsion of the ordinary solid rod in which the angle of twist varies from zero on the axis to a maximum at the outer surface. In striving to reach the various desiderata, the apparatus went through a process of development, and results were published in three installments.

[1] This chapter is based on material contained in the first NDRC report, the second Watertown report, and B14.

The basic idea was the same throughout. The strained region was confined to *two* narrow bands machined on the cylindrical specimen. The entire specimen was compressed longitudinally and the two ends kept from rotating, and torsional strain imposed on the two bands simultaneously by rotating the part of the cylinder between the bands against the two fixed ends, as suggested in Fig. 128. This arrangement has the obvious advantage of eliminating the necessity for thrust bearings and

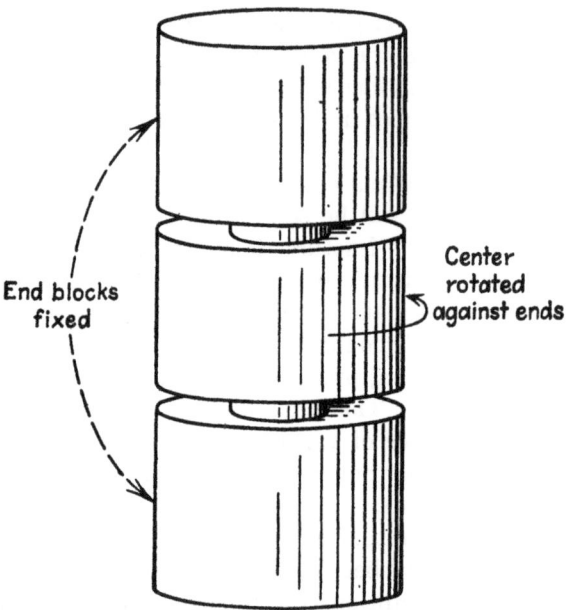

FIG. 128. General scheme of specimen in which torsion is combined with simple longitudinal compression. The plastic yield in torsion is confined to the two narrow isthmuses. The design gives stability under the compressive thrust.

thus does away with error from friction. Furthermore, since virtually two experiments are performed simultaneously, the error is less than if only a single region had been strained.

The first experiments were performed on solid cylinders and so could give only the shearing stress for the average of a strain which ranged from zero to maximum. These first experiments were performed on my own responsibility in my own laboratory. The application to the problem of armor penetration was suggestive, however, and in the first NDRC report the effects are described and the results of measurements given on an armor plate and on cast iron; the latter showed a dramatic increase of torsional strength under simple compression. It was soon realized even in the first experiments that an essential was to mount the specimen in such a way that the clamps which prevent rotation of the ends do not at the same time prevent longitudinal displacement. The reason is that simultaneously with the shearing strain in the bands there is a longitudinal strain also; if the latter is inhibited the former is affected.

The next stage in the development was to make the cylindrical specimen hollow so that the twisted band is confined in a thin cylindrical shell, and strain is therefore approximately homogeneous. At the same time the necessary stability against buckling was given by inserting a solid guiding pin in the center of the cylinder. This pin could be proved to introduce no error by friction. The band machined out of the specimen had a fixed shape, being made always with the same tool. The contour of the tool was rounded at the end, as indicated in Fig. 129. The reason for the rounded contour was to avoid stress concentrations at corners, and so an irrelevant source of fracture.

With specimens prepared in this way, comparative studies may be made of the behavior of different varieties of steel and of different compressive loadings, in both the plastic and the fracture regions. A number of such comparative studies were made as part of the work under contract for the Watertown Arsenal and presented in the second Arsenal report. The computation of any precise strains with rounded contours was obviously difficult, so that these experiments did not allow a quantitative comparison of the strain-hardening effects produced by torsion (or shearing) with the effects which had already been found by straining by tension. Since a major interest in the measurements consisted in this comparison, the method was further improved to permit a reasonably precise calculation of the strains. The latter work was again done on my own responsibility and was published in the *Journal of Applied Physics* in 1943. The essential modification in the latest work was to give sharp edges to the band subjected to twist. To my surprise these sharp edges exerted no predisposition to fracture, which occurred indifferently in any part of the band. On reflection, this does not appear so strange, because fracture by shearing slip would seem to have little connection with the sort of stress discontinuity introduced by a sharp corner of this kind. At the same time, the best dimensions for the specimen were studied by empirical cut-and-try methods, and the dimen-

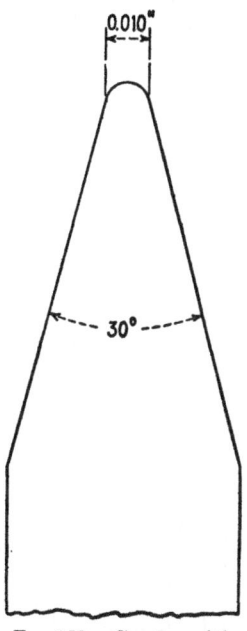

FIG. 129. Contour of the tool with which the isthmus was cut in the first series of torsion experiments.

250   OTHER TESTS INVOLVING LARGE DEFORMATIONS

sions indicated in Fig. 130 adopted as best for the final measurements. It appears from the figure that the region of most intense twist is confined to a band 0.010 in. long and 0.010 in. wide in the form of an annulus 0.277 in. ID and 0.297 in. OD. The conditions inside this annulus must be very close to two-dimensional and little affected by the curvature.

The method adopted in all the work was a slow dynamic method as distinguished from a static method. A predetermined angular velocity of twist was impressed on the specimen and the torque measured which was necessary to do this. The reason for this procedure was that slow drift of the angle of twist continues so long after every change of the loading torque that an impossibly long time would have been consumed in an experiment if equilibrium had been waited for. Furthermore, even if equilibrium had been waited for, the significance of the results would have been obscure, being complicated by the various spontaneous recovery effects which are known to occur. The important thing was to attain reproducible conditions. By a study of the effects of varying speed over a wide range it was established that in the region of speeds employed any speed effects were negligible. It would seem that apart from questions of convenience the slow dynamic method is better than a static method, unless a complete and much more elaborate analysis is made of all the different sorts of time effect concerned.

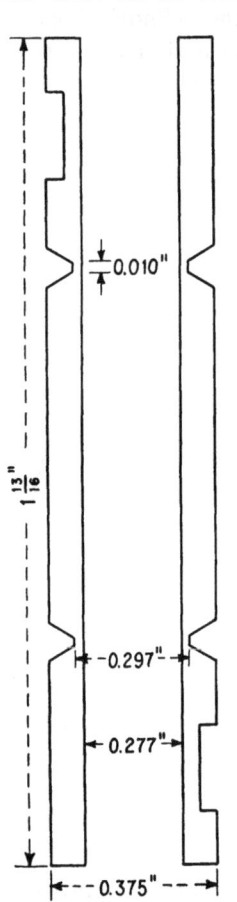

FIG. 130. Dimensions of the final torsion specimens.

The Apparatus. The apparatus consists of a part in which the specimen may be clamped while a torque and simultaneously a longitudinal compression are applied to it, a part by which the specimen may be twisted at a number of different speeds, and a part in which the twisting force is measured and automatically recorded photographically as a function of the angle of twist. The measurements are concerned with the region of plastic flow; the apparatus is not sensitive enough to give good measurements in the elastic region. The middle portion and the ends of the specimen shown in Fig. 128 or 130 are clamped to other parts of the apparatus by square wedges driven into seats milled on the specimen. The two ends are clamped to crossbars so attached to the rest of the apparatus that the ends cannot rotate, while the central part is clamped to a wheel to which a torque may be applied. The experiment consists in rotating the central portion against the ends and recording the torque as a function of the angle of twist. The twist in the specimen is confined almost entirely to the region at the bottom of the notch. Since the notch is narrow, high shearing deformations may be obtained with small angles of twist.

The specimen is mounted between the platens of a miniature hydraulic press, the ram of which is only 1 in. in diameter. Longitudinal compression is applied during the twisting by applying pressure to the ram of the press. The packing of the ram is so constructed as to move with negligible friction. Pressure is applied by means of a hand pump connected through a dead-weight piston gauge; any desired compression is applied by selecting the proper weights, and the compression is held constant during the experiment by keeping the piston of the gauge floating. The pressures used ranged up to a maximum of 1,500 psi on the ram.

The specimen must be so mounted that it is free to expand or contract longitudinally under the longitudinal compressive load or to permit the natural longitudinal motion which accompanies torsion even in the absence of a longitudinal load. To this end the crossbar at the upper end to which the specimen is wedged carries at its ends two holes which engage two heavy pins in the upper platen of the press in such a way that the crossbar can slide longitudinally along the pins. The bottom crossbar is rigidly bolted to the bottom platen.

Torque is applied to the central portion by means of a single cable running in a groove on the periphery of the wheel bolted to the central part of the specimen. The diameter of the wheel is approximately 52 cm. The bottom notch diameter of the specimen is about 0.6 cm. This means that the translational drag exerted on the specimen by the cable is negligible compared with the torque.

In Fig. 131 is shown the arrangement by which the torsion part of the apparatus is actuated. The cable from the torsion wheel makes a right-angle turn over a ball-bearing pulley mounted at the end of a freely swinging arm 2 m long hanging from the ceiling. The cable is wound on a drum actuated by a worm, driven by a 1.4 hp motor through one or two reduction boxes and V driving belts and pulleys of various sizes by which a wide range of speeds may be obtained.

The force in the cable is measured by attaching the freely swinging arm to a stiff flat steel spring (heavy hack-saw blade) which gives the force by its deflection. The spring is so stiff that departure of the swinging frame from the vertical is negligible over the entire range of force, so that the only effective force acting on the spring is the force in the horizontal section of the cable from the pulley to the torsion wheel. The spring is about 5 cm long. Its stiffness may be varied by changing its length with a suitable clamp. The motion of the spring is magnified by a light wooden arm 60 cm long attached to it. This arm carries at its end a source of light, a 6-volt bulb, mounted with a simple lens system so as to give a point of light on a sensitive paper (Eastman, P.M.C. No. 1, normal bromide paper) mounted on a rotating drum. The arm is damped with a vane in a dashpot with heavy oil. The drum is connected by belting through variable pulleys to the drum on which the cable is wound. What is obtained on the paper is therefore a record of force in the cable (torque) against angle of rotation. The light is pro-

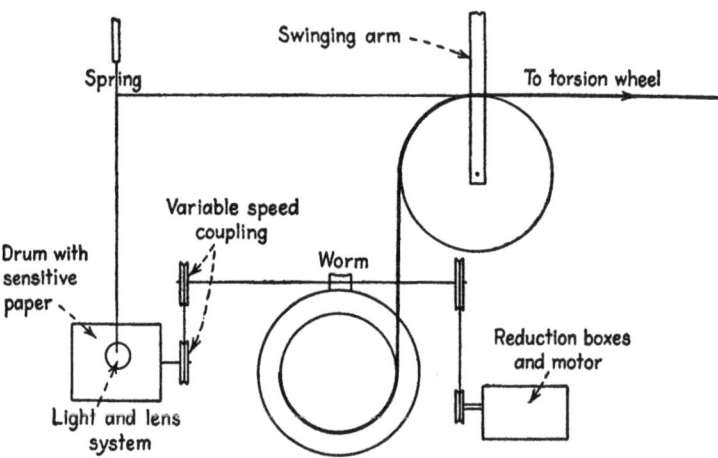

FIG. 131. General scheme of torsion and recording mechanism.

vided with a rheostat and ammeter so that the intensity of the light may be chosen suitably for the speed. Arrangements are made for putting various fiducial marks on the paper. The wheel may be clamped and force applied without motion of the specimen, thus correcting for the small deflection of the spring; the wheel may be entirely disconnected and known weights applied to the cable in its place, thus calibrating the deflection of the spring for force; and finally, the rotating wheel itself carries a graduated circle, which makes it possible to snap the light off and on at various angular displacements and thus calibrate the photographic drum for angle of twist of the specimen. Because of slight stretch in the cable and lack of perfect rigidity in other parts, the photographic paper gives the record of force against angle in oblique instead of rectangular coordinates. The various fiducial marks make it possible to correct back to rectangular coordinates. The photographic parts are contained in a lighttight box, so that the room need not be darkened.

There is a heavy red window in the box by which the position of the light may be found, which permits a control of the experiment, as when it is desired to discontinue the twisting after the force has reached its maximum.

Methods of Calculation. In order to evaluate the characteristic properties of the material, the shearing stress and the shearing strain in the material corresponding to any observation should be determined. The mean shearing stress can be found at once from the torque and the dimensions. If it is assumed that the shearing stress S is constant across the section, which must be very approximately true under the conditions, the following equation holds:

$$\text{Torque} = 2S \int_{r_i}^{r_0} 2\pi r^2 \, dr$$

or

$$S = \frac{3}{4\pi} \frac{\text{torque}}{r_0^3 - r_i^3}$$

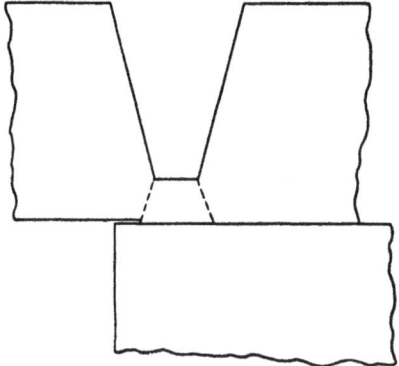

Fig. 132. Enlarged detail of the notch of the torsion specimen with part of the central pin.

where r_i and r_0 are the internal and external radii of the notch. The factor 2 before the integral arises from the two sections of twisting.

The mean shearing strain demands a knowledge of the angular displacement and the width of the zone of plastic flow. This width is far from constant, but the zone spreads out from the bottom of the notch to the inner surface in a manner indicated qualitatively in Fig. 132. One can see that the spreading must be an accompaniment of the strain-hardening. At the surface of separation of the elastic from the plastic zone the flow stress is at its minimum, corresponding to zero plastic distortion and zero strain-hardening. In the interior of the flow zone stress is greater because of strain-hardening. Equilibrium demands that the area of action of the minimum flow stress be greater than the area of action of the greater flow stress. Spreading of the flow zone and extension of the plastic-elastic boundary provide this greater area of action for the smaller stress. There seems to be no method by which one can calculate with confidence the amount of spreading to be expected. One might perhaps hope to get it experimentally by polishing and etching a section of the twisted specimen, but an attempt to do this did not yield definite results. It turns out that the spreading can be obtained by scratching the surface of the untwisted specimen with suitable straight lines and measuring the displacement of these scratches after twisting. By running a fine scratch down the sides of the notch and across its bottom it was possible to establish in the first place that at the corner of the notch the zone of flow does not extend perceptibly out of the notch into the surrounding material. After twisting, the scratch continues to run straight down the sides of the notch, with no deviation as it approaches the corners. At the cylindrical surface of the notch the scratch takes a sharp break in direction and runs diagonally across the cylindrical face of the notch, still in a straight line, indicating uniform shearing strain across the width at the outside of the flow zone.

The inside surface was marked with six longitudinal scratches engraved with a special broach pushed through the hole. After the conclusion of the experiment the specimen was split in two lengthwise with a thin milling cutter; the general appearance of the displaced scratches is indicated in Fig. 133. The circumferential displacement a and the angle ϕ were measured. Unlike the outer surface, the edges of the flow zone at the inside are not sharply marked, as shown by the rounding of the corners in Fig. 133. However, in the central two-thirds or three-quarters of the distance a the scratches remained straight, indicating constant shearing strain in this part of the flow zone. The effective width w_i of the flow zone at the inside was taken to be $w_i = a \tan \phi$. For one series on a 1045 steel w_i was found to vary from 48 to 77 per cent greater than the initial width of the notch bottom at the outside surface. The lower figure, 48, is for twist with no longitudinal load, and the larger, 77, for twist with the highest load and an angular displacement three times as great. Because of longitudinal flow the notch itself decreases in width under twist with compression, the decrease being greatest for the greatest loads. The resultant of the increase in width at the inside and the decrease at the outside is a mean width that is approximately independent of the compressional load within limits of error and is about 20 per cent greater than the initial notch width. The dimensions of the specimen were such that the mean width is roughly equal to the mean breadth as determined by the difference between inside and outside radius at the notch. The decrease of width of the bottom of the notch under the action of compressional load may be determined from measurements before and after twisting. The corners of the notch remain sharp, and precise settings are possible both before and after. In the case of the

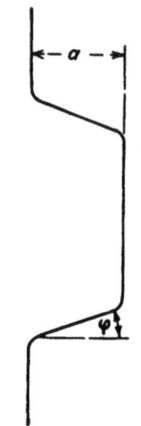

FIG. 133. Enlarged schematic view of the interior longitudinal scratch as distorted from a straight line by the rotation of the central part against the ends.

1045 steel mentioned above the notch under the greatest compressional load flowed down to 67 per cent of its initial width.

The mean shearing strain was taken as the mean circumferential displacement divided by the mean width of the flow zone.

In addition to the longitudinal flow under combined twist and compression, decreasing the notch width, there is radial flow, increasing the diameter at the bottom of the notch. This radial flow is nearly uniform across the entire notch width, any tendency to assume a barrel shape being slight. There is a tendency to radial flow in both directions, out at the outer surface, and in at the inner surface, but the inner guide pin prevents the inward flow. The increase of outer notch diameter was measured at the conclusion of twisting, before slitting. For the steel mentioned above the increase of notch diameter under the largest compressional load was 2.2 per cent. The diameter enters the formula for the mean shearing stress through the difference of two cubes, so that the

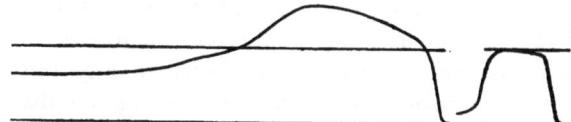

FIG. 134. Photographic record of angle of twist (abscissa, reading from right to left) against torque for cold-rolled steel. The short graph on the right was obtained with zero compressional load, the long graph on the left under an average compressional load at the isthmus of 81,000 psi. The total angular displacement of the long graph is 85°.

effect of a change of diameter is much magnified in the final result. Thus above, the effect of a 2.2 per cent increase in notch diameter is a 32 per cent correction on the shearing stress.

In reducing the observed torques and twists to mean shearing stress and strain at intermediate points below the maximum it was assumed that the over-all longitudinal and radial flow were distributed in proportion to the angle of twist. This is not quite correct, as will appear later, but the error so introduced is negligible.

The various measurements just outlined cannot be made unless the specimen hangs together during the splitting subsequent to twisting. Since the specimen becomes increasingly fragile with increasing twist, in most cases twisting was stopped shortly after the torque had passed its maximum. Under these conditions the specimen retained sufficient strength, even when twisted without load, to permit such handling as was involved in splitting it longitudinally.

Typical experimental results are shown in Figs. 134 to 137. In Figs. 134 and 135 are reproductions of the photographic record of torque against angle of twist for two varieties of steel. Figure 136 shows a transcribed photographic record with attached coordinates, for another steel. Figure 137 shows cast iron, with and without compressive load.

Effect of Varying Compressive Load at Constant Speed. Measurements were made on five specimens of 1045 steel supplied by the Arsenal, $\frac{5}{16}$ in. OD and 1.8 in. ID. The steel was heat-treated at the Arsenal to a Rockwell C hardness of about 34. The compressive loads varied from zero to 102,000 psi on the initial cross section at the bottom of the notch.

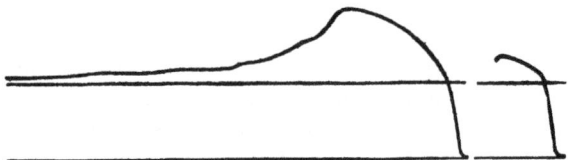

Fig. 135. Like Fig. 134, except that it is for "drill rod" instead of cold-rolled steel.

The constant speed for all five specimens was an angular displacement of 1.35° per second. The curves of torque against angular displacement for the first and fifth specimens, that is, for zero compressive load and the maximum load, 102,000 psi, have been shown in Fig. 136. With zero compressive load, rupture is complete with an angular displacement of 20°, whereas with compressive load an angular displacement of 85° is

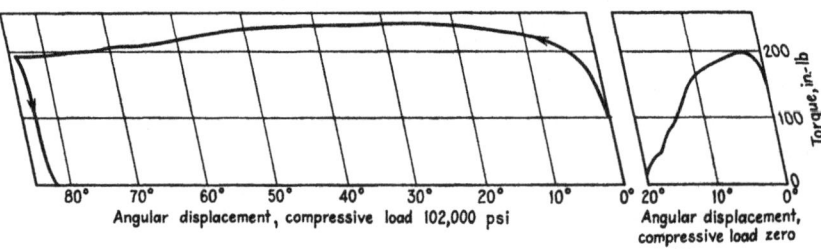

Fig. 136. Torque against angular displacement, transcribed from the photographic record for a 1045 steel from the Arsenal.

supported without rupture, and in fact with a decrease of torque only to 80 per cent of its maximum value. The maximum of the curve for no load is comparatively sharp and is reached at an angular displacement of 2.7°, whereas the maximum of the curve with load is very broad and reached at an angular displacement of perhaps 28°; that is, with compressive load strain-hardening is spread out through 10 times the angular displacement without the load. The maximum torque without load is 197 in.-lb, and with load 244 in.-lb. The effect of load on the maximum is thus perfectly definite but is much smaller in magnitude than the effect in increasing the distortion before rupture. The ratio of the torques at the maximum appears somewhat too large because of the effect of radial flow under compressive load. It is possible to apply a rough correction for it in this case by measuring the diameter of the end of the ruptured stump. The diameter was thus found to have increased because of radial flow by a total of 0.0045 in. If it is assumed that the radial flow is proportional to the angle of twist, for which justification has been found, then the increase of diameter due to radial flow at the point of

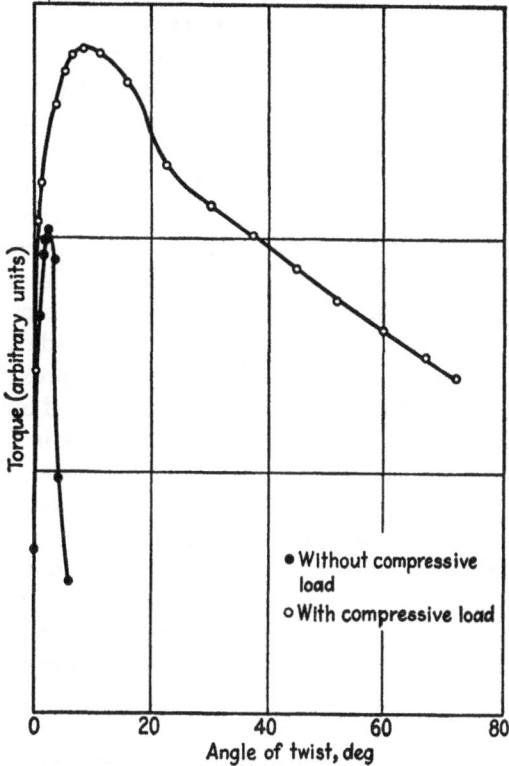

Fig. 137. Torque versus angular displacement for cast iron, with and without compressive load.

maximum torque is 0.0015 in. Applying a correction proportional to the difference of the cubes of the external and internal diameters, this brings the figure 244 above down to 235; that is, the effect of a compressive load of 102,000 psi is to increase the maximum shear stress in the ratio of 235 to 197, or by 19.3 per cent, over that with no load. The angular displacement to effect this increases 10-fold. This means that the strain-hardening produced by an increase in the strain of 10-fold is only 19.3 per cent. This is much less than the strain-hardening in tension produced by an increase in linear stretch of 10-fold.

Detailed figures will not be given for the other three specimens at intermediate compressive loads. The results lie between the extremes, spaced roughly proportionally to the compressive load. This applies both to the angular displacement and to the torque at the maximum.

Effect of Speed at Constant Compressive Load. A good deal of attention was given during the war in other laboratories to the question of the effect of speed of deformation in connection with the problem of the penetration of armor plate. Since it was so easy to vary the speed of the present apparatus, this question was examined in some detail. Two grades of steel were used: the 1045 steel heat-treated as above, and a 1020 steel, dead soft. An extreme variation of speed from 0.016 to 45° per second was used, that is, a range of 2,800-fold. At the highest speeds the effect of the viscosity of the dashpot oil becomes quite appreciable and must be corrected for before the absolute values can be obtained for the stress-strain curve. The maximum is so flat, however, that there should not be appreciable error in the maximum stress because of damping.

The first set of measurements was with five samples of 1045 steel and was mostly by way of orientation. One run was made with a speed of 0.12° per second, one with 1.75°, and three with approximately 42° per second. The results with the two lower speeds differed only slightly, but the variation between themselves was not consistent in direction. The three runs at higher speed gave more consistent results and were consistent in sign with respect to the mean of the runs at lower speeds. From the average of the two low-speed runs and the three high-speed runs it appears that at higher speeds both the maximum torque and the angular displacement at the maximum torque are increased. For an increase of speed of 40-fold the maximum torque is increased about 7 per cent, and the angular displacement at the maximum torque by some 1.8-fold. The latter figure is too large because of the effect of damping; the former figure should be essentially correct. These runs were all made with the maximum compressive load of 102,000 psi in order to stretch the curves out to the maximum amount, making the manipulations easier and incidentally reducing error from damping.

More extensive measurements were made on the 1020 steel. Eleven specimens were used with six speeds ranging from 0.016 to 36° per second. Except for one of the intermediate speeds two runs were made for check at each speed. The compressive load of all these runs was the same, 68,000 psi, less than with the 1045 steel because of the lower compressive strength. The results are shown in Figs. 138 and 139, in which the maximum torque and the angular displacement at the maximum torque are plotted against log speed, speed being measured in degrees per second. The points scatter, but not too much to obscure the trend. Both the maximum torque and the angular displacement at the maximum torque increase as speed increases. Assuming a linear dependence on log speed, and drawing the best straight line through the scattered points, it appears that the maximum torque increases about 4 per cent for an increase of speed of 3,200-fold, and the angular displacement at the maxi-

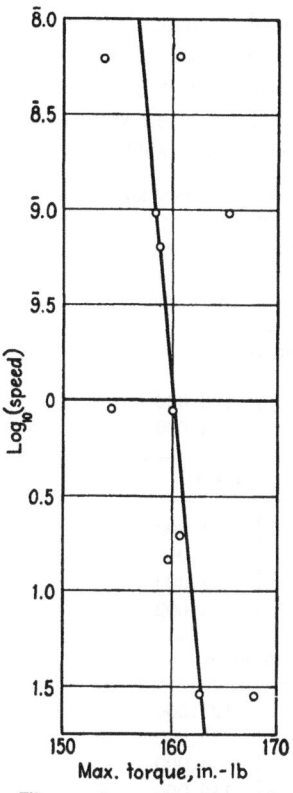 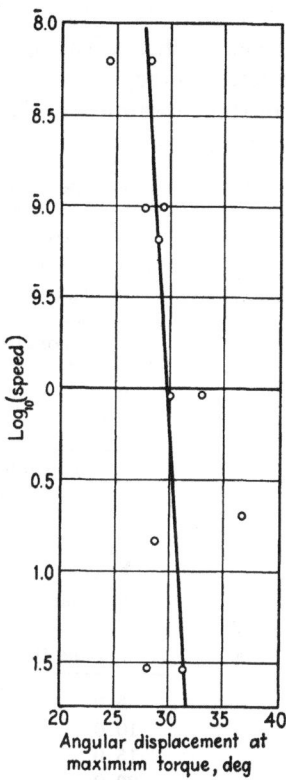

Fig. 138. The maximum torque as a function of speed of twisting for 1020 steel twisted under a longitudinal compressive stress of 68,000 psi.

Fig. 139. The angular displacement at the maximum torque of 1020 steel twisted under a longitudinal compressive stress of 68,000 psi.

mum torque by about 14 per cent for the same increase of speed. Exactly what is the significance of this in terms of fundamental properties of the material is not easy to say, the maximum torque being probably reached when the weakening due to initiation of new local focuses of fracture overtakes the strain-hardening distributed more or less uniformly throughout the mass. Roughly, the significance would seem to be that the strain-hardening increases somewhat with speed and also the amount of deformation before fracture. Probably, however, neither effect is so large that it will ordinarily have to be taken into account, nor will conclusions have to be essentially altered which are obtained with measurements made with a single moderate speed. In most of the following only a single speed was used, of the order of 1° per second.

It should be possible by a mathematical analysis to obtain information about what I called "primary creep" in discussing two-dimensional compression from the data for the effect of variation of speed just presented. The mathematics appears to be complicated, however, and the experimental accuracy is not high enough to tempt me to the effort.

Effect of Heat-treatment. The 1045 steel was tried with three different heat-treatments: (1) quenched into water from 1575°F and drawn back

TABLE XVI. RESULTS FOR STEEL 9-2, (UNCORRECTED FOR RADIAL FLOW)

Angular displacement, deg	Average shearing strain	Average shearing stress, psi
Longitudinal load zero		
2.5	0.42	112,000
4.0	0.68	114,000
Longitudinal load 70,000 psi		
2.5	0.42	114,000
5.0	0.85	122,000
7.5	1.28	128,000
10.0 (max)	1.70	132,000
Longitudinal load 105,000 psi		
5.0	0.85	123,000
10.0	1.70	135,000
15.0	2.55	144,000
20.0	3.40	148,000
22.1 (max)	3.76	148,000

to 800°F, (2) quenched into salt at 800°F from 1575°F, and (3) quenched into salt at 1100°F from 1575°F. Three specimens of each heat-treatment were used, at longitudinal compressive loads of zero, 70,000 and 105,000 psi. All were twisted at a speed of 0.15° per second. The outside diameter of all these specimens was 5/16 in., the inside diameter 0.197 in., and the notch diameter 0.225 in. The width of the notch was about 0.010 in., but the corners were somewhat rounded so that the exact width of the zone of flow at the bottom of the notch was somewhat indefinite. These specimens were the first in which the procedure was

TORSION COMBINED WITH SIMPLE COMPRESSION

used of making longitudinal scratches on the inside, splitting the specimen longitudinally after the run, and measuring the displacements of the internal scratches in order to get the width of the flow zone at the inside. The zone of flow was found in this way to be about 85 per cent wider at the inside than at the bottom of the notch. It was possible therefore to obtain for the first time both shearing stress and shearing strain (average).

TABLE XVII. RESULTS FOR STEEL 9-3 (UNCORRECTED FOR RADIAL FLOW)

Angular displacement, deg	Average shearing strain	Average shearing stress, psi
Longitudinal load zero		
2.5	0.42	80,000
3.9 (max)	0.66	81,000
Longitudinal load 70,000 psi		
2.5	0.42	76,000
5.0	0.85	89,000
10.0	1.70	100,000
15.0	2.55	106,000
18.4 (max)	3.13	108,000
Longitudinal load 105,000 psi		
5.0	0.85	92,000
10.0	1.70	105,000
15.0	2.55	112,000
20.0	3.40	118,000
25.0	4.25	123,000
30.0	5.10	128,000
34.3 (max)	5.83	129,000

These results, however, were not corrected for the effect of radial flow due to the compressional load. The effect of such a correction, as already explained, would be to lower the apparent strain-hardening, that is, to lower the computed shearing stress, at the larger angular displacements.

The results are given in Tables XVI, XVII, and XVIII. A plot is shown in Fig. 140 at the highest compressional load. The order of the results with respect to heat-treatments is what would be expected from the behavior in tension.

Comparison may be made with the tensile figures for these same heat-treatments (page 56). The true stresses at fracture at atmospheric pressure under tension were 306,000, 215,000, and 191,000 psi, respectively. This is the order for the maximum shearing stresses in torsion. The maximum shearing stress in a tension specimen is one-half the tensile stress. We may therefore expect some correspondence between half the tensile figures above and the maximum shearing stress in torsion. This comparison gives 153,000 against 148,000, 107,500 against 129,000, and 95,500 against 124,000. The first two are nearly equal, probably within

TABLE XVIII. RESULTS FOR STEEL 9-4 (UNCORRECTED FOR RADIAL FLOW)

Angular displacement, deg	Average shearing strain	Avearge shearing stress, psi
Longitudinal load zone		
2.5	0.42	68,000
4.4 (max)	0.75	71,000
Longitudinal load 70,000 psi		
2.5	0.42	72,000
5.0	0.85	83,000
10.0	1.70	93,000
15.0	2.55	100,000
20.0	3.40	105,000
23.0 (max)	3.91	106,000
Longitudinal load, 105,000 psi		
5.0	0.85	82,000
10.0	1.70	97,000
15.0	2.55	107,000
20.0	3.40	114,000
25.0	4.25	120,000
30.0	5.10	123,000
33.9 (max)	5.76	124,000

experimental error. In the last two pairs the shearing stress reached during torsion is higher than that in tension by an amount beyond probable error.

It is to be considered what the effect of strain-hardening is in this comparison. The principal elongations corresponding to different shearing strains may be computed by well-known methods, as given, for example, in Love's "Elasticity." The elongation gives immediately the reduction of area in the tensile specimen. The maximum shearing strains reached with the three heat-treatments were, respectively, 3.76, 5.83, and 5.76. These correspond to ratios of area in tension tests

(A_0/A) of 4.02, 6.03, and 5.96, respectively. The ratios of areas at fracture in tension of the three heat-treatments were 2.42, 2.52, and 2.43, respectively. The shearing strains reached in torsion under load are thus materially greater than the corresponding strains at fracture in tension under atmospheric pressure. The difference between the stresses in the two cases is to be ascribed to the greater strain-hardening in torsion, which is made possible by the longitudinal compression. The excess of shearing stress over one-half tensile stress is so much larger for the two softer steels because they support greater strain without fracture and are

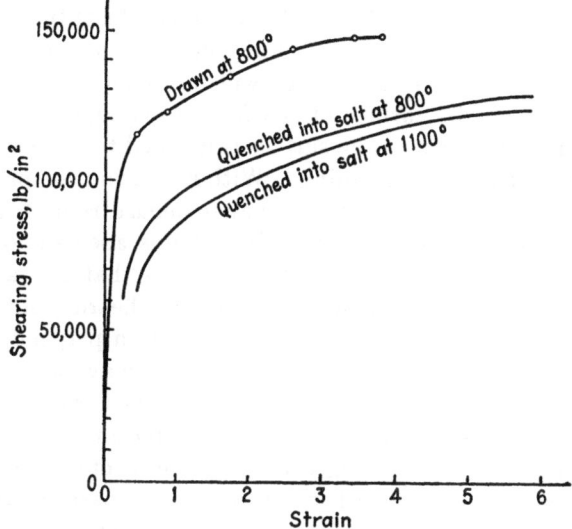

FIG. 140. The shearing stress as a function of shearing strain for several heat-treatments of 1045 steel twisted under a longitudinal compressive load of 105,000 psi.

therefore carried farther into the region of strain-hardening. Fracture in torsion without longitudinal compressive load occurs at a stress and strain distinctly less than in tension at atmospheric pressure.

A more significant comparison between the tensile and the shearing experiments is obviously to be obtained by using data for the tensile specimens pulled under hydrostatic pressure, when it was possible to reach strains equal to those reached in torsion under load. From the curves given earlier (page 64) it may be found that one-half the tensile stress in the tension experiments at the elongations equal to those at the maximum in the torsion experiments are, respectively, 195,000, 163,000, and 150,000 psi. These are all higher than the corresponding shearing stresses and in the same order. This brings out the fact, for which there are other lines of evidence, that distortion in shear is less effective in strain-hardening than the same distortion in tension. The ratios of the three figures just given to the corresponding shearing stresses, that is, $195/_{148}$, $163/_{129}$, and $150/_{124}$, are 1.32, 1.26, and 1.21, respectively. These ratios give a rough measure of the relative effectiveness of strain-hardening by tension and by shear; the ratios are higher for the harder and stronger steels.

Fracture. In no case, even with no compressional load, is fracture the catastrophic affair that it is in simple tension. Fracture starts in one or more localized spots as gaps on the planes perpendicular to the longitudinal axis, such that light can be seen through the gap. As twist proceeds, the fracture spreads peripherally until the entire circumference is involved. If twist is suspended at any intermediate stage, as at 15° in Fig. 132, parts of the circumference will be found to cohere. At zero compressional load the surface of the fracture is approximately a plane perpendicular to the axis. The surface is inclined to have a mat appearance, with little trace of burnishing. With increasing compressional load the fracture shows an increasing tendency to depart from the plane, sometimes with a distinct radial pattern like the escapement on some old-fashioned clocks. The probable explanation is that slip always has a tendency to occur on the planes of maximum shearing stress and that these planes are rotated from their initial location perpendicular to the axis by the compressional load. The fracture may be situated indiscriminately over the notch width; if it runs into the corner it does not run along the corner indefinitely but snaps back after an interval, giving the escapement-like pattern. Along with the increase of compressional load goes an increase in the burnished appearance of parts of the fracture, evidently due to rubbing under compression. Sometimes slivers of metal appear to get broken from the corners of the teeth of the escapement and are rolled between the opposing surfaces. Sometimes the metal balls up in places, and little nubbins of apparently harder material plow through the mass of softer metal.

Since fracture is such an indefinite thing and may be thought to be involved at least to some extent everywhere beyond the maximum torque, in most of the experiments twist was not carried beyond the point of maximum load. The point of maximum could be established by watching through the heavy red glass covering the recording paper and the rotation stopped at the beginning of decrease of torque. In this connection it appeared to be worth while to examine the reproducibility of the maximum. Tests were made on eight samples of 1045 steel, heat-treated to a Rockwell C of 34, four each at two different speeds, 0.12 and 40.0° per second. The longitudinal load on all eight was 102,000 psi. The results are shown in Table XIX. Three of each group of four are consistent within limits somewhat closer than were found in the study of the effect of speed in the last section. There is, however, one in each group that diverges considerably from the mean. In each case the maximum torque is abnormally high; but in one case, that at the lower speed, the corresponding angular displacement at the maximum is abnormally low, and in the other case abnormally high. The explanation is probably different in the two cases. The fracture of the low-speed specimen was abnormal in that it did not run uniformly around the circumference in the approximate center of the notch, but it was continually trying to escape from the notch, running off at an angle. This means that it would tend to twist up against the longitudinal load and thus would demand a greater

TABLE XIX. REPRODUCIBILITY OF RESULTS WITH DIFFERENT SPECIMENS

	Maximum torque, in.-lb	Angular displacement at maximum torque, deg
Speed 0.12° per second	248	22.3
	243	19.5
	232	14.2
	293	12.0
Speed 40° ± per second	249	34.5
	250	35.4
	255	38.5
	294	66.6

torque. The abnormality of the other was probably connected with some abnormality in the heat-treatment. Rockwell measurements disclosed at least one soft spot in which the C hardness was only 25 against an average for other parts of the specimen and for the other specimens of 32 to 36. A soft spot would mean greater angular displacement without fracture and, if it gave rise to slipping at an angle to the circumference, would mean a greater torque to overcome the effect of the longitudinal load.

It may be concluded that one may expect deviations of the maximum torque from the mean of not more than 3 or 4 per cent, and of the angular displacement at maximum torque of somewhat more. Abnormal cases may probably be thrown out on the basis of some other discoverable abnormal feature.

It is especially to be emphasized that the maximum marks no abrupt break in any of the physical properties. It would be natural to think that the maximum marks the beginning of fracture, but examination discloses nothing that could be described as an onset of fracture on passing the maximum. An experiment that might be expected to be especially informative with regard to fracture is the following. If, after the specimen has been twisted with longitudinal load, twist is released, longitudinal load released, and then twist reapplied without load, it would be expected that fracture would occur at less than the normal torque if microscopic fractures had been initiated by the previous twisting under load. As a matter of fact, if the initial twisting has been through an angle as much as the angle at the maximum torque in the absence of compressional load, the torque which can be supported on the second application after strain-hardening may be considerably more than the normal maximum torque without load. This holds whether the twisting under load stopped short of the maximum or was carried somewhat beyond it. As twisting under load is carried to greater extents beyond the maximum there is a progressive deterioration, until finally the torque which can be supported the second time sinks below that for the virgin specimen without load. But there is no change in the trend to mark passing of the maximum.

Up to the maximum the curves have good reproducibility; beyond the maximum increasing divergences may be expected between runs on different specimens. In general, beyond the maximum the curve drags out to a long tail, approaching an asymptote. The abruptness of the drop beyond the maximum depends greatly on the grade of steel. The drop to the asymptote may be punctuated by episodes in which the rate of fall is temporarily accelerated, or even sometimes by brief reversals. Even after deterioration is extreme and the specimen is full of visible wide-open cracks, there is still some cohesion between the different parts of the specimen if the load is great enough. The cohesion is doubtless a cold welding as the high parts are dragged past each other under intense pressure on small local areas. It is generally recognized that cold welding of metals would be expected to occur if perfectly clean metal surfaces could be brought into contact, even without considerable normal pressure. Such welding does not occur under usual conditions because the surfaces are kept out of immediate contact by a film of absorbed air or other impurity. In these experiments some of the fractures must occur under such conditions that the surface film does not have a chance to form, so that cold welding would be expected.

In Fig. 135 the asymptotic torque corresponds to a coefficient of friction between the parts, assumed perfectly free from each other, of about 0.7.

On the long tail of the torsion curves beyond the maximum when there is a heavy compressional load the metal is in a highly abnormal condition. It is still fairly strong for torsion, but for other sorts of stress it may be very weak, and it has lost nearly all capacity for further plastic flow. If a torsion specimen is removed from the apparatus after having been carried over into the long tail of the curve, it will be found still hanging together, but it may be snapped like a pipe stem in the fingers by bending.

If at any stage of the twisting process the torque is removed, keeping compressional load constant, and torque is then reapplied, twisting flow will not be resumed until the previous torque has been reached. If torque is removed, then compressional load removed, and then torque reapplied without compressional load, it is possible, if the preceding strain has not been too severe, to apply a torque greater than the maximum which can be applied to a virgin specimen free from compression. Under these conditions, however, very little further plastic flow takes place, and the specimen breaks very much more abruptly than when in the virgin condition.

Under such conditions the concept of "fracture" becomes hazy. The metal is highly disorganized from the original lattice; just how much or what kind of disorganization there must be to justify being called a fracture does not appear. In fact one would expect cold welding to occur to a certain extent, so that open fractures might under proper circumstances be partially welded together again. Many examples of such cold welding were found in my previous work on shearing under more extreme conditions of load and distortion.[1] Incidentally under these conditions strain-hardening in shear reached an asymptotic limit.

In general, two types of fracture are to be distinguished, tensile and shearing fracture. Of these the tensile fracture is much more clean-cut and definite, and it is no accident that by far the greater part of the speculation in the literature has been concerned with tensile fractures. The difference is primarily one of geometry. In tensile fracture the direction of the forces is such as to remove the fractured segments from the reach of mutual atomic attraction; when the separation is beyond the domain of atomic forces the process becomes irrevocable and catastrophic. In shear the primary process is one of tangential slip, and the direction of motion does not remove the atoms from mutual reach. Fracture is a secondary and incidental process under such conditions. Slip is accompanied by local disorderings; strain-hardening is one evidence of this disordering. As disordering progresses the displaced parts can fit together less perfectly in their displaced positions, until eventually the cohesive forces between the high spots become unable to check the momentum imparted by the external forces and slip becomes catastrophic. One would expect under such conditions that fracture would not be

[1] P. W. Bridgman, *Phys. Rev.* **48**, 825–847, 1935; *Proc. Am. Acad. Arts Sci.*, **71**, 387–459, 1937.

clean-cut, and one would be prepared for the large effects of normal pressure which are actually found.

If the onset of fracture in shear is gradual, there must be an interaction with the strain-hardening mechanism. It is this interaction doubtless that is responsible for the maximum in the strain-hardening curve. In tension, on the other hand, fracture terminates the strain-hardening curve abruptly, with no warning. It is much more appropriate to speak of distinct mechanisms for strain-hardening and fracture in the case of tension than in the case of shear.

Fracture without Load of Specimens Twisted under Load. It is of interest to know how the physical properties of the steel have been altered by strain-hardening in shear. One might expect to find something from a microscopic examination; this has been attempted, but as already mentioned, the zone of flow is not so well distinguished from the unstrained metal as would be desirable. More work should be done here, but in the meantime a direct measurement of the physical properties is in order. The simplest of these is to determine how the behavior in torsion is affected. To this end a series of eight runs was made, counting the control. Specimens of 1045 steel, loaded to 102,000 psi, were twisted at a constant speed of 1.27° per second through successively larger angles from 10 to 105°, the load released, and then the specimen twisted to fracture at zero load. Although these results anticipate the subject of Part III of this book, behavior as affected by various sorts of prestrain, it would seem that the immediacy of the application justifies inclusion here.

The results are shown in Table XX, which gives also the figures for the control, fractured immediately at zero compressive load. The control reaches its maximum torque at 2.7° angular displacement, passes through a period of progressive disintegration with decrease of torque from 197 to 157 in.-lb at an angular displacement of 11°, and here abruptly fractures. The specimens twisted under longitudinal load are thereby strengthened for a subsequent twisting without load, in the sense that a greater torque is required to produce fracture. This is shown by the figures in column 7 of the table. The maximum torque without load is, however, less than the previous maximum torque under load, as shown by the figures in column 6. Between 40 and 55° of twist under load, a marked deterioration sets in, as shown by the drop in the figures in column 7. In all cases, after twisting under load, the character of the subsequent twist without load is entirely different from that of the virgin specimen, as shown by a detailed inspection of the data. The virgin specimen passes through a period of pronounced plastic yield and strain-hardening before the maximum is reached at 2.7°, and after the maximum there is a period of relatively slow decline before catastrophic fracture occurs. The specimen previously twisted under load, however, shows no period of further plastic deformation when twisted without load, but torque rises with no appreciable plastic flow to a sharp peak, from which it abruptly falls with fracture. The material has lost its capacity for further strain-hardening, but its absolute strength is increased. Such a material would obviously be very dangerous in a machine part.

It does not appear from these results whether the capacity for plastic flow would be lost if angles of twist under load of less than 10° had been

TABLE XX. FRACTURE WITHOUT LOAD OF SPECIMENS TWISTED UNDER LOAD

Under 102,000 psi longitudinal compressional load				Fractured without load, max torque, in.-lb	Ratio col. 5 to col. 2	Ratio col. 5 to control
Max angular displacement, deg	Torque at max angular displacement, in.-lb	Max torque, in.-lb	Angular displacement at max torque, deg			
(1)	(2)	(3)	(4)	(5)	(6)	(7)
10	226	...	....	202	0.89	1.03
20	244	...	....	216	0.89	1.10
30	242	245	21.3	197	0.81	1.00
40	252	256	32.0	210	0.83	1.07
55	218	235	20.3	154	0.71	0.78
70	203	235	24.1	121	0.60	0.61
105	169	248	22.5	79	0.47	0.40
Control* (zero load)........	197		2.7			

* Abrupt rupture occurs at 157 in.-lb and 11°.

used; this would have to be determined by special examination. It seems unlikely, however, that at angles less than 10° any materially greater capacity for plastic flow would be produced by twisting under load than that possessed already by the virgin specimen, or any materially greater strength. In this respect these results for shear differ markedly from those found in tension. In general it appears that shearing distortion is much less effective in changing the properties than distortion by stretching.

Although distortion in stretch may be more effective than distortion in shear, nevertheless the effect of severe shearing distortion is striking enough. Thus above, under load a shearing distortion was supported without fracture of 10 times that which produces complete fracture without load, and with a strength of 0.4 of the initial strength.

Experimental Results in Terms of Stress and Strain.

Results for one series on a 1045 steel are shown in Fig. 141. Figure 142 gives the results for annealed drill rod, said to be a 1.25 C steel. These curves were

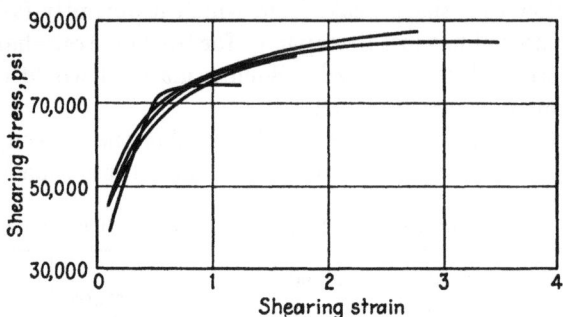

FIG. 141. Shearing stress against conventional shearing strain for a 1045 steel twisted under various longitudinal compressive loads. The loads were zero, 33,000 psi, 62,000 psi, and 81,000 psi, respectively, in the order of the length of the curves.

obtained by calculation by procedures already indicated from curves similar to Figs. 130 and 131. The calculations in Figs. 141 and 142 were not carried beyond the point of maximum torque. It is a consequence

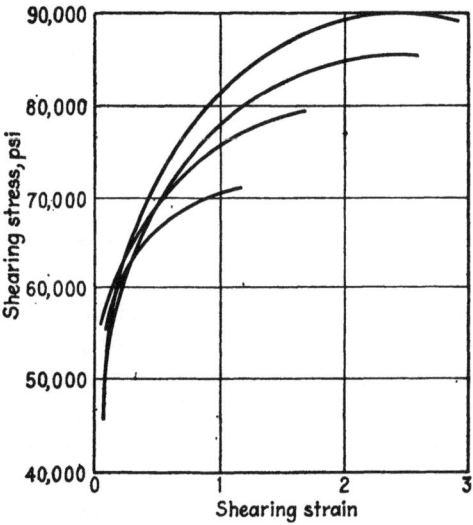

FIG. 142. Same as Fig 141, but for "drill rod" instead of 1045 steel.

of the radial flow during twist that the maximum torque of Figs. 130 and 131 does not always correspond to a maximum *stress* in Figs. 141 and 142.

The figures show marked differences between the two steels. Up to the maximum the shape of the stress-strain curve for the 1045 steel is little affected by the compressional load beyond a low initial effect. The difference in the curves is mostly in the magnitude of the strain which can be reached at the maximum shearing stress; this is a strong function of the compressional load. It is probable that below the maximum the apparent variation with load is within experimental error. With the 1.25 C steel, on the other hand, the shape of the curve below the maximum depends markedly on the compressional load, and the shearing stress at maximum torque increases by more than 50 per cent with compressional load in the range covered. If the strain at maximum torque is plotted against compressional load it will be found that the curve is concave upward for both steels, that is, the strain at maximum torque increases more rapidly than the compressional load.

Discussion. In the light of the existence of a maximum the significance of the strain-hardening curve requires special consideration. It is common experience that the strain-hardening curve, that is, the curve of plastic flow against stress, usually rises, as shown by the very use of the term "hardening." The extent to which the curve can be followed is limited by the fracture of the material. Under ordinary conditions fracture intervenes at a fairly early stage in the straining process; but, if fracture can be postponed as here, the "hardening" curve can be followed much farther. The fact that under such conditions the curve passes through a maximum cannot but help to throw light on the mechanism of strain-hardening in general. This aspect of the matter will be discussed again later.

Correlation between Tension and Shear. A question that has been much discussed is the nature of the connection between strain-hardening curves for different types of deformation, and in particular the connection between plastic flow in simple tension and in simple shear. The general impression seems to be that for small strains there is a connection but that for larger strains the simpler correlations, at least, that have been commonly discussed may be expected to fail by progressively larger amounts. The new evidence which we have here, in which the strain-hardening curve in simple shear has been carried beyond a maximum, pretty well throws out of court the possibility of any extensive correlation at all, because, as we have seen, the strain-hardening curve in tension rises linearly in terms of natural strain to magnitudes of strain much in excess of those reached here for shear. However, even below the maximum of the curve, the shearing strains reached here are considerably in excess of those usually considered, so that it is perhaps of some interest to inquire into the possibility of correlating the curves for tension and shear in the region below the maximum.

One method of correlation between tension and shear is in terms of the "octahedral" strains and stresses. The thesis is that, if the octahedral shearing strains and shearing stresses are computed, for either tension or shear, it will be found, for low strains, that the strain is the same function of stress for both tension and shear.[1] The formulas for the conversion of ordinary shearing stress and shearing strain into the corresponding octahedral values have already been given in the discussion of two-dimensional compression. The numerical values were computed for three of the points on the run on the 1045 steel under a mean compressional load of 81,000 psi and will be plotted presently. This steel was used for the comparison because the strain-hardening curve in tension has already been determined.

The octahedral components may be obtained easily for tension. If the strain in tension is designated as usual in terms of the natural strain at the neck, namely, by $\log_e (A_0/A)$, then the conversion formulas for the octahedral components will be found to be

$$\text{Octahedral shearing strain} = \sqrt{2} \log_e \frac{A_0}{A}$$

$$\text{Octahedral shearing stress} = \frac{\sqrt{2}}{3} \times \text{true tensile stress}$$

Since these are both relations of simple proportionality, it follows from the already known linear relation between the ordinary components that for tension octahedral shearing stress is a linear function of octahedral shearing strain. Two points therefore determine the line. In Fig. 143 the straight line for tension is drawn through two of the previously determined points reduced to octahedral components by the formulas just given. In the same figure are plotted the three points above from the torsion experiments. The figure leaves no question as to the wide divergence of the curves for shear and tension. It is probably not ruled out that within experimental error the two curves may extrapolate to the same value for infinitesimal strains.

The same sort of thing is shown by some other methods of correlation which are as a rule less successful than the coordination by octahedral components. The conclusion is that in general the strain-hardening curve for simple shear rises less rapidly than the curve for simple tension. In speculating as to the reason for this the different character of distortion in shear and in tension is to be kept in mind. At low strains there is an illusive similarity, but at the strains of the magnitude of those of interest here, the differences are accentuated. It is instructive to con-

[1] The "octahedral" strain or shearing stress is the strain or shearing stress across one of the "octahedral" planes, that is, the plane whose normal is equally inclined to the three axes of principal stress.

sider in detail the atomic kinetics of tensile and shearing strain. In tension, the principal axes are lines of fixed direction and also lines of maximum elongation and contraction. Since in the virgin material the atoms are already practically in contact, an elongation or contraction of a line of atoms can be accomplished only by pushing new atoms into the line or expelling them from it. The process is not uniform or continuous as it is in the macroscopic equations. Extensions of more than 20-fold are easily reached in tension combined with pressure; this means that along one axis at least 19 new atoms have been forced into line for every original atom, and along the other axis only one out of at least 20 atoms remains in place. The paths by which the atoms find their way into their new positions cannot be simple, and the entire mass of the material

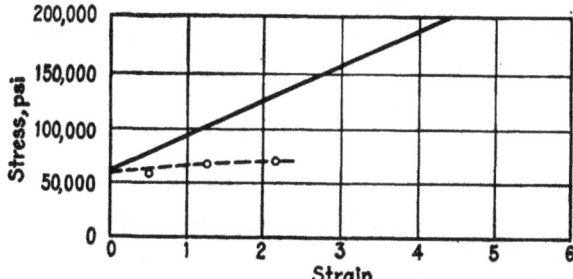

FIG. 143. "Octahedral" stress against "octahedral" strain for tension and shear on a 1045 steel. The upper straight line is from tension tests under hydrostatic pressure; the lower dotted curve is from the longest of the four curves of Fig. 141.

must be a field of great atomic disorganization. Expressed otherwise, in simple tension, the planes of shear envelop a cone and hence interfere with each other.

In simple shear, on the other hand, any plane of shear retains its direction and all the atoms in the plane retain their separation. From the idealized point of view, simple shear, no matter how great in amount, consists merely in the sliding of intact planes of atoms over other intact planes. It is natural to think that in any actual case of shear the total disorganization must be less than in tension. Furthermore, from the idealized atomic point of view, after a shearing displacement of one plane with respect to its next atomic neighbor by one atom has been reached, the initial configuration is recovered and the phenomenon repeats. In other words, from the idealized point of view, any strain-hardening there may be in shear should be complete for shears less than unity; beyond this the strain-hardening curve should be flat. Actually, of course, there is interference between slip in the different elementary domains, and the idealized state of affairs is smeared out. There should be enough residue of the idealized situation in any actual situation, however, to justify the expectation that strain-hardening in shear would be less than in tension and that it would flatten off earlier. This is the state of affairs which we have found and which is indicated in Fig. 143.

Agreement with the Equations of Plasticity. We next inquire to what extent the phenomena described here conform to the conventional mathematical formulations of plasticity theory. The experiments, however, were not primarily designed for the study of flow, and results bearing on this topic were obtained only incidentally. The experiments in the literature most similar to these in their bearing on flow are probably those of Hohenemser and Prager[1] on torsion combined with tension. The details of their experiments were, however, quite different; they maintained the angle of twist constant and observed the decay of torque when the tension was manipulated.

The most generally used flow equations, written in terms of principal strains and stresses, are

$$\dot{\epsilon}_1 = \alpha[\sigma_1 - \tfrac{1}{2}(\sigma_2 + \sigma_3)]$$
$$\dot{\epsilon}_2 = \alpha[\sigma_2 - \tfrac{1}{2}(\sigma_3 + \sigma_1)] \qquad (15\text{-}1)$$
$$\dot{\epsilon}_3 = \alpha[\sigma_3 - \tfrac{1}{2}(\sigma_1 + \sigma_2)]$$

In addition to the flow equations a condition on the stress is required to ensure that the material is in the "plastic" condition. We assume the form of von Mises, namely,

$$(\sigma_1 - \sigma_2)^2 + (\sigma_2 - \sigma_3)^2 + (\sigma_3 - \sigma_1)^2 = k \qquad (15\text{-}2)$$

The "constant" k may conceivably vary during the flow to express the progress of strain-hardening.

In applying these equations to the problem above, the first step is to find the principal stresses. Conventional cylindrical coordinates will be used: r along the radius, θ circumferential, and z along the axis. Under torque with no longitudinal load the only nonvanishing stress component in the cylindrical system of coordinates is $\Theta_\theta = Z_\theta = S$, where S is the ordinary shearing stress. The directions of principal stress are the two directions inclined at 45° to the axis in the $z\theta$ plane, and the direction of the radius. Denoting the first two principal stresses by σ_1 and σ_2 and the radial component by σ_3, the numerical values of these principal stresses are

$$\sigma_1 = S \qquad \sigma_2 = -S \qquad \sigma_3 = 0$$

Under longitudinal compression an additional component $Z_z = -C$ is added, where C is the compressional load per unit area. What are the new principal stresses? If we seek to determine these in terms of the

---
[1] K. Hohenemser and W. Prager, Z. angew. Math. Mech., **12**, 1–14, 1932.

boundary conditions at the free faces, we find that they are underdetermined. On the free faces perpendicular to the radius there is no force; on the faces perpendicular to z there is a normal component C and a shearing stress S, but the surfaces perpendicular to θ are not free because of the geometry, the hoop closing in on itself, and other considerations are necessary to fix the θ components of stress. The simplest assumption is that the Θ_θ component remains zero even after the addition of longitudinal load. In this case the two nonvanishing principal stress components are rotated in the $z - \theta$ plane, the axis of the compressional stress being rotated toward the cylinder axis. The amount of rotation and the altered values of the principal stress may be computed by elementary methods. The important feature is that in virtue of the rotation the direction of maximum shearing stress will no longer be the direction of plastic flow. Because of the geometry, the direction of plastic flow is constrained to be circumferential, and the maximum shearing stress always bisects the angle between greatest and least principal stress.

The principle of maximum shearing flow in the direction of maximum shearing stress can be maintained if one supposes that the longitudinal load generates a suitable Θ_θ component of stress. It is natural to expect the generation of a certain amount of such stress. It is a matter of observation that the compressional load produces radial flow. It would then also tend to produce circumferential flow, which would build up a Θ_θ component of stress because there is no free surface across the θ direction as there is across the radial direction. Any net Θ_θ component of stress would obviously be held in equilibrium by elastic stresses in the bulk of the material beyond the notch. If the Θ_θ component resulting from the compressional load is equal to the compressional stress, then the net result of adding the load is to add a two-dimensional hydrostatic stress to the original stress system. The direction of the principal stresses will be unaltered and therefore the direction of maximum shearing stress, so that plastic flow still has the direction of maximum shearing stress. The numerical values of the principal stresses are now $\sigma_1 = S - C; \sigma_2 = -(S + C); \sigma_3 = 0$.

The actual stress system is probably somewhere between the two extremes just considered. Further evidence on this point is afforded by the character of the fractures, which will be discussed later. The effect is in any event not of prime importance, for in the extreme cases in the experiments reported here the direction of flow cannot differ by more than 10° from the direction of maximum shearing stress. Let us assume for the present that flow is always along the direction of maximum shearing stress, and inquire what the equations have to say about the effect of adding longitudinal load to twisting force.

1. When twist takes place without longitudinal load we have

$$\sigma_1 = S \quad \sigma_2 = -S \quad \sigma_3 = 0$$

and the condition of von Mises (15-2) becomes

$$6S^2 = k$$

2. When twist occurs with load, assuming flow in the direction of maximum shearing stress, we have for the stresses

$$\sigma_1 = S' - C \quad \sigma_2 = -S' - C \quad \sigma_3 = 0$$

The condition of von Mises now gives

$$6S'^2 + 2C^2 = k$$

Eliminating k, we have

$$S'^2 = S^2 - \tfrac{1}{3}C^2$$

In other words, less shearing stress is required to produce plastic flow in torsion when a longitudinal load acts simultaneously than when no load acts. The form of the result shows incidentally that tension as well as compression reduces the torque necessary for torsional flow.

This conclusion might at first appear to contradict the experimental findings above. The results thus far emphasized apply, however, only to large strains. At smaller angles of twist there are effects in the opposite direction. This is shown very definitely by the crossing of the curves at the low strains in Fig. 141, and careful examination of the photographic records of Figs. 132 and 134 shows the same thing. The apparatus was not sensitive enough to give good numerical values for the effects. In fact, the effect was usually noticeable only for the larger loads; this is to be expected because the effect involves the square of the load.

What about the other components of flow? In Eqs. (15-1) take the direction of $\dot{\epsilon}_3$ as that of radial flow. In the first stress system above, shearing without longitudinal load, the equations make $\dot{\epsilon}_3$ zero, and there should therefore be no radial flow. But in the second stress system, twist with load, the equations give $\dot{\epsilon}_3 = \alpha C$. There should, therefore, be radial flow under any compressional load, no matter how small. The flow equations give

$$\frac{\text{Radial velocity}}{\text{Velocity of shear}} = \frac{C}{3S'}$$

Accompanying the radial flow there is an equal longitudinal flow as demanded by the condition of constancy of volume. The total radial and longitudinal flows may be measured at the termination of the experiment, or if special methods are adopted, they may be measured during the progress of the twist. The longitudinal flow may be determined either from the over-all length or from the notch width. Individual measurements of diameter or notch width after torsion are likely to be somewhat irregular, because of local irregularities in the flow, particularly after the first beginnings of rupture. The over-all length, however, smooths out the local irregularities and may be taken as the best measure of the secondary flow.

The longitudinal flow satisfies the same equation as the radial flow, that is,

$$\frac{\text{Velocity of longitudinal flow}}{\text{Velocity of shear}} = \frac{C}{3S'}$$

The experimental setup was such that C remains constant but S' increases during progress of twisting because of strain-hardening. Hence the total longitudinal shortening, plotted against the total angle of twist, would be expected at first to increase more rapidly, and then more slowly.

The longitudinal flow was studied with six series of specimens, 16 specimens altogether, and with three grades of steel: a commercial cold-rolled, an annealed drill rod (carbon tool steel, about 1.25 C), and the heat-treated 1045 steel already referred to. At constant compressional load the change of length for the cold-rolled steel was found to be roughly proportional to the angle of twist, but with a tendency to become less proportionally at the greater angles. The strain-hardening of cold-rolled in torsion is comparatively slight. Six measurements on one sample of drill rod gave a markedly greater falling off from linearity at the larger angles, corresponding to greater strain-hardening. Measurements with different compressive loads up to 82,000 psi gave for the shortening per degree of twist at the point of maximum torque a perfect straight line through the origin when plotted against load for the drill rod. For the 1045 steel, on the other hand, there was appreciable departure from linearity, the value at half final load being 58 per cent of the final value. These results are qualitatively in accord with the flow equations and are therefore not inconsistent with an approximate isotropy of flow and strain-hardening under the conditions.

Isotropy after Plastic Flow. Bearing more immediately on the question of isotropy another sort of experiment was performed. A piece of drill rod, cut to the regular dimensions of the torsion specimen, was loaded in compression without twisting. A maximum compressional stress at the isthmus of 194,000 psi was applied, and under this the total shortening was 0.0065 in., or about 33 per cent of the total notch width. Compressional flow was confined to the notches. It was then twisted through 15° under a compressional load of only about one-third as much, 62,000 psi, and during the twisting received an additional shortening of 0.0012 in. The curve of angle of twist against torque for this specimen was indistinguishable from that of a virgin specimen. The longitudinal flow was, however, greatly affected by the previous cold-working in compression. A virgin specimen twisted through the same angle under a compressional load of 62,000 psi shortens 0.0044 in., or 3.5 times as much as the cold-worked specimen. It would seem then that under certain conditions strain-hardening and flow may be far from isotropic.

CHAPTER 16

SHEARING COMBINED WITH APPROXIMATELY HYDROSTATIC PRESSURE[1]

Introduction. In the experiments described in Part I the material under test was exposed to a hydrostatic pressure exerted by a true liquid, so that the conditions of the test were simple and well defined. The upper limit of the pressures reached there was a more or less naturally imposed limit in that any practical liquid freezes solid at higher pressures at room temperature. This natural limit was about 450,000 psi. If one is willing to put up with less well defined conditions by transmitting the stresses through a solid it is possible to reach much higher pressures and obtain information which is useful in its range of inferior precision. During the last 15 years I have made a number of studies in this range. Certain of these studies are pertinent to our present concern with plastic flow and fracture and will be briefly summarized here in order to present as complete a picture as may be of this entire domain. These studies have been presented extensively in journals which are more or less readily accessible so that an elaborate reproduction of the results would not be justified.

The experiments to be described are concerned with the shearing of various materials under average hydrostatic pressures running up to 700,000 psi in most of the published work but which have been extended in unpublished work for a number of substances to twice as much.

The Apparatus and Method. The material on which the tests are made is in the form of thin disks subjected to normal pressure between hardened steel (or carboloy) surfaces, which are simultaneously rotated in opposite directions about an axis normal to the plane of the disk, thus dragging the two faces of the disk circumferentially in opposite directions, and producing in the material of the disk shearing slip on planes parallel to the faces. Figure 144 shows the essential scheme of the apparatus. The disk of material A is compressed between the steel block C, called the "anvil," and a short boss on a cylindrical block of steel B, called the "piston." The anvil is rotated about the axis of the piston while the piston remains stationary. In order to avoid error from friction from end thrust, the experiment is doubled by using a second disk and piston

[1] This chapter is based on material contained in B6, B7, B8, B9, B11, B18, and B26.

on the other side of the anvil, the anvil being rotated between the two stationary pistons. Pressure on the disks is produced by pushing the pistons together between the platens of a hydraulic press. The force required to produce rotation is measured as a function of the normal pressure and curves of one plotted against the other. Knowing the dimensions of the apparatus the rotating force may be converted into mean shearing force on the surfaces of the disks and the total force pushing the pistons together into mean hydrostatic pressure over the faces of the pistons. The curves of directly observed forces were thus converted into curves of mean shearing stress against mean pressure. The range of hydrostatic pressures in nearly all the published experiments in which the apparatus was constructed of steel was 700,000 psi; but, as already stated, in unpublished work with similar apparatus constructed of carboloy the pressures have been pushed to 1,400,000 psi. As high a value as 700,000 with steel apparatus was made possible by the use of special steels and by the design, which is such that the highly compressed bosses are supported by a large surrounding mass of steel in which the mean stress is much lower.

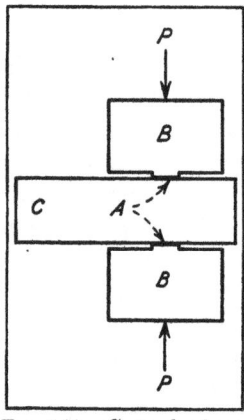

FIG. 144. General scheme of apparatus for combining large shearing strains with approximately hydrostatic pressure.

As normal pressure between the pistons is first increased beyond the yield point of the material of the disk, the material extrudes in the form of a sheet from between piston and anvil until the thickness has been so much reduced that further extrusion is prevented by the radial drag of the friction of the flat faces. It is obvious that, if the friction is finite, any normal pressure whatever can be held in equilibrium if only the thickness becomes small enough. In these experiments the limiting thickness at 700,000 psi was between 0.001 and 0.004 in., usually nearer the former than the latter. Detailed calculations, for which reference must be made to the original papers, show that if the disk is thin enough the effective friction which prevents further extrusion is confined to a relatively narrow annulus at the edge of the disk, and the great central bulk of the material of the disk is in a state of approximately hydrostatic pressure, equal to the mean pressure exerted by the pistons. Superposed on this effective hydrostatic pressure is a shearing stress acting radially in the planes of the faces during rotation.

In most cases the anvil was rotated manually, at convenient speeds up to a maximum of the order of 1 radian in 5 sec from a minimum of perhaps one-tenth as much. The maximum speed was so low that the thermal effects are negligible, as may be shown by a simple calculation. Rotation was almost always back and forth through an angle of about 60°; the rotating force plotted against pressure was the mean of the force for the two directions of rotation, which did not differ except for extraneous effects connected with asymmetries in the apparatus. The regular run was not begun until the material had been subjected to a seasoning process by subjecting to half the final pressure, and by rotating the anvil back and forth until a steady state had been reached. The mere fact that a steady state was always reached, and that after a comparatively small number of rotations, was itself one of the major results of the experiments which perhaps could not have been predicted in the light of the indefinite strain-hardening of steel in tension. For a steady state means that, at any fixed mean hydrostatic pressure, the plastic shearing strength of the material reaches a limiting value for sufficiently increased strain. That this is the case will appear more definitely from the discussion, in the immediately following, of the nature of the strains.

The magnitude of the shearing distortion at the edge of the disk corresponding to a rotation of 60° is 130 radians in a disk 0.001 in. thick (the diameter of the face of the piston is 0.25 in.). This was the usual magnitude of the strain imparted by a single oscillation of the anvil. Since strain of this character is largely irreversible, the effect of many oscillations must have been to impart a total strain of approximately the sum of the individual contributions so that effectively the distortions extended into the thousands of radians. By a simple rearrangement of the apparatus, however, it was possible to give the anvil, if desired, a continuous rotation in the same direction rather than an oscillation and thus subject the material to distortions of thousands of radians all in the same direction. No essential difference was introduced by this procedure. The steady value, after initial seasoning, was usually reached after a rotation of 10° or less.

The Qualitative Nature of the Results. The curves of shearing stress against mean pressure consist of two essentially different parts. The initial part of the curve corresponds to ordinary frictional slip at the surface between surface of the piston and the material, and the force required to rotate gives at once the usual coefficient of friction. Accompanying the tangential frictional drag on the surface of the disk there is of course an internal shearing stress within the body of the disk on planes parallel to the faces. In the initial stages this internal shearing stress is below the elastic limit of the material of the disk. As pressure on the faces increases, the frictional drag, and therefore the internal shearing stress, increases proportionally to the pressure if the coefficient of friction is constant, or more rapidly than the pressure if the coefficient of friction increases somewhat with pressure, as it usually does. Eventually a pressure will be reached at which the internal shearing stress reaches the plastic flow point, and from here on the mechanism of rotation is by an internal slip of the material of the disk on planes parallel to the faces, the faces of the disk being frozen by friction to piston and anvil.

The change from surface slip to internal slip is reflected by a change in the direction of the curve. If the shearing stress at plastic flow (which will be called hereafter simply the shearing "strength") were independent of mean hydrostatic pressure, then the curve of shearing stress would run horizontally beyond the point where internal slip begins. Although there are experiments in the stress range of ordinary engineering practice that would suggest that the shearing strength does not vary importantly with normal pressure, and hence that the actual curves would be approximately horizontal after the surfaces had been frozen by friction, our previous acquaintance in this book with the effects of higher hydrostatic pressures would prepare us to find that the curve does not remain horizontal but rises by appreciable amounts. This does indeed turn out to be the case, and for most metals the shearing strength rises by rather large amounts for pressures up to 700,000 psi.

The transition from surface slip to internal yield is never sharp but is spread out into a "knee" which separates the initial part of the curve corresponding to external slip from the latter part corresponding to internal slip. The reason for this is easy to see in the failure of the conditions inside the disk to be homogeneous, the strain varying directly as the radius. The measured curve is thus a mean over the radius, and transition from external to internal slip takes place at different radial distances at different times. Beyond the knee, in most cases the curve remains concave toward the pressure axis for the balance of its course.

The stress distribution within the disk obviously depends on the thickness. If the disk is thin, friction on the faces is confined to a comparatively narrow annulus around the edge, and most of the interior is under approximately hydrostatic pressure. The thicker the disk the wider the annulus required to contain the internal pressure. If the disk is initially thick enough, the frictional annulus will eventually eat its way to the center as pressure is increased. When this point is reached the disk becomes unstable, and further increase of pressure is accompanied by explosive expulsion of the material of the disk. In the early work a number of such explosive expulsions were encountered, the explanation of which did not at once appear. It was found that if the disk was made initially not more than 0.005 in. thick there was practically no danger of such an expulsion, whereas if the thickness was 0.015 in. or more expulsion might be expected. This is, however, a matter of the individual material;

by no means all materials can be made to exhibit explosive expulsion under any circumstances, but give only smooth extrusion.

It will be noticed that the geometry of the apparatus is entirely different from that associated with the testing of a tension specimen. Even if the material of the disk should "break" across the entire extent of some plane, the broken fragments are not separated by the forces from each other but continue within their mutual sphere of action. This, in conjunction with the pressure, brings it about that, even if fracture should momentarily occur, the fracture may be self-healed by restoration of the molecular bonds. This self-healing is obviously enormously facilitated by pressure. It is an obvious disadvantage of the method that it is not possible to follow the process of fracture and self-healing continuously into the ordinary region of zero pressure. The reason is that if one goes to too low pressures one passes from the region of internal to the region of external slip.

Fracture with self-healing results in all sorts of differences in the detailed behavior of different substances. Most metals, particularly those crystallizing in one or another of the common cubic systems, rotate smoothly, with no suggestion of any fracture phenomena at all. But there are many substances which, instead of rotating smoothly, rotate only with protest, usually noisy, of one sort or another. Some substances chatter, others squeak, others make a grinding noise, and one has been found to hiss. The most important of these variants of behavior is doubtless snapping. The rotation of many substances is interspersed with snapping at more or less regular intervals; the snapping is accompanied by more or less violent elastic reactions in the handle and other parts of the rotating mechanism and consequently a drop in the rotating force. The simplest way of describing the material of the disk is that it lets go to get a fresh hold. The actual angular displacement of the disk while it is getting a fresh hold is usually very small. After it has got its fresh hold, the force climbs back to its former value, and the phenomenon presently repeats itself. If the apparatus is so arranged that the material cannot get a "fresh hold," as sometimes when the rotating handle is driven by a constant weight instead of being pushed manually, then when the first fracture has occurred as indicated by a snap, the apparatus runs away to the end of its permitted rotation. Under these conditions failure is catastrophic. Once the motion has ceased, however, so that a fresh hold may be taken, the material self-heals, and the reverse rotation may be made as usual.

This phenomenon of snapping is usually superposed on the ordinary phenomena of plastic flow; in the intervals between snaps, yield may be perfectly smooth, and there is usually no way of anticipating when the snap will occur. The angular interval between snaps often depends on the speed of rotation, the slower the rotation the greater the interval. In the case of some substances it is possible to rotate so slowly that the snapping entirely disappears, while with other substance varying speed through a wide range gives no indication that snapping would ever disappear even at much slower rotations. On the other hand, there are a few substances, of which graphite at the highest pressures and quartz glass are examples, which do not yield at all by plastic flow but in which rotation is accomplished only by a succession of jumps and snaps.

The initial seasoning process required to get the virgin specimen into a steady state may continue for a very long time. Thus in the case of lead the state had not yet become entirely steady even after 100 double rotations through 60°, that is, after a displacement of one atomic plane tangentially with respect to the next through 30,000 atomic steps. After the initial seasoning has been accomplished there are various transient effects on each reversal of rotation. These transient effects, as already stated, usually disappear in a rotation of not more than 10°. The transients may be of various sorts. The final steady value of force may be reached from above or from below or after two or three oscillations of diminishing amplitude.

There is a very definite effect of the speed of rotation on the shearing strength; this effect is a strong function of the material. If the body were in the ideal plastic condition as defined by the von Mises function the force required to produce plastic flow would be independent of the speed, in sharp contrast to a liquid where the speed is proportional to the force. To test this matter the apparatus was arranged so that the speed of rotation could be varied within a range of some 10,000-fold. Special means was required to find the effect, because within the limits of manual manipulation of the rotating lever it is not possible for most metals to establish any variation with speed at all, so that it was obvious that the description of von Mises is a very much closer approximation than the analogy to a viscous liquid. Even when the speed is varied within much wider limits, the variation of force with speed for most metals is not large. The metals which showed the largest variation with speed were the soft metals, lead and tin. Figure 145 shows the variation of force in arbitrary units for different speeds plotted on a logarithmic scale over a range of 10^4. For 10^4 variation in speed the flow stress of copper varies from 5.5 to 7.5, arbitrary units, that is, by 36 per cent. Tin under a mean pressure of 350,000 psi shows a 10-fold variation of force for the same 10^4-fold variation of speed, and tin under 640,000 psi a somewhat smaller variation. Mica, on the other hand, shows no measurable variation of force for the same variation of speed. It would seem obvious that the yielding of a material like mica would depend on the boundary conditions in an entirely different way from the yielding of the conventional viscous liquid, and one cannot help wondering what would be the effect on geophysical speculations of treating the earth as a truly plastic substance instead of the conventional highly viscous liquid. The fact that the metals which show the largest speed effects are the soft metals, lead and tin, is perhaps not surprising in view of their known capacity for recrystallization at room temperature. If there is some readjustment of the internal structure during the process of flow, and if this readjustment proceeds at its own characteristic pace, then the effect of the readjustment would be expected to be greatest at the lowest speeds.

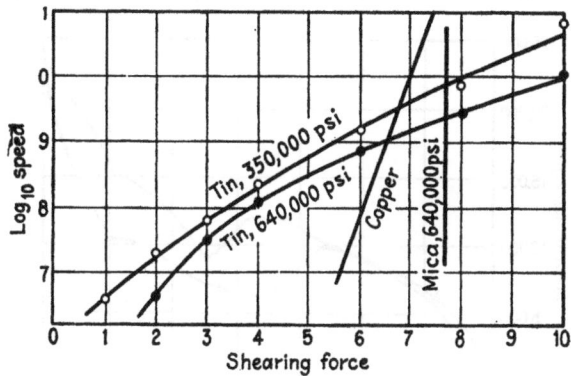

FIG. 145. Shows the variation of plastic shearing strength as a function of speed for several substances.

Many substances show special features in their behavior under high shearing distortion. Perhaps the most common of these is the occurrence of breaks in the curve of shearing strength against mean pressure. These breaks usually correspond to a polymorphic transition in the material produced by the pressure. In actually carrying out the experiments some of these breaks were found at pressures where it was already known from previous work that polymorphic transitions took place, and in other cases the break was subsequently found to correspond to previously unknown transitions, the check being made in other forms of apparatus made for the purpose. In fact, because of the ease with which the shearing experiments could be performed, one of the principal uses made of the shearing method in the early stages was as a tool of exploration for finding previously unknown polymorphic transitions. Figure 146 shows the shearing curve for bismuth, a substance with several polymorphic transitions.

The shearing strength may jump either up or down on passing through a transition. The controlling factor here is probably the nature of the lattice change at the transition. Other things being equal, one would expect the lattice with the greatest number of natural cleavage planes to have the lowest shearing strength. A change from a crystal system of low symmetry to one of high symmetry therefore might be accompanied by a decrease of shearing strength; and, conversely, since there is no universal rule with regard to the change of type of crystal lattice that is brought about by pressure, but many examples of both sorts occur, it would be expected that examples of both sorts of break in the shearing-strength curve would be found.

The mere fact that polymorphic transitions occur under these conditions is itself informative. One might perhaps anticipate that the lattice

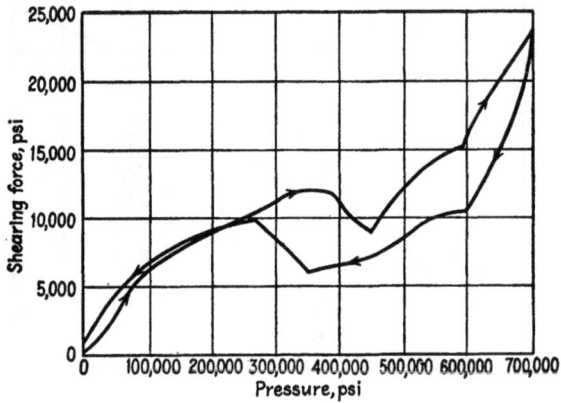

FIG. 146. Shearing strength of bismuth as a function of pressure.

would be so distorted by these extreme shearing displacements that it would no longer be capable of exhibiting polymorphic transition. On the contrary, it appears that polymorphic transition may be assisted rather than hindered under these conditions. Examples have been found in which the transition takes place at room temperature under shearing stress, although without the shearing stress the transition will not run until the temperature has been raised by as much as 200°C to overcome the internal viscosity. The fact that polymorphism persists under such drastic conditions indicates that the lattice structure essentially persists. What the nature of this persistence is should be the subject of further study. It is conceivable that the crystal might be broken up into a mosaic of smaller and smaller blocks, with little alteration of structure within the block, or that the distortion is spread more or less homogeneously through the entire structure. The question could be approached systematically by X-ray analysis. I have had unsystematic analyses made in a few cases. It was found, for example, that under intense shearing the crystal structure of copper is so far broken down that only a single very diffuse characteristic X-ray line can be detected, meaning a grain size probably less than 10 A on a side.

Besides polymorphic transitions, many substances show chemical effects under these conditions. In the early work a number of cases were found of apparently explosive chemical decomposition which were later traced to mechanical instabilities and the accompanying high-temperature effects which may be the result of too great initial thickness in the sample. After this factor was recognized, however, a number of examples remain in which genuine chemical changes occur. Some of these could be established by X-ray analysis of the products. Thus examples were found of change of valence and reduction of the oxides. Chemical synthesis may also be accomplished, as, for example, synthesis of the sulfides by shearing together the component elements. In addition to the examples of chemical change, the nature of which could be fully established by X-ray analysis, there were other examples in which color changes made it probable that some chemical change had occurred, but in which the quantities of material were too minute for X-ray analysis. It is evident that some of these changes must be self-limiting. If the material is transformed first on the surface, as seems probable, and if the new material has a lower shearing strength than the original material, then plastic shear will be confined to the infinitesimally thick surface film which has once formed, and there will be no mechanism by which the change can spread through the bulk of the material.

The shearing curves of noncrystalline substances, or those in which the crystal structure is not dominant in determining the properties, as in many organic substances, may differ qualitatively from those of crystalline substances in that the shearing curve is concave upward, which means that the limiting shearing strength increases more rapidly than the hydrostatic pressure. This may mean some interlocking effect between the complicated molecules, of somewhat the same nature as the interlocking which seems indicated to explain the enormous effect of high pressure on the viscosity of liquids. Under these conditions the molecules of long polymeric chain compounds are doubtless pulled apart by the mechanical stresses. Rubber, for instance, is converted into an entirely different sort of thing, like horn.

Many ordinarily brittle substances lose their capacity for fracture under these conditions. For example, borax glass emerges after subjecting to high-pressure shearing in a single coherent piece with few internal cracks, which may very well have occurred during release of stress. Furthermore, the borax glass is welded to the piston and anvil just as if it had been fused on. In fact, welding to the steel parts is a very common phenomenon for all types of material. Some glasslike materials do not self-fuse or weld; quartz glass, for example, falls apart into an impalpable powder after it is subjected to such conditions, with no trace of welding to the steel.

Quantitative Results. Measurements up to 700,000 psi have been made on several hundred substances. Reference must be made to the original papers for the numerical details. In most cases, the limiting shearing strength is given as a function of pressure at five equally spaced pressure intervals, together with a qualitative statement as to the nature of the shearing process, whether smooth or with internal fracture, as evidenced by snapping or chattering or what not. The materials embrace elements and various types of compound, from simple salts to minerals and glasses. Metals are presumably of most interest to us here. In Table XXI results are given for 16 metals. In this table the limiting shearing strength is given at top (700,000 psi) pressure and half top pressure, and for comparison the half tensile strength at atmospheric pressure. The latter is the shearing stress at the fracture point under ordinary tensile conditions. It is not a limiting shearing strength but is the shearing strength at that degree of strain corresponding to tensile fracture. It will be noticed that in all cases except that of magnesium, where the atmospheric data are almost certainly grossly incorrect, the half tensile strength at atmospheric is much less than the limiting shearing strength at 350,000 psi. This is as would be expected. The significance of the comparison of the high pressure with the atmospheric results is obscured for cadmium, calcium, and thallium by the fact that these metals have low-pressure polymorphic transitions.

There is not sufficient basis for the attempt at a more exact calculation of what might be expected, in the first place because the shearing strength at atmospheric is not a limiting strength for indefinitely large distortion, and in the second place because the shearing deformation in the tensile test is more complicated than the essentially one-dimensional shear of the high-pressure experiments.

Inspection of the table shows that in most cases the shearing strength at 700,000 psi is markedly less than twice the strength at 350,000 psi. Detailed inspection of the original curves will show that it is true in general that the curve is concave toward the pressure axis after the knee has been passed. Extrapolation of the curves to zero pressure is not significant because of the transition from internal to external slip that occurs at the lower pressures. The degree of concavity toward the pressure axis is a function of the material. Magnesium is a metal with little strain-hardening, whereas in iron the strain-hardening is not far from linear in the pressure, something which we might have anticipated from the tension experiments.

Only the beginning has been made at extending these results to 1,400,000 psi with the carboloy apparatus, and measurements have been made for some dozen substances. The results for copper and several other common metals have been checked in the range common to the earlier measurements. It would appear probable in general that instances will be common in which the curve of limiting shearing strength becomes convex toward the pressure axis at the upper end of the range instead of

TABLE XXI. LIMITING SHEARING STRENGTHS

Substance	Strength under 700,000 psi	Strength under 350,000 psi	One-half tensile strength at atmospheric pressure
Al	44,000	26,000	4,300
Pb	9,700	5,200	1,300
Cd	27,000	16,000	4,600
Ca	25,200	16,300	3,500
Co	89,000	46,000	17,500
Fe	170,000	94,000	17,500
Au	64,000	34,000	10,000
Cu	70,000	43,000	15,000
Mg	12,400	10,400	14,000(??)
Ni	124,000	57,000	35,000
Pd	81,000	57,000	15,000
Pt	82,000	57,000	13,500
Ag	67,000	37,000	9,200
Tl	16,000	6,200	650
Zn	26,000	15,000	12,500 ±
Sn	11,000	7,200	1,800

being concave as usual in the lower range. This perhaps is not strange, the higher stress bringing about the same interlocking effect as shown by many noncrystalline substances at lower pressures.

Discussion. These experiments invite the attempt to understand the nature of the mechanism that may be responsible for the effects. The big broad requirement that any mechanism has to meet is to explain the qualitative difference between flow in a liquid and flow in a plastic solid. In a liquid the rate of flow is proportional to the shearing stress; in a solid plastic flow does not occur at all until a critical stress has been reached, and once this stress has been reached flow varies from too slow to measure to too rapid to measure in a narrow range of stress. This is the broad picture, on which may be superposed secondary modifications. The mechanisms in solid and liquid must therefore be essentially different. The mechanism in a liquid is fairly well understood. It is a kinetic mechanism; shearing stress is the result of the transfer of a tangential component of momentum across the planes on which slip occurs. A solid surface moving tangentially with respect to the liquid is bombarded by molecules which have a net component of relative velocity parallel to the surface in virtue of the motion of the surface. Since this net relative component is proportional to the velocity, the momentum transfer, and so also the force, is proportional to the velocity.

In order to see what may happen in a solid, consider in the first place the ideal case of a perfect lattice, with slip taking place along one of the crystallographic planes, as indicated in Fig. 147. Tangential slip of one plane past the next is at first resisted by the ordinary elastic forces up

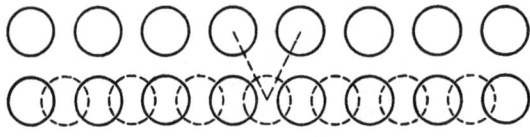

Fig. 147. Idealization of the arrangement of atoms in a lattice to illustrate the forces arising as a result of slip along one of the crystallographic planes.

to a maximum displacement of one-half an atomic step. At this displacement the net tangential force vanishes by symmetry. During the final half of the displacement by a whole atomic step, considerations of symmetry show that the tangential component is exactly the reverse of that at the corresponding stage in the first half; so that, if there is no permanent distortion in the lattice, no net work at all is required to displace the lattice by one atomic step. This means that, if the lattice can once be got into a steady state of slip, it will continue permanently in that state of slip with no force to maintain it. We have here a new state of matter, the "state of slip." In order to get matter into the state of slip a force must be applied sufficient to overcome the maximum elastic force, but the force then reverses sign.

This ideal state of slip of course does not occur; in understanding why it does not occur I think we find the clue to the phenomena of flow in solids. Flow is essentially an irreversible phenomenon; the energy input required to maintain flow is continually being dissipated. This demands some microscopic irreversibility in connection with the relative atomic displacements, and accordingly some modification of the picture. It is natural to connect the irreversibility with imperfections in the lattice structure. Two atomic slip planes in the final steady state must contain numerous atomic roughnesses, in the form of atoms projecting above their surroundings. When slip occurs, these will encounter other roughnesses and will push them aside. If the amount of distortion necessary to allow the projecting roughnesses to slip past each other is too great, the pushing aside will produce a permanent displacement of the projection. The result will be that the second half of a displacement through one atomic step is not symmetrical with respect to the first half; so that the work done during the first half, which would have been entirely compensated if the system had remained elastic, is no longer compensated, and there will be a net outstanding expenditure of work for the complete atomic step. It is this outstanding net difference between the two halves of the displacement that determines the shearing strength, the shearing force being the average force over the entire slip plane.

When the steady state is reached, as many new roughnesses are being created as are being destroyed by the pushing aside of other roughnesses. These new roughnesses will carry on the function of the roughnesses destroyed. Since the creation and destruction of roughnesses are irreversible phenomena, we have here a dissipative mechanism that results in an outstanding contribution to the force in the direction resisting slip.

The net force, the resultant of the forces in the two halves of the atomic step, is obviously to a first approximation determined by the geometry of the situation and is independent of the speed of slip. Whatever part may be played by the natural atomic vibrations cannot depend to any important extent on the speed of slip, for the highest speeds reached in any of these experiments are very low compared with the atomic vibrations, something of the order of 10^6 atomic vibrations occurring during one atomic step. We thus have an explanation of the most conspicuous difference between a liquid and a solid.

If the irreversible pushing aside of the roughnesses is followed by some sort of recovery process which proceeds at a speed comparable with the speed of atomic slip, then we may expect the force required to maintain continuous slip to vary with the speed. We have seen that this effect does occur to a certain extent, particularly in the low-melting metals like tin and lead where recrystallization proceeds with measurable speed at room temperature. If, on the other hand, there are no such phenomena of readjustment, it would be expected that the shearing force would be determined predominantly by the geometry and hence would be insensitive to temperature. This was found to be the case by special experiments on copper. An increase of temperature from room temperature to 150°C resulted in only an unimportant decrease of shearing strength. In a liquid, on the other hand, viscosity is a very strong function of temperature.

Any satisfactory theory of these effects must eventually account numerically for the increase of shearing strength at high pressures. Doubtless an important part of this is connected with the general increase in the intensity of atomic forces when the atoms are pushed into closer contact. There must be a connection with the almost universal decrease of compressibility at high pressure. However, a close connection between the two phenomena would be difficult to evaluate, because of the difference in the type of atomic displacement in the two cases. In general, the increase of shearing strength is markedly greater than the falling off of compressibility. There are doubtless other factors in increasing the shearing strength in addition to an intensification of the atomic forces. For instance, any slipping past each other of lattice irregularities will become more difficult at higher pressures because of the simple geometrical fact that such irregularities occupy a larger proportional part of the volume at the low volumes produced by high pressure.

Part III
PLASTIC FLOW AND FRACTURE AFTER PRESTRAINING

One of our fundamental problems is to find into what physical condition a substance has been thrown by various sorts of large plastic deformation. A partial answer to this question can of course be given by microscopic or X-ray analysis, and these may be very informative. But a more direct method, if it is desired to know how this or that property has been affected by the large deformation, is to measure that property directly. If one is interested in the alteration of mechanical properties by large plastic deformation, one measures the mechanical properties of the deformed specimens; that is, one makes tension or torsion or other tests on specimens cut from previously strained material. Apart from the practical interest in knowing how plastic flow affects such mechanical properties, there is theoretical interest, because some of the theories of plastic flow make specific assumptions about what the alterations will be, and it is desirable to know the range of validity of these assumptions. In particular, most elementary theories assume that the material remains isotropic after plastic flow. This assumption can be tested qualitatively by very simple tests. For example, if a simple compression test is made on a cylinder cut in any orientation from a block of previously strained material, then the cross section of the cylinder should remain circular when it is shortened if the material was isotropic. If the cross section becomes elliptical, as it often does, the departure from circularity offers a measure of the deviation from isotropy.

In the following, tests in simple tension, simple compression, and torsion have been made on specimens prestrained in simple tension (usually under pressure), simple compression, and two-dimensional compression.

CHAPTER 17

SIMPLE TENSION AFTER PRESTRAINING IN TENSION[1]

Introduction. Tests of this sort were made as part of the Arsenal contract and may be found in the first, third, and seventh reports. The experiments in the first and third reports were made on specimens which had been pulled under pressure to various degrees of necking and then were repulled at atmospheric pressure after the neck had been refigured by filing to a predetermined diameter and radius of curvature R. This involved in all cases an increase of R. The purpose of this refiguring was to make the conditions of the second testing more nearly comparable by starting with smaller a/R values. The cure, however, proved worse than the disease. It is obvious that because of the rapid variation of diameter in the neighborhood of the neck the strain-hardening produced by the first pulling varies rapidly along the longitudinal axis at the neck. Accordingly when the diameter in the neighborhood of the neck is made more nearly uniform by increasing R, opportunity is given for the new neck on the second pulling to be formed in a displaced position at some point where the decreased strain-hardening inherited from the first pulling more than compensates for the larger diameter. Hence the total strains on the second pulling, calculated by adding to the original strain at the neck the additional strain at the new displaced neck, will be in error, and false conclusions will be drawn about the strain-hardening curve. This source of error was suggested by Dr. Zener and Captain Hollomon at the Arsenal as the way out of one or two paradoxical conclusions from the results of the first and the third reports.

To avoid error from displacement of the position of the neck, repetition of the experiments was made with no refiguring. The regularity of the results indicated that no displacement of the neck took place under the new conditions. It is natural to expect this, in view of the well-known fact that a tension test may be resumed at atmospheric pressure with no break in the stress-strain relation. A break would be expected if the neck on the second pulling were displaced. Probably a mathematical proof should be possible that under these conditions a new neck is not formed, but any such proof would involve assumptions about the stress-strain relations on the second pulling, and it was just these that were in

[1] This chapter is based on material contained in the first, third, and seventh Watertown reports.

question. In any event, the empirical evidence for no new second neck seems adequate. Misgivings as to errors introduced by the varying initial degrees of necking were considerably allayed, when the decision not to refigure was made, by accumulating experience in applying the various corrections for stress distribution at the neck. This was reinforced by an experimental study of the justifiability of the assumption of uniformity of strain across the neck, which was specially made before the new studies were initiated. This study has already been described; the experimental evidence is illustrated in Figs. 4 and 5.

Apparatus. If the first pulling under pressure was to only a moderate reduction of area, the second pulling at atmospheric pressure might have been continued in the pressure apparatus, merely omitting to fill the pressure vessel with the pressure-transmitting liquid. However, if the area had been much reduced by the first pulling, the tensile loads on the second pulling were too small for accuracy by this method. This led to the construction of a miniature testing machine, which was then used in the testing of all specimens. In this machine, tensile load was applied with a stiff spring and the load was determined in terms of the deflection of the spring, which was measured with an Ames 0.0001-in. gauge. One end of the specimen was coupled to the loading spring. The other end was coupled to a graduated screw by which the specimen could be extended. The procedure was to extend the specimen by the desired amount with the screw and to observe the load as it crept back to a steady value after the extension. The final value was reached in a few minutes, and this was taken as the reading. No consistent study was made of the creep phenomena, although qualitative observations were made. The measured extension of the specimen played no part in the final results, being merely a parameter which facilitated the manipulations, just as during the measurements under pressure. Corresponding to each equilibrium load the neck diameter was determined, measured with a reading microscope suitably attached to the testing machine. What is more, the microscope was so mounted that the diameter could be measured at various positions along the axis, and so the radius of curvature R of the contour could be determined. Complete stress-strain data could thus be obtained at all intermediate points for the pullings at atmospheric pressure, instead of only at the terminal point, as was the case under pressure.

The Tests. The tests were made on two steels, tempered pearlite and tempered martensite, the data for which under pressure have already been given in Table VI under the designations 9-7 and 9-8. To the data for pulling under pressure are now added the data for the second pulling at atmospheric pressure. This second pulling was always to fracture. The

296   PLASTIC FLOW AND FRACTURE AFTER PRESTRAINING

TABLE XXII. TENSION DATA

Designation of specimen	Data for pulling under pressure			Data for pulling at atmospheric pressure	
	Pressure of pulling, psi	$\log_e \frac{A_0}{A}$	Flow stress, psi	$\log_e \frac{A_0}{A}$ (total)	Flow stress, psi
(1)	(2)	(3)	(4)	(5)	(6)
9-7-1	Virgin	Virgin	Virgin	1.040	193,000(f.)
9-7-2	100,000	1.099	197,000	1.129 1.166 1.240 1.396	188,000 192,000 193,000 210,000(f.)
9-7-3	189,000	1.252	210,000	1.332 1.401 1.432 1.634	197,000 201,000 203,000 224,000(f.)
9-7-4	290,000	1.147	187,000	1.208 1.263 1.317 1.354 1.400 1.600	190,000 192,000 195,000 197,000 198,000 226,000(f.)
9-7-5	385,000	1.250	222,000	1.332 1.421 1.482 1.554 1.739	195,000 203,000 203,000 206,000 255,000
9-7-6	Virgin	Virgin	Virgin	0.134 0.229 0.363 0.482 0.587 0.634 0.701 0.824 1.003	135,000 143,000 146,000 153,000 160,000 166,000 169,000 173,000 183,000(f.)
9-7-7	365,000	2.190	270,000	2.262 2.475	239,000 268,000(f.)
9-7-8	370,000	2.766	316,000	2.786 2.973(f.)	254,000 ....
9-7-9	352,000	2.367	285,000	2.580 2.670(f.)	238,000 ....
9-7-10	355,000	1.732	246,000	1.755 2.117(f.)	207,000 ....
9-7-11	365,000	2.146	278,000	2.176 2.344(f.)	225,000 ....
9-7-12	369,000	3.289	325,000	3.317 3.340(f.)	264,000 ....

SIMPLE TENSION AFTER PRESTRAINING IN TENSION

FOR TEMPERED PEARLITE

		Fracture data			
Nature of fracture (7)	Ratio of area of tensile break to total area (8)	Additional strain to fracture at atmospheric (9)	Extrapolated flow stress at fracture, psi (10)	Hydrostatic tension on axis at fracture, psi (11)	⅓Σ at fracture, psi (12)
Cup-cone	0.33	1.040	193,000	85,000	149,000
Cup-cone	0.33	0.297	200,000	93,000	160,000
Cup-cone	0.30	0.382	212,000	110,000	181,000
Cup-cone	0.31	0.453	208,000	105,000	174,000
Cup-cone	0.31	0.489	214,000	113,000	184,000
Cup-cone	0.34	1.003	180,000	62,000	122,000
Cup-cone	0.34	0.285	245,000	154,000	237,000
Fibrous	....	0.207	260,000	165,000	252,000
Cup-cone with star	0.34	0.303	240,000	149,000	229,000
Cup-cone	0.35	0.385	226,000	124,000	199,000
Cup-cone	0.39	0.198	235,000	142,000	220,000
Fibrous	....	0.051	265,000	186,000	274,000

TABLE XXIII. TENSION DATA

Designation of specimen	Data for pulling under pressure			Data for pulling at atmospheric pressure	
	Pressure of pulling, psi	$\log_e \frac{A_0}{A}$	Flow stress, psi	$\log_e \frac{A_0}{A}$ (total)	Flow stress, psi
(1)	(2)	(3)	(4)	(5)	(6)
9-8-1	Virgin	Virgin	Virgin	1.235	209,000(f.)
9-8-2	110,000	1.686	241,000	1.778	246,000(f.)
9-8-3	211,000	1.666	245,000	1.754 1.856	223,000 228,000(f.)
9-8-4	307,000	1.564	241,000	1.662 1.744 1.895	219,000 220,000 227,000(f.)
9-8-5	400,000	1.721	266,000	1.820 2.000	227,000 241,000(f.)
9-8-6	Virgin	Virgin	Virgin	0.202 0.327 0.497 0.645 0.792 0.895 1.026 1.203	146,000 151,000 160,000 172,000 178,000 179,000 185,000 201,000(f.)
9-8-7	365,000	2.219	302,000	2.237 2.385(f.)	225,000 ....
9-8-8	367,000	3.153	330,000	3.174 3.217(f.)	279,000 ....
9-8-9	112,000	2.092(f.)	....	....	....
9-8-10	100,000	1.300	210,000	1.371 1.47 ± 1.49	199,000 217,000 176,000(?)
9-8-11	196,000	1.408	231,000	1.576 1.787(f.)	219,000 ....
9-8-12	281,000	1.376	246,000	1.509 1.589 1.739(f.)	214,000 220,000 ....
9-8-13	390,000	1.109	227,000	1.182 1.258 1.361 1.605(f.)	198,000 203,000 209,000 ....
9-8-14	380,000	2.945	326,000	3.016(f.)	....

FOR TEMPERED MARTENSITE

		Fracture data			
Nature of fracture (7)	Ratio of area of tensile break to total area (8)	Additional strain to fracture at atmospheric (9)	Extrapolated flow stress at fracture, psi (10)	Hydrostatic tension on axis at fracture, psi (11)	⅓Σ at fracture, psi (12)
Cup-cone	0.35	.....	195,000	88,000	153,000
Fibrous	....	0.092	222,000	118,000	192,000
Fibrous	....	0.190	225,000	122,000	197,000
Fibrous	....	0.331	227,000	123,000	199,000
Cup-cone Fibrous	0.34	0.279	232,000	128,000	205,000
Cup-cone	0.35	.....	195,000	87,000	152,000
Fibrous	....	0.166	248,000	144,000	227,000
Fibrous	....	0.064	280,000	175,000	268,000
Cup-cone	0.19				
Single longitudinal split					
Combination	0.47	0.379	222,000	118,000	192,000
Combination	....	0.363	220,000	116,000	189,000
Cup-cone	0.33	0.496	215,000	110,000	182,000
Fibrous	....	0.071	273,000	168,000	259,000

flow stress at fracture as directly measured was not reliable because of the distortions of the neck produced in the act of fracture. A more reliable flow stress at fracture is given by extrapolating the flow stress versus strain curve to the strain of fracture, the total cross section at the contour being affected to a minor degree by those errors which affect the value of a/R and so the corrected flow stress.

Tables XXII and XXIII give the various observed and calculated data for the second pullings at atmospheric pressure. These data have already been shown, directly plotted, in Figs. 23 and 24, pages 67 and 68. Another aspect of the data is plotted in Figs. 148 and 149.

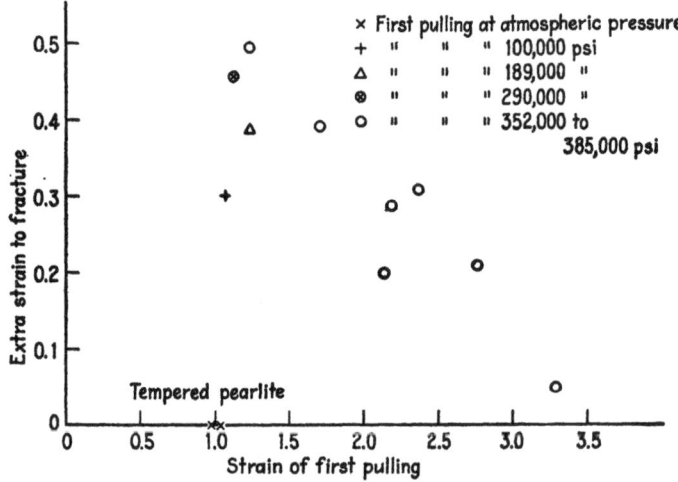

Fig. 148. The "extra strain to fracture" as a function of the strain of first pulling under various hydrostatic pressures for tempered pearlite.

The tables and figures establish various tensile properties of the prestrained material. In the first place, every specimen which was pulled under pressure, when pulled again at atmospheric pressure, suffers further plastic deformation before fracturing. This is in spite of the fact that under pressure the specimen may have been pulled to a strain much beyond the normal fracturing strain at atmospheric pressure, or to a flow stress much beyond the normal fracturing stress. It is true that some of the specimens pulled to the greatest strains under pressure do not permit much additional deformation at atmospheric pressure, but there is always a measurable amount. It is probable that if the pulling under pressure had been pushed to the very verge of fracture the additional deformation tolerated at atmospheric pressure would have become vanishingly small, but no special effort was made to reach these conditions.

In the second place, the stress-strain points on the second pulling at atmospheric pressure fall on a single line, irrespective of the pressure or the strain of the first pulling. This fact, together with the fact mentioned in the preceding paragraph, means that it is possible to extend the stress-strain curve for atmospheric pressure to much higher strains than can be reached without fracture in normal tests on virgin material at atmospheric pressure. In fact the curve may be extended to any strains that can be reached under pressure.

The stress-strain curve for pulling under pressure lies higher than the stress-strain curve for atmospheric pressure, the amount by which it lies higher being greater the higher the pressure. This means that the stress-

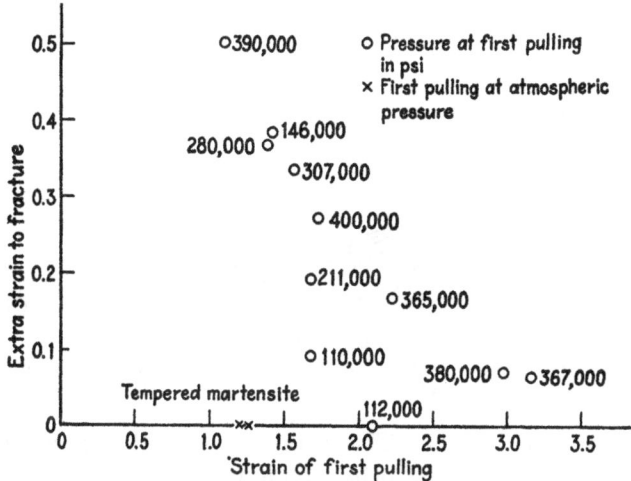

FIG. 149. The "extra strain to fracture" as a function of the strain of first pulling under various hydrostatic pressures for tempered martensite.

strain curves for different pressures are different. The matter has already been discussed on page 69, and the general nature of the dependence of the stress-strain curve on pressure suggested in Fig. 25. The differences in the curves arising from different pressures of pulling are not large and in most of this work could be and were disregarded. The magnitude of the effect would seem to be a function of the steel, and the probability is that it is rather larger than usual for the two steels studied in detail here. For the 9-8 steel, the flow stress at a natural strain of 3 is about 20 per cent higher under a pressure of 400,000 psi than at atmospheric pressure.

This comparatively small rise of flow stress with the pressure of pulling is by no means sufficient to obscure the important gross qualitative fact that most of the enhancement of strength imparted by a pulling under pressure is retained at atmospheric pressure. What is not retained at atmospheric pressure is the further ductility that would have been exhibited if pulling had been further continued under pressure. There is nevertheless some remanent ductility at atmospheric pressure in most cases.

Although the flow stress versus strain curves at atmospheric pressure are independent of the pressure of the previous pulling, the fracture phenomena are not. A specimen pulled at atmospheric pressure after prepulling under pressure follows smoothly along the unique stress-strain curve until it bumps suddenly, without warning, into fracture. At least there is no warning within the accuracy of the measurements. It could not be maintained that if measurements were made with sufficiently increased precision some warning of fracture might not be found from the course of the stress-strain curve. There was no evidence for such an

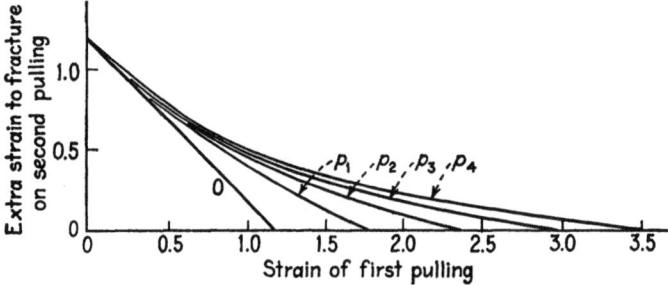

FIG. 150. The "extra strain to fracture" on the second pulling as a function of the strain of first pulling under successively higher pressures.

effect in these experiments. The distance which the specimen follows along the stress-strain curve before bumping into fracture is a strong function of the pressure of the previous pulling as well as of the strain of previous pulling. In general, the additional strain which is tolerated at atmospheric without fracture is less the greater the strain of the previous pulling and greater the greater the previous pressure. These statements may be substantiated by an examination of Figs. 148 and 149.

Figure 150 suggests qualitatively what may be the nature of the connection between the strain of first pulling, the pressure at which the first pulling was made, and the extra strain required to fracture on the second pulling. It is obvious that the line for atmospheric pressure in Fig. 150 must be inclined at 45° to the two axes because of the experimental result that a pulling may be interrupted and resumed with no change in the final result. It is also obvious that the lines for all pressures cut the axis of extra strain at the same point because of the experimental result that a mere exposure to hydrostatic pressure (pulling to zero strain under pressure) produces no permanent alteration in properties. Furthermore, if the pressures on the different curves are spaced with a uniform increase, the intercepts on the strain-of-first-pulling axis will be spaced uniformly because of the known linear connection between pressure and strain at fracture. The numerical values represented in Fig. 150 indicate what is roughly to be expected for steel 9-7, the upper pressure being 420,000 psi, and the pressures, accordingly, being spaced at intervals of 105,000 psi.

Data already presented in Fig. 8 of Chap. 1 have shown that a/R does not depend on the pressure of pulling. Data in Tables XX and XXI show furthermore that when the specimen is first pulled under pressure and the pulling is then resumed at atmospheric pressure the a/R values at atmospheric pressure continue to lie on the same curve, a function of the strain of pulling only. The fact that a/R is a function only of the strain has most important implications with regard to fracture. Consider two different specimens pulled at atmospheric pressure, both at the same strain, but one pulled previously at a higher pressure than the other. Both these specimens have the same strain, the same a/R, the same mean load, and, because they have the same a/R, the same detailed distribution of stress across the section. But the specimens are not equivalent as regards fracture, for the one which was pulled at the higher previous pressure may be pulled by the greater additional amount before it fractures; that is, these two specimens have identical strains and stresses but different fracturing characteristics. Here then is peremptory proof that fracture cannot be determined even by any function of stress and strain taken in combination, to say nothing of there being a fracture function of strain alone or of stress alone.

The feature by which the two specimens above differ is their past history, so that in general fracture must be determined by past history as well as by instantaneous stress and strain. Or put another way, because no physicist would accept the idea that a really *complete* description of the present condition does not determine present behavior (disregarding quantum effects), it must be that a specification of both stress and strain does not afford a complete enough description of present condition to determine fracture. This appears reasonable enough. Fracture may be initiated at instabilities so deep down in the atomic structure as to be beyond the averages which constitute the stress and strain of the conventional analysis of the theories of elasticity and plasticity.

Fracture. Certain data for fracture are given in Tables XX and XXI. In general, two types of fracture predominate. One is the conventional cup-cone fracture, and the other is a fracture in which an appreciable part of the surface of separation is contributed by axial planes, the fracture switching irregularly back and forth between planes perpendicular to the axis and longitudinal planes passing through the axis. This fracture may be described as "fibrous," perhaps a not very good name for it.

It is closely related to the "star" fracture illustrated in Fig. 22 of Captain Hollomon's Watertown Arsenal Report No. WAL 732/10-3. This, under extreme conditions, degenerates into separation on a single axial plane, as illustrated by his Fig. 23.

If the type of fracture is plotted as a function of the strain at which fracture occurs, as in Fig. 151, a definite regularity is manifest. The cup-cone fractures occur at low strains and the fibrous fractures at high strains. There is an intermediate region in which both types of fracture occur together. The transition region from one to the other type occurs at a definitely lower strain for tempered martensite, with a strain in the neighborhood of 1.8, than for tempered pearlite, with a strain in the

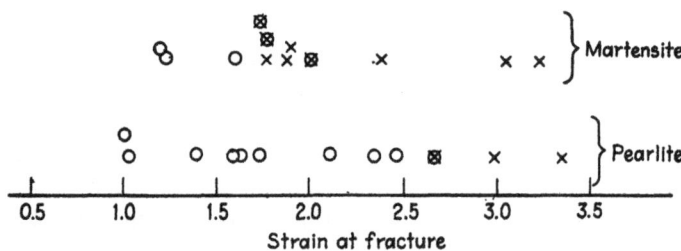

FIG. 151. The type of fracture as a function of the strain at fracture for tempered pearlite and tempered martensite.

neighborhood of 2.7. This is practically the only particular under the conditions of these experiments in which there is a significant difference between the behavior of tempered pearlite and tempered martensite.

One special case, specimen 9-8-10, merits special mention. This specimen failed by a split along a single jagged longitudinal axial plane extending completely through the neck. The maximum width of the crack was 2.2 per cent of the neck diameter and the length 52 per cent of the neck diameter. At the moment of the appearance of the fissure there was a sharp drop in the total load. This drop in load again constitutes evidence for the reality of the hydrostatic tension on the axis of a necked specimen; at the instant of fissuring this hydrostatic tension must largely vanish, and with it a component contributing to the support of the load. A precise explanation for this particular fracture does not appear, but it seems plausible to suppose that this, and the other fibrous fractures also, are connected in some way with the transverse component of tension on the axis arising from the necking. At the value of a/R associated with the strain at which the fissure described above appeared, this transverse tension was not far from 100,000 psi. Tables XX and XXI show values for the hydrostatic tension on the axis at fracture ranging roughly between 100,000 and 200,000 psi. Under normal conditions of fracture of virgin specimens at atmospheric pressure the hydrostatic tension does not reach values nearly so high because of the lower values of a/R. Since in general a hydrostatic tension works in the direction of embrittlement, it is natural to expect a change in the character of the fracture under these conditions.

The ratio of the area of the tensile part of the break, that is, of the bottom of the cup, to the total neck area is given in Tables XX and XXI for those fractures which are of the cup-cone type. The ratio varies little throughout the tables. This constitutes an extension of results previously found. It has been previously noted that the cup-cone disappears if the fracture occurs above a certain pressure, and up to that pressure the ratio is approximately linear in the pressure at which fracture occurs. To this is now added the fact that the ratio is the same for all fractures at atmospheric pressure, independent of the amount of previous pulling or the shape of the neck as given by a/R, or of the pressure of previous pulling. We can hardly expect to understand this striking phenomenon until we understand what it is that determines the extent of the tensile part of the break in general. It would seem to be fairly certain that fracture starts as a tensile break on the axis and travels outward. When it reaches a critical distance from the axis it switches and travels on a shear surface up the sides of the cup. Under present conditions there is compensation of various tendencies, so that the over-all result is that the ratio of the areas of the two sorts of fracture is a function only of the hydrostatic pressure at which fracture occurs. A detailed evaluation of the effect of the various factors is obviously of extreme complication. In particular, consideration must be given to the propagation of stress at the instant of fracture, and this is a highly complex thing in a plastic metal strained to the high degree of nonuniformity of the metal at points not actually in the neck itself.

General Considerations. Certain very general considerations make it not unplausible to expect some of the phenomena just discussed. Plastic deformation of the magnitude of that dealt with here is, from the atomic point of view, a highly discrete process. Lines of atoms are crumpled and atoms originally in the line are forced out, and other lines of atoms are extended and new atoms forced into line. Such forcible changes of position must mean the opening of free spaces between atoms of at least the order of magnitude of atomic dimensions. The mean dimensions of these interatomic spaces will be constrained to be smaller the higher the hydrostatic pressure prevailing during the plastic flow, and after the termination of flow the number of atomic cavities of exceptional size which persist will be smaller the higher the pressure during flow. The first effect means a higher mean force between atoms during flow, thus being consistent with the rise in the stress-strain curve of Fig. 25 with higher pressure. The second effect means a greater strain to fracture on the atmospheric pressure curve of Fig. 25 the higher the pressure of prior pulling.

CHAPTER 18

SIMPLE TENSION AFTER PRESTRAINING IN SIMPLE COMPRESSION[1]

Introduction. The material for these tests was 9 of the 10 heat-treatments of the steel which supplied the tests already described on simple compression. The tenth heat-treatment, that tempered at 250°C, was too hard to be used in the tension tests. It will be recalled that two specimens of each treatment were subjected to simple compression. The compression of one of these specimens was carried through only three stages, to a shortening from 1.5 to 0.42 in., or to a natural compressive strain of 1.25. At this stage further simple compression was suspended, and the specimen was cut up to provide specimens for the other sorts of test. Three specimens were cut out altogether; one was used in the tests to be described presently on simple compression, and two were used in the present tests on tension. One of these tension specimens was cut with the longitudinal axis along the previous axis of compression, and one with its length transverse to the previous compressional axis, that is, along the line of previous extension.

All the tests to be reported in this chapter are thus tests in tension after a prestraining in simple compression to a single value of strain, namely, 1.25. An exhaustive study of the matter would obviously demand that tests be made on specimens prestrained by varying amounts in compression. The present tests were, however, adequate to establish certain features of what may be anticipated to be the general situation.

Since the specimens were small, the same miniature testing machine for tension was used that was described in the previous chapter.

The Results. The data for the tension specimens are shown in Figs. 152, 153, 154, and 155. These figures reproduce the results with sufficient accuracy so that tables of numerical values are not necessary. The flow stresses plotted in the figures have all been corrected for stress distribution at the neck according to the procedure described in the first chapter, the correction being determined by the reduction of area in accordance with the empirical curves. It is to be considered whether this procedure was justifiable under these conditions, since the fracture of the prestrained specimens did not have the well-defined cup-cone character of

[1] This chapter is based on material contained in the eighth Watertown report and B21.

most of the fractures which formed the basis of the previous empirical determination of the correction in terms of the strain. In order to answer this question the ratio a/R was determined for all the fractured specimens by projecting onto paper an image magnified 14-fold and determining the best radius of curvature of the contour by fitting circles of various

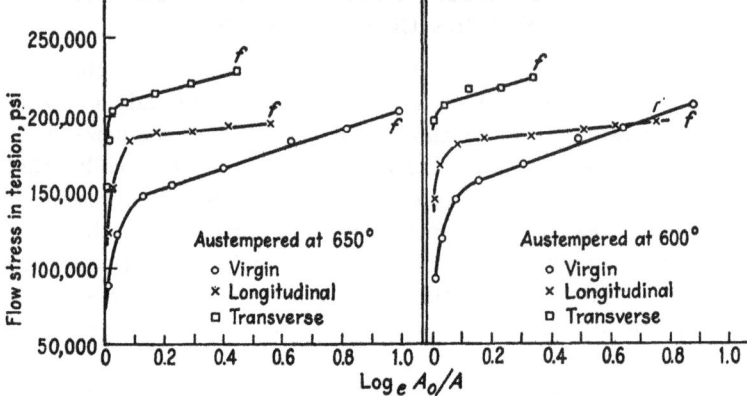

FIG. 152. The flow stress in tension as a function of natural strain for steel austempered at 650 and at 600°C. The three curves refer to virgin specimens and specimens cut in the longitudinal and transverse directions from specimens previously strained in simple compression.

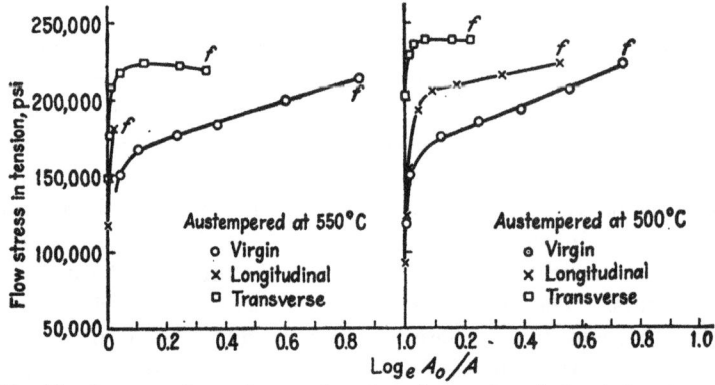

FIG. 153. The flow stress in tension as a function of natural strain for steel austempered at 550 and 500°C. The curves refer to virgin specimens and specimens cut in the longitudinal and transverse directions from specimens previously strained in simple compression.

diameters. The values of a/R so determined lay on the same empirical curve as that formerly found connecting a/R with the strain, with no more scatter than formerly, except for a few specimens for which the determinations were difficult and in doubt. These exceptional cases were disregarded and a correction uniformly applied on the same basis as before.

Figures 152 to 155 show that the curves for the prestrained material in all cases lie higher than the curves for the virgin material; that is, the metal has been hardened for tension by the previous straining in compression. We are, therefore, here beyond the range of the Bauschinger effect.

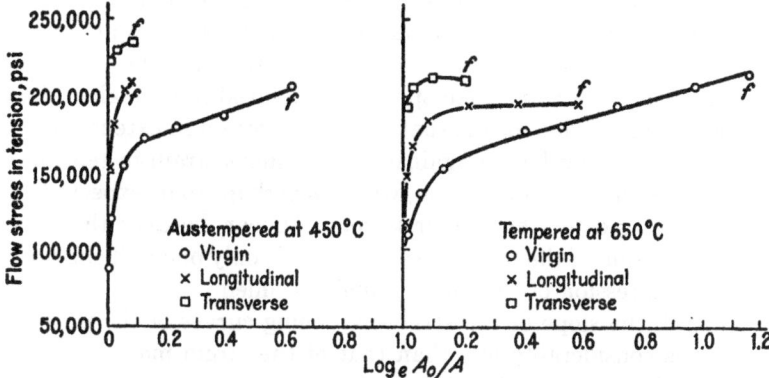

FIG. 154. The flow stress in tension as a function of natural strain for steel austempered at 450°C and tempered at 650°C. The curves refer to virgin specimens and to specimens cut in the longitudinal and transverse directions from specimens previously strained in simple compression.

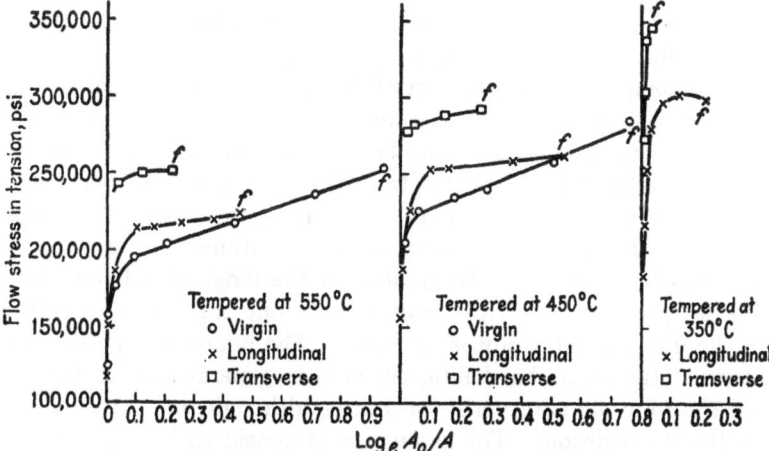

FIG. 155. The flow stress in tension as a function of natural strain for steel tempered at 550, 450, and 350°C. The curves refer to virgin specimens and to specimens cut in the longitudinal and transverse directions from specimens previously strained in simple compression.

In all cases there are marked differences between the curves for the longitudinal and transverse directions. This shows that the prestrained material is far from isotropic, an effect which would appear to be almost inevitable in a material exhibiting strain-hardening, although the effect is neglected in the elementary formulations of the plasticity conditions.

The prestrained metal always breaks at a lower strain than the virgin metal. The precise strain at which fracture occurs is subject to considerable irregularity, and it would appear that flaws are more likely to develop in the prestrained metal; a notable exception is the longitudinal specimen austempered at 550°C. Discounting the effect of flaws, it would appear that the general tendency is for the stresses at fracture of the prestrained metal to be somewhat higher than for the virgin metal. The stresses for the transverse specimens invariably lie higher than for the longitudinal. This means that a previous strain in extension is more effective in strain-hardening against a subsequent strain in tension than is a previous strain in compression, which is perhaps to be expected. The discrepancy is greater than at first would appear, because the previous strain in extension of the transverse specimen is only one-half the previous strain in compression of the longitudinal specimen.

In general, the slope of the strain-hardening curves of the prestrained specimens is considerably less than that of the virgin material. This is probably to be understood in terms of a reversible component of plastic flow. Plastic flow is essentially an irreversible phenomenon as far as most physical properties are concerned, but from the geometric point of view it is reversible because a plastically deformed metal can be pushed back into its original shape. If the effect were determined by the gross geometry alone, then an application of tension to the specimen prestrained in compression should carry it back to the condition of the virgin metal, with a strain-softening at a linear rate instead of a strain-hardening. The phenomena are far from being of this nature, but a small part of the reversibility to be expected from the gross geometry may well get back into the microscopic domain, accounting for the smaller slope of the strain-hardening curve under these conditions. In general, this effect would be expected to be greater for the longitudinal than for the transverse specimen, the latter having only one transverse direction in which the direction of strain is reversed. The figures show that in fact the slope of the strain-hardening curve is in general less for the longitudinal than for the transverse specimens, although this is by no means true without exception. The same sort of consideration indicates that the strain at fracture of the longitudinal specimen should be greater than for the transverse specimen, which again is true in general, although not without exception.

The fractures of all the specimens were in general of coarse, granular, irregular character, with the cup-cone character less prominently developed than usual.

CHAPTER 19

SIMPLE TENSION AFTER PRESTRAINING IN TWO-DIMENSIONAL COMPRESSION[1]

Small tension specimens were cut from two of the blocks described in Chap. 13 which were strained in two-dimensional compression by shortening to two-thirds the original dimension in one direction, expanding to three-halves the original dimension in one direction at right angles and with no change of dimension in the third perpendicular direction. These

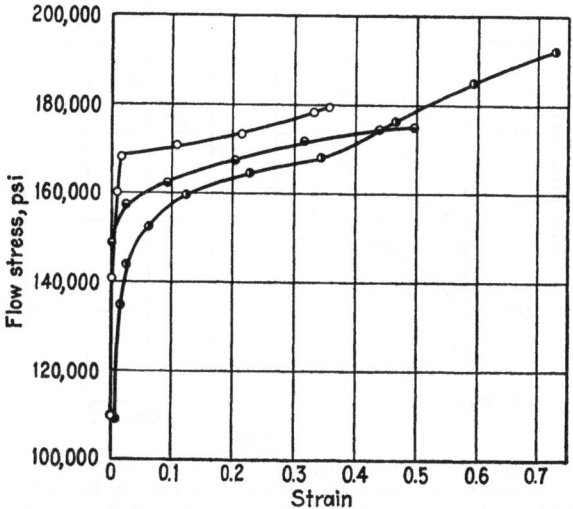

FIG. 156. Stress-strain relations in tension for small specimens of Solar steel cut in different orientations from a block previously strained in two-dimensional compression. Reading from the top down, the first curve is for the specimen with axis along the previous x direction, the second for the previous y direction, and the third for the previous z direction. Fracture occurred at the final points of all the curves.

were tested at atmospheric pressure in the small tension-testing machine as already described, and the results are sufficiently reproduced in Figs. 156 and 157.

Again the marked failure of isotropy is manifest. The effect furthermore is a strain-*hardening* for all directions, whether or not the prestrain in that direction was in the same or the opposite direction from the second strain. The strain-hardening on second tension is least in the direction of previous compression. The strain at fracture on the second pulling is greatest in the direction of previous compression. Both these

[1] This chapter is based on material contained in B21.

effects seem natural and would suggest that there is a small reversible component in the effects of plastic flow, a suggestion for which support was also found in the previous chapter. The other orientations do not show such uniformity; the strain-hardening is greatest for the X orientation of one of the steels and for the Y orientation of the other.

It would be expected that the circular section of a tension specimen would go out of round if the prestraining in different transverse directions had been different. Failures of circularity were in fact found, rising in the extreme case to an excess of the largest diameter over the smallest of

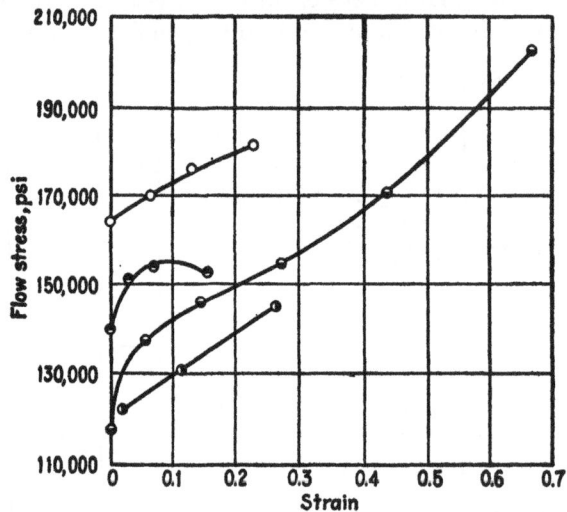

FIG. 157. Stress-strain relations in tension for small specimens of an annealed 1035 steel. The lower curve is for a virgin specimen. The other curves are for specimens cut from a block previously strained in two-dimensional compression. Of the three other curves, reading from the top down, the first is for the specimen with its axis along the previous y direction, the second for the previous x direction, and the third for the previous z direction. Fracture occurred at the final points of all the curves.

as much as 15 per cent. There was no obvious correlation between the ellipticity and the prestraining, and it would appear that many more data would be needed to establish a significant difference between the asymmetries in the prestraining experiments and the failure of circularity normally found in virgin tests of ostensibly homogeneous material. The effects of prestraining would appear to be liable to be dominated by the large plastic flow associated with the normal necking process.

Much work evidently remains to be done on this subject. Especially to be mentioned are examination of the effect of different degrees of prestraining, and of the degree of ellipticity in the sections of specimens for which the prestrain across the section normal to the tension axis had different signs in different orientations.

CHAPTER 20

SIMPLE COMPRESSION AFTER PRESTRAINING IN TENSION UNDER PRESSURE[1]

An experiment which should be tried some day is to find to what extent a tension specimen can be pushed back into its initial cylindrical figure after it has started to neck. Such an experiment would give information about the degree of reversibility of plastic deformation, an effect which several lines of evidence have suggested may sometimes play a part. There will obviously be difficulties in such an experiment from instability of alignment, which is perhaps why the experiment does not seem to be in the literature. However, failing full restoration of the original cylindrical figure, a certain amount of information may be obtained with regard to the behavior in simple compression of the strained material at the neck of a tension specimen. The following experiments were made in connection with the Watertown Arsenal contract.

Because the region of approximate homogeneity in the neighborhood of the neck is small, any such experiments must be made on small specimens. The small size made impractical an elaborate study by multiple compressions with refiguring, such as was used in studying simple compression, and a more simple procedure was adopted which, although not yielding complete information, was capable of qualitative and comparative information.

The material of these tests was the 9-0 steel of Table VI. Eight specimens were studied in all; seven of these were prepulled in tension at various pressures and by various amounts. The behavior under pressure is described in Table V under specimens 9-0-18 to 9-0-24. The prepulling followed the regular pattern. Complete stress-strain curves, which need not be reproduced here, were taken during the prepulling. The range of natural strain of the prepulling was from 0.422 to 2.204. The pressure of the prepulling varied (except for the specimen at atmospheric pressure) from 220,000 to 363,000 psi, maximum. The stress-strain curve for the prepulling falls within experimental error on the same line as the previous results for the 9-0 series.

The neck of the tension specimen was machined after prepulling for the subsequent compression tests as shown in Fig. 158. All the specimens were machined with the same grinding of the same carboloy tool. The

[1] This chapter is based on material contained in the third Watertown report.

exact dimensions after machining were measured with a micrometer microscope. The diameter fluctuated within plus or minus 3 per cent about 0.105 cm, and the length was 0.127 cm. The final results were corrected for fluctuations in the initial dimensions.

Longitudinal compression was applied to the specimens in the same miniature testing machine that was used for the tensile tests already described. Tension was converted into compression by a simple yoke arrangement which need not be described in detail. The over-all length of the sample was measured by two feeler rods bearing directly against the ends of the specimen through small holes at the center of the platens by which the compressive force was applied to the specimens. The differential motion of the two feeler rods was transferred through a yoke to an Ames 0.0001-in. jeweled gauge. Compression was applied in small steps, and readings were made both with increasing and decreasing compression, usually about 25 readings in all. In the plastic part of the range no fixed time schedule was followed, but after each increment of load the reading was not made until creep had sensibly stopped. This might demand something of the order of 1 min; creep diminished rather abruptly.

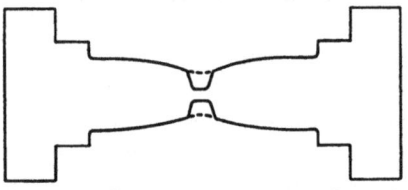

FIG. 158. The method of figuring the neck of a tension specimen for subsequent testing in simple compression.

The compressive load on all specimens was increased to such a value that the shortening of all specimens was approximately the same, about 18 per cent of the initial length. The specimen becomes barrel-shaped. After release of stress, the length and largest diameter of the isthmus were measured, and the final diameter at the waist used to obtain the maximum true compressive stress at the waist, that is, the final compressive load divided by the final waist area. The increase of section at the waist varied between 28 and 35 per cent, corresponding to a natural strain of from 0.25 to 0.30. The average increase of section as computed from the measured decrease of length was 22 per cent. The discrepancy between this and the increase at the waist indicates the amount of barreling. It is obvious that a precise solution of the problem here would demand a determination of the stress distribution at the waist, analogously to the problem of the neck of a tension specimen. It is probable, however, that in view of the smallness of the strain, any effects of stress distribution are small.

It is not necessary to give all the results in detail, but it will suffice to show in Fig. 159 curves of shortening against compressive stress, measured on the original area, for three specimens distributed over the range of prestrain. The curves for the other specimens fit into the same diagram, spaced according to the amount of prestrain; it would only confuse the diagram to try to give all the data. The most striking result shown by the figure is that prestrain in tension decreases the subsequent plastic flow in compression. It was not at all certain before trying the experiment that this would be the case, the existence of the Bauschinger effect suggesting the possibility of the opposite sign.

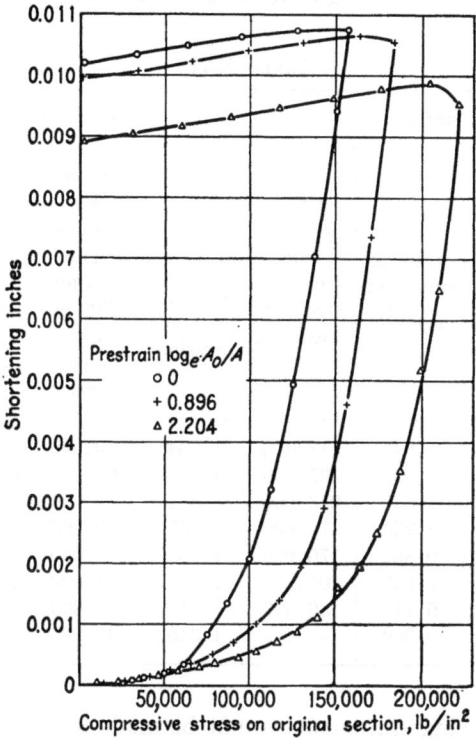

FIG. 159. The shortening against compressive stress of three specimens previously strained in simple tension to the indicated amounts.

In order to represent the change in the plastic properties by a single parameter, we may compare the stresses calculated on the original area at some definite shortening, the same for all specimens. In Table XXIV the compressive stress is listed for a total shortening of 0.009 in., which is not far below the maximum shortening of all the specimens. The stress-strain curve is running so steeply near the end, and the maximum shortenings of all specimens are so nearly the same, that nearly as satisfactory a parameter to represent the relative plastic behavior is simply the maximum compressive stress of the experiment. By listing this maximum one has the advantage that the true stress may be calculated in terms of the measured final dimensions. In Table XXIV the maximum compressive stresses are also listed.

In Fig. 160 are plotted the two parameters of Table XXIV, true compressive stress at the maximum and the stress on the original area at a shortening of 0.009 in., against the natural prestrain. Both parameters are linear in the natural prestrain within experimental error. The pressure range of the prestraining is not so great as in some of the other work,

TABLE XXIV

Designation of sample	Prestrain data			Data for compression			
	Pressure of prestrain, psi	Natural strain of prestrain $\log_e \frac{A_0}{A}$	Corrected true tensile stress of prestrain, psi	Ratio of final to initial section for compression	Compressive stress on original area for 0.009-in. shortening, psi	Compressive stress at max		
						On original section, psi	On final section, psi
9-0-18	363,000	1.541	213,000	1.304	200,000	210,000	161,000
9-0-19	302,000	1.637	235,000	1.308	203,000	211,000	161,000
9-0-20	252,000	1.530	213,000	1.307	192,000	202,000	154,000
9-0-21	266,000	2.204	270,000	1.280	219,000	221,000	172,000
9-0-22	221,000	0.896	170,000	1.350	177,000	182,000	135,000
9-0-23	260,000	1.649	201,000	1.277	201,000	204,000	160,000
9-0-24	Atmos.	0.422	130,000	1.314	153,000	156,000	119,000
9-0-25	Virgin	0.000	....	1.346	147,000	156,000	116,000

so that the following statement cannot be made with as great accuracy as might be obtained. However, within the limits of this work there seems to be no correlation with the pressure of prestraining, and both parameters are functions only of the amount of prestrain, not of the pressure at which it was imparted. Plotted also in Fig. 160 is the stress-strain line for the prestraining. Its slope is more than twice as great as the slopes of the lines for compressive stress in the same diagram. This means that at high prestrains in tension the subsequent flow stress in compression at a compressive strain of roughly 0.3 is a smaller fraction of the previous flow stress in tension than it is at smaller prestrains; that is, the greater the strain-hardening by prestrain in tension, the smaller the fraction of it which is retained as hardening for compression.

The scale of Fig. 159 is too small to show the displacements for small stresses. The initial part of the stress-strain curves is shown on an enlarged scale in Fig. 161 for five of the specimens distributed over the range of prestrain. The curves cross; the curve for the virgin specimen starts lower and ends higher than for the other specimens. With increasing prestrain the curves become progressively flatter. At first, with increasing prestrain, the initial slope increases, but presently it passes through a maximum and at the highest prestrain the initial slope

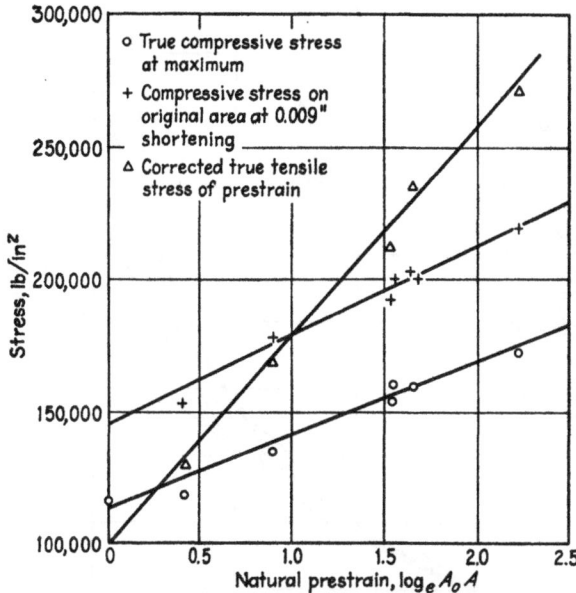

Fig. 160. Two compressive stresses, as indicated, as a function of the natural strain of the previous straining in tension. Also shown is the stress-strain line for the first straining in tension.

approaches that of the virgin material. This initial region where the curves cross is the region of the Bauschinger effect—a decrease of the plastic yield in compression produced by a previous raising of the plastic yield in tension. The range of the Bauschinger effect thus appears to be limited in two dimensions; it is limited to a certain range of compression and limited to a certain range of prestrain in tension.

Figure 161 shows, as is well known, that there is no well-marked yield point in compression. One might be tempted to describe the varying initial slopes by saying that the initial elastic constant in compression had been altered by the prestrain in tension. This description is, however, probably not legitimate. It is probable that the initial region is a region in which accommodation has not yet been achieved, so that hysteresis and other nonelastic effects are to be looked for, making it illegitimate to speak of an elastic constant.

The unloading curves may be similarly plotted on an enlarged scale. Near the upper end the curves are concave toward the stress axis, which may be explained by a continuation of the previous creep even after unloading begins. For considerably more than half the stress range on unloading, however, all the curves are sensibly straight and with the

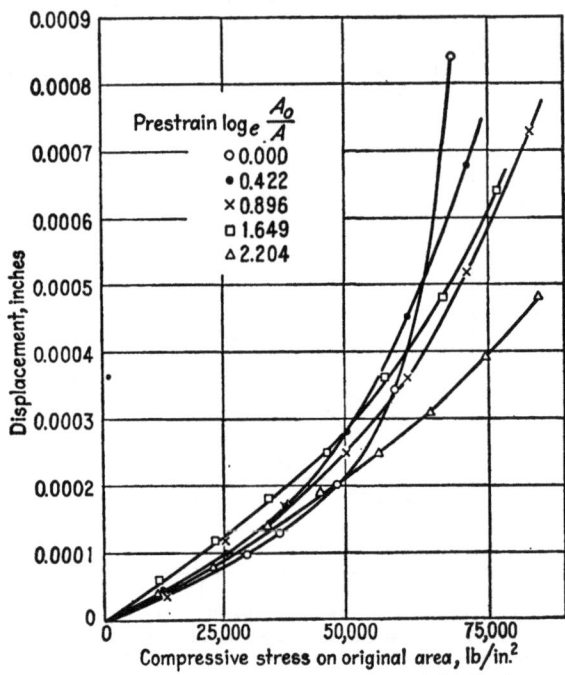

FIG. 161. Similar to Fig. 159; enlarged plot at low stresses of the displacement against compressive stress for five specimens previously strained in tension to the indicated amounts.

same slope. Since there is no special feature to be brought out, the enlarged unloading curves will not be reproduced here.

These results for simple compression after tension have been carried to strains in compression of roughly only 0.3. We have already seen that the stress-strain curve for simple compression does not begin to straighten out and become linear to a first approximation much before strains of 0.5. It would have been interesting if the compression measurements could have been carried to higher strains in order to find whether there is any effect of the magnitude of the previous strain in tension on the ultimate slope of the compression curve. If one may judge by Fig. 161 it is probable that any such ultimate effect is small.

CHAPTER 21

SIMPLE COMPRESSION AFTER PRESTRAINING IN SIMPLE COMPRESSION[1]

If the direction of the second simple compression is the same as that of the first compression, we have essentially merely a simple compression test resumed after unloading. Such tests have already been discussed in connection with the attainment of large strains in simple compression by refiguring and reloading, and need not be considered further here. We shall here discuss only tests in which the specimen for the second test is cut in the direction transverse to that of the first test, so that the new direction of shortening is the direction of previous transverse lengthening. The new transverse section now contains one direction of previous shortening and one direction, at right angles, of previous lengthening.

The tests were made under the Arsenal contract. The material of the tests was the same as that used in the tests on simple compression with refiguring, already described on page 182. The transverse specimens were cut from the second series of the two series of specimens described there, which had been prestrained in simple compression by three or four stages of compression to strains in the neighborhood of 1.2. The data for the prestraining have already been given.

The lateral specimens were compressed in the same apparatus as was used for the first compression, which needs no further description. The data are presented in Table XXV for the tempered and the austempered series. For the sake of clearness the data for the first compression are repeated. The results for the two compressions are plotted in Fig. 162.

If the strain-hardening were isotropic it would be expected that the deformation of the transverse specimen would be indistinguishable from that of the longitudinal specimen at the same stage of compression. This means that the compressive stresses, when plotted against cumulative strain, will lie on a smooth curve. Figure 162 shows that, on the contrary, the stress for the tranverse specimen is certainly never higher and is usually lower than that of the previous longitudinal application by an amount certainly beyond experimental error. We may therefore conclude that the strain-hardening by simple compression is definitely not isotropic.

Parallel with the decrease of compressive stress under transverse com-

[1] This chapter is based on material contained in the eighth Watertown report.

TABLE XXV

Specimen	Compression in first orientation			Subsequent compression of laterally oriented specimen				
	Strain $\log_e \frac{l_0}{l}$	Compressive stress, psi	Rockwell C hardness	Additional strain	Compressive stress, psi	Hardness	Ratio of diameters	Ratio of transverse strains
Tempered at 650°C	0 0.395 0.801 1.205	94,000 149,000 178,000 191,000	21.5 27.8 31.6 33.7	0.417	191,000	33.8	1.078	1.46
Tempered at 550°C	0 0.325 0.734 1.139	135,000 196,000 206,000 250,000(?)	32.7 33.1 36.8 38.0	0.402	216,000	37.6	1.097	1.63
Tempered at 450°C	0 0.223 0.622 1.023	189,000 219,000 236,000 262,000	40.6 39.4 41.4 42.5	0.412	243,000	42.0	1.116	1.71
Tempered at 350°C	0 0.393 0.702	213,000 279,000 290,000	47.7 47.8 48.6	0.414	283,000	48.2	1.115	1.74
Tempered at 250°C	0 0.406 0.821	260,000 346,000 371,000	51.9 52.9 53.9	0.751	349,000	55.3	1.206	1.75
Austempered at 650°C	0 0.409 0.829 1.241	56,000 155,000 171,000 182,000	13.8 25.2 29.7 31.5	0.409	179,000	31.1	1.126	1.84
Austempered at 600°C	0 0.400 0.818 1.227	66,000 159,000 174,000 185,000	16.9 26.0 30.7 32.4	0.404	186,000	32.0	1.087	1.58
Austempered at 550°C	0 0.409 0.826 1.194	87,000 174,000 188,000 195,000	23.8 28.7 32.5 34.3	0.413	192,000	33.6	1.105	1.64
Austempered at 500°C	0 0.391 0.795 1.203	92,000 176,000 193,000 211,000	26.3 30.5 34.5 36.0	0.401	203,000	35.6	1.111	1.74
Austempered at 450°C	0 0.389 0.798 1.195	90,000 178,000 196,000 215,000	26.3 30.8 34.6 36.8	0.414	209,000	36.6	1.116	1.72

pression there goes a slight decrease of hardness. This also appears natural in view of the directions of prior strain that are involved in a measurement of hardness by the usual techniques.

Failure of isotropy is also shown by the failure of circularity of the transverse section in the second orientation. In all cases the ellipticity is marked, and in all cases the long axis of the ellipse is in the direction

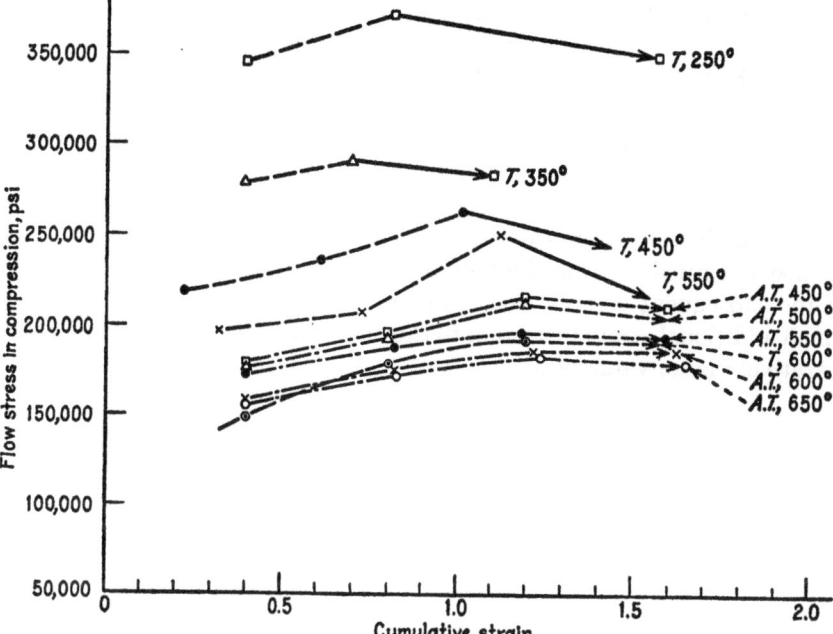

FIG. 162. The flow stress in simple compression as a function of the cumulative arithmetic strain for specimens cut in the transverse direction from blocks previously strained in simple compression.

of the previous compressive strain. In other words, in the final transverse section strain-hardening for extension is greater in the direction of previous extension than in the direction of previous compression. This is what might be expected.

Departure from the circular figure, that is, the ratio of maximum to minimum diameter, does not afford a proper measure of the relative degree of strain-hardening, because this becomes relatively greater the greater the absolute distortion. A more suitable measure is the ratio of maximum and minimum transverse strains, which may be written

$$\frac{\log_e (d_{\max}/d_0)}{\log_e (d_{\min}/d_0)}$$

where d_{max}, d_{min}, and d_0 are the maximum and minimum diameters after strain and the original diameter before strain, respectively. It is this ratio which is given in the final column of Table XXV; it varies from 1.46 to 1.84, and thus indicates a considerable failure of isotropy. With the conspicuous exception of the specimen austempered at 650°C, there is in general a definite tendency for the anisotropy to be greater for the heat-treatments giving greater hardness.

Returning to Fig. 162, although the stresses for the final compression lie below those reached in the final stage of the precompression at right angles, they are in all cases higher than in the next but one stage of precompression, and therefore all the more higher than in the first stage of precompression; that is, the previous extensional strain along the reoriented compression specimen has raised the flow stress for subsequent compression. This is the opposite of the Bauschinger effect, which we have already seen occurs only in the region of small strains. It is, of course, not ruled out that the Bauschinger effect would be found here also if the subsequent compression after prestrain had been made to only a small additional strain instead of to the approximately constant and comparatively large value of 0.4.

CHAPTER 22

SIMPLE COMPRESSION AFTER PRESTRAINING IN TWO-DIMENSIONAL COMPRESSION[1]

A number of qualitative experiments of the indicated kind were made during the very extensive preliminary work incidental to getting the apparatus for the two-dimensional compression functioning satisfactorily. These experiments consisted sometimes in subjecting a rectangular block to simple compression along either one of the three different directions of the preceding two-dimensional compression, and sometimes in reorienting the block in one of the several possible ways for a second two-dimensional compression. In these early experiments the great difficulty was slight inequalities of strain on the first compression which resulted in marked geometrical asymmetries on the second application. For this reason none of the preliminary experiments yielded results of sufficient value to justify presentation here or numerical calculation. Qualitatively, none of the early results were inconsistent with what appears to be the major generalization from most of the work on prestrained specimens, namely, that the yield stress on a subsequent straining is raised by the greatest amount in that direction in which the original straining was in the same direction as the subsequent straining, and by the least amount in that direction in which the two directions of straining are opposite. In the latter case the second yield stress may be actually diminished.

Only two quantitative experiments were made after the apparatus had been got to functioning properly. One of these was on an annealed Cr-V steel, and the other on a stainless steel of composition unknown from the stock of the laboratory machine shop, and designated as "free-machining." These were first strained in two-dimensional compression, each in six stages, to the final strain given in Table XXVI, that is, shortening to roughly one-half the initial height. From each rectangular block thus strained a cylindrical specimen was cut for subsequent straining in simple compression. The longitudinal axis of the cylinder was along the X direction of the original block, that is, along the line of zero original strain or along the axis of intermediate stress. The axis of original zero stress or original free extension is taken as the Y axis, while the Z direction of the original block is the direction of maximum shortening or of maximum compressive stress. The transverse section of the specimen for second

[1] This chapter is based on material contained in B21.

compression thus contained one direction in which the original strain was the maximum shortening and the direction at right angles in which the original strain was an equal lengthening.

The stresses and the strains of the first two-dimensional compressions are shown in Table XXVI. The conventional strain-hardening is

TABLE XXVI
Cr-V *Block*

Stress			Strain		
X_x, psi	Y_y, psi	Z_z, psi	e_x(nat)	e_y(nat)	e_z(nat)
First Compression					
−57,000	0	−104,000	0	+0.105	−0.105
−62,500	0	−119,000	0	0.223	−0.223
−75,000	0	−122,000	0	0.322	−0.322
−78,000	0	−130,000	0	0.440	−0.440
−94,000	0	−139,000	0	0.562	−0.562
−87,000	0	−145,000	0	0.702	−0.702
Second Compression					
−142,000	0	0	−0.167	(+0.702) +0.072	(−0.702) +0.095
Stainless Steel First Compression					
−48,500	0	−87,500	0	+0.068	−0.068
−62,500	0	−123,000	0	0.149	−0.149
−79,000	0	−149,000	0	0.250	−0.250
−108,000	0	−199,000	0	0.368	−0.368
−198,000(?)	0	−228,000	0	0.476	−0.476
−161,000	0	−243,000	0	0.604	−0.604
Second Compression					
237,000	0	0	−0.121	(+0.604) +0.067	(−0.604) +0.054

exhibited. The table also shows the results of the second straining in simple compression. These results were unique in that no strain-hardening was shown by either specimen, but both shortened over the entire range of stress under a practically constant simple compressive *stress*, the small increase of *load* during the compression being almost entirely that contributed by the increase of cross section during compression. Only the final strain of the second straining is indicated in the table for this reason. The constancy of stress was established for the Cr-V specimen by seven equally spaced readings and for the stainless specimen by five equispaced readings. In fact, the actual readings, taken at their face value, indicate a slight strain-softening rather than a strain-hardening. The stress at initial yield, which was marked with unusual sharpness for both these specimens, and in this constitutes a second point of difference between the behavior in these tests and normal behavior in compression, was actually somewhat higher than the recorded final yield stress. The initial values are indicated in the table in parentheses. It is not certain that they differ from the final values by more than experimental error.

The values for the strains on second compression in the table show that there was deviation from the circular cross section. This indicates failure of isotropy. According to this measure of anisotropy, the Cr-V specimen failed to be isotropic by 32 per cent (0.095/0.072 = 1.32), and the stainless specimen by 24 per cent (0.067/0.054 = 1.24). The deviation from isotropy is in the anticipated direction for the Cr-V steel, the direction of minimum extension on the second straining being the Y direction, that is, the direction of maximum extensional strain on the first straining, and therefore the direction in which the strain-hardening for extension would be expected to be a maximum. The expected relations are reversed for the stainless steel, however, the direction of *maximum* extension on second straining being the Y direction. It will be recalled that a stainless steel has already been found to be quite out of line with other steels in the slope of the strain-hardening line in simple tension under hydrostatic pressure. In comment on the abnormal relation shown by this stainless steel it is to be remarked that the simple anticipation is based on an oversimplified picture, since flow in one direction is affected by conditions at right angles, and therefore strain-hardening in one direction cannot be a function only of the previous strain-hardening in that direction. The full situation is of great complication and involves the functional dependence of tensors on each other.

The data in Table XXVI allow a check on the adequacy of the octahedral stress as a criterion of flow. It will be found that for the Cr-V steel the stress function $(X_x - Y_y)^2 + (Y_y - Z_z)^2 + (Z_z - X_x)^2$ at the last point on first compression is 3.20×10^{10} against 4.15×10^{10} (psi)^2 at the first yield point on subsequent simple compression with reorientation. For the stainless steel the corresponding figures are 9.17 and 11.23×10^{10}. The discrepancy in both cases is in the same direction and considerably more than the experimental error.

CHAPTER 23

TORSION AFTER PRESTRAINING IN TENSION UNDER PRESSURE[1]

The specimens were all cut from the same bar of 9-0 steel furnished by the Arsenal. Nine specimens were prestrained in tension. These were divided into three groups of three each. Three were strained at atmospheric pressure by different amounts, three at a pressure of approximately 230,000 psi, and three at approximately 370,000 psi. In addition to these nine, a virgin specimen was also tested in torsion. At atmospheric pressure the maximum prestrain was limited by the danger of rupture; previous experiments had shown that the natural strain at rupture of this steel is 0.78. The greatest prestrain given to any of the atmospheric specimens was 0.66. At the higher pressures the maximum prestrain was not pushed so close to the rupture point under pressure; if this degree of strain had been approached too closely there was danger that the conditions would be too inhomogeneous at the neck.

The prestraining offered no special features; it followed the canonical procedure described in Chap. 2. The entire stress-strain curve was determined. The terminal points of these curves have already been given in Table V under the designations 9-0-9 to 9-0-17. In correcting for inequality of stress distribution across the neck the value of a/R was not directly measured but was taken from the mean curve for many experiments given in Fig. 7. It is probable that the tensile stress for specimen 9-0-17, the specimen with the greatest prestrain, 1.165, is considerably too high. It seems to lie above the best straight line through the other points and is also much higher than the line given by previous measurements on this series, which were determined over a range of strain nearly 3 times more extensive, and which therefore are doubtless better.

After the prestraining in tension an isthmus was turned at the neck in preparation for the torsion tests. This isthmus was similar to that described on page 314, Fig. 158, for the compression tests, except that the corners were not sharp. The isthmus was finished with the same carboloy tool for all specimens, so that its shape was identical for all specimens except for slight variations in diameter. The diameter was measured with a micrometer microscope; the fluctuation through the series was plus or minus 1.2 per cent. The final torsion results were corrected to

[1] This chapter is based on material contained in the third Watertown report.

the mean diameter. Preliminary investigation had been made to find the best shape for the isthmus. The fillet at the edge of the isthmus proved to be necessary; if the corner of the isthmus is left sharp, fracture in torsion always occurs at the sharp corner. This is unlike the results already described for torsion combined with longitudinal compression where sharp corners proved admissible. With the fillet the fracture occurs at various points indifferently along the length. The radius of curvature of the fillet was of the order of 0.025 cm, the straight length of the isthmus 0.075 cm, and the diameter of the isthmus 0.102 cm. By making a light longitudinal scratch along the isthmus it was possible to establish that the plastic twist is distributed uniformly along the uniform part of the isthmus and ceases nearly abruptly at the fillet.

The torsion tests were made in a specially constructed miniature testing machine. Torque was applied through ball bearings by means of the elastic twist of a steel piano wire 0.13 cm in diameter. The yield point in torsion was reached on the average at a total twist of the wire of about 50°. This was read from a pointer on a scale divided to 1.0°, reading to tenths by estimation. The wire was calibrated with weights on a lever arm. The twist of the specimen was measured by the relative angular displacement of the two ends, determined by arms directly attached to the specimen. One of the arms carried a scale graduated to 0.5° and the other a pointer. The position of the pointer on the scale was read with a micrometer microscope giving 0.005° during the initial elastic part of the twist, and with a hand glass giving by estimation 0.05° during the plastic part of the twist.

In order to apply torsion to the specimen, flats were machined on the large end, which were keyed into the shafts of the testing machine. One of the ends was so mounted as to permit longitudinal motion, so that no longitudinal stress was applied during the twisting, an important precaution.

After reaching plastic yield, readings were made on a time schedule at intervals of approximately 1 min. An increment of angular displacement of approximately 10° was applied to the torque wire in an interval of a few seconds by rotating the far end. After this rotation, the position of the far end was maintained constant by the action of a bearing designed to have friction greater than the maximum torque required to rupture the specimen. At the conclusion of 1 min the difference of angular displacement between the two ends of the torque wire was read, giving the torque, and also the difference of angular displacement of the two ends of the specimen. Although sensible creep had not entirely ceased at the end of 1 min, it had so much slowed down that no important error in the readings is to be anticipated. The important feature of any time schedule is that it should be the same for all specimens; in this way comparable results should be obtained. There is, of course, a great deal of detail within the region of creep. A detailed investigation of this, however, was felt to be outside the scope of this work.

Rupture usually occurred at a total twist of the order of 100°; sometimes it occurred during the twisting process and sometimes during the 1 min of rest.

Typical results for the torque against angle of twist are shown in Fig. 163 for specimen 9-0-13. The initial elastic part of the curve is not well shown on the scale of this figure.

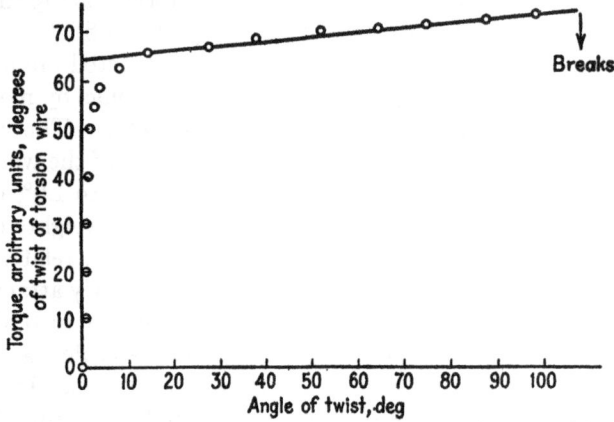

FIG. 163. Torque against angle of twist of a specimen previously strained in tension under pressure.

If the elastic part of the curve is plotted on a large scale it will be found that there is no significant correlation between the amount of prestraining and the slope of the curve; that is, the average shearing modulus is not affected by prestraining in tension. The accuracy of this statement is about 3 per cent. In other words, at the maximum torque before obvious departures from linearity began, the corrected elastic strain for all specimens varied by not more than 3 per cent above and below the mean, with no correlation with the amount of pressure or of prestrain.

The data for the plastic part of the deformation in torsion are contained in Table XXVII. For convenience of reference the data for the prestraining in tension are repeated in Table XXVII from Table V. The fifth column of Table XXVII was obtained from diagrams like Fig. 163. This diagram shows that in the plastic part of the range the relation between angle of twist and torque is approximately linear. Diagrams like Fig. 163 were constructed for all the tests, the best straight lines drawn, and the intercepts on the torque axis tabulated in column 5. This

TABLE XXVII

Designation of specimen	Prestrain data					Twisting data				
	Pressure of prestrain, psi	Max prestrain $\log \frac{A_0}{A}$	Max "true stress" of prestrain		Torque intercept at 0° twist, arbitrary units	Slope of stress-strain curve for torsion, arbitrary units	Torque at 70° twist, arbitrary units	Shearing stress at torque intercept, psi	$\frac{4}{3}\frac{T_1 - T_0}{T_0}$	Angle of twist at fracture, deg
			Uncorrected, psi	Corrected, psi						
9-0-9	Atmos.	0.378	137,000	128,000	62.3	0.087	68.4	71,000	0.175	100
9-0-10	Atmos.	0.453	149,000	138,000	64.3	0.080	70.0	72,000	0.156	92
9-0-11	Atmos	0.655	163,000	145,000	66.8	0.090	73.2	76,000	0.169	89
9-0-12	155,000	0.275	139,000	133,000	62.1(?)	0.080(?)	67.8	72,000	0.162	103
9-0-13	153,000	0.462	157,000	144,000	65.2	0.093	72.0	74,000	0.179	109
9-0-14	162,000	0.718	187,000	163,000	69.6	0.108	77.4	79,000	0.194	92
9-0-15	375,000	0.236	142,000	137,000	58.0	0.093	64.0	66,000	0.201	108
9-0-16	360,000	0.513	186,000	169,000	65.8	0.097	72.1	75,000	0.186	110
9-0-17	365,000	1.165	318,000	262,000	71.0	0.168	82.7	81,000	0.274	86
9-0-18	Virgin				48.9	0.124	58.1	56,000		154

intercept gives a rough indication of the plastic yield point in torsion. There was usually little doubt as to the best straight line, particularly if one confines oneself to the central part of the curve. The last point before rupture almost always lies somewhat below the curve; the true curve is doubtless gently concave downward instead of straight. The deviation from linearity was not great enough to give rise to important doubt with regard to the intercept except for the virgin specimen and specimen 9-0-12. Column 6 gives the slope of the line whose intercept is given in column 5, that is, the slope of the strain-hardening line in torsion. In column 7 is given the torque at an angle of twist of 70°, well below the

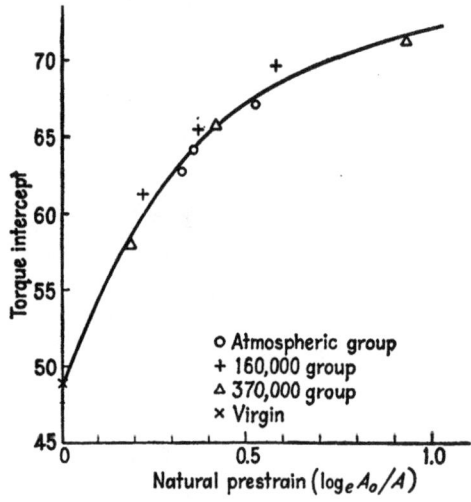

FIG. 164. Extrapolated torque at zero twist against prestrain of specimens previously pulled in tension.

rupture point of any of the specimens, and far enough above the yield point to show the effect of strain-hardening in torsion. This column gives a composite of the effects of raising the yield point and strain-hardening. Columns 8 and 9 will be described later. Finally, in column 10 is given the angle of twist at rupture. Because of the irregular time effects these angles are probably not significant within 5°.

The data of Table XXVII are exhibited in several figures. In Fig. 164 is shown the extrapolated torque at zero twist (column 5 of the table) plotted against the natural prestrain, and in Fig. 165 the torque at 70° twist against the natural prestrain. In each figure the tests are grouped into sets of three, depending on the approximate pressure at which the prestrain was imparted. The points in Figs. 164 and 165 lie approximately on single curves. This means that the torque extrapolated to zero twist and the torque at a fixed angle of twist are functions only of the amount of prestrain, not of the pressure at which the prestrain was imparted.

The curves of both Figs. 164 and 165 rise; that is, the material is hardened for torsion by prestrain in tension. The rise, however, is not linear, but the curves are concave downward. Large prestrains in tension are proportionally less effective in hardening in torsion than smaller prestrains. The phenomena for torsion are thus unlike those for tension. Hardening in tension by prestrain in tension was found to be linear in the prestrain.

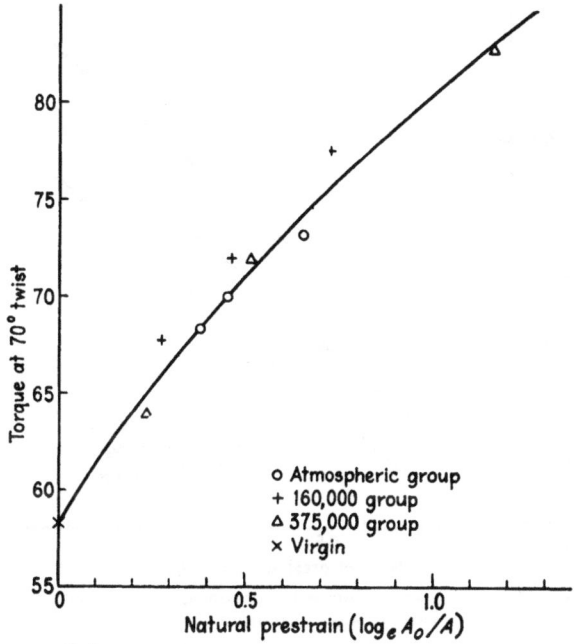

FIG. 165. Torque at 70° twist against prestrain of specimens previously strained in tension.

In Fig. 166, column 6 of Table XXVII, that is, the slope of the stress-strain curve in torsion, is plotted against prestrain. There is much scattering of the points in Fig. 166, and the point for the virgin specimen falls entirely out of line with the others. To the extent that it is justified to draw lines through these points it would appear that the slope rises in general with prestrain; that is, strain-hardening in torsion increases more rapidly the greater the prestrain in tension. The effects here are qualitatively different from those for tension; the slope of the strain-hardening curve for the second pulling in tension was found to be constant independent of the prestrain in tension.

There also seems to be a tendency for the slope to be a function of two parameters, that is, of both the amount of prestrain and of the pressure at which the prestrain was imparted. For a given prestrain the slope is greater the greater the pressure of the prestrain. This is unlike the state of affairs shown in Figs. 164 and 165.

In Fig. 167 the angle of twist at fracture is plotted as a function of the natural prestrain. The greater the prestrain the smaller the angle of twist at fracture; this is to be expected. Again there is much scattering, but again there seems to be a tendency for the angle of twist at fracture to be a function of both the amount of prestrain and of the pressure at which

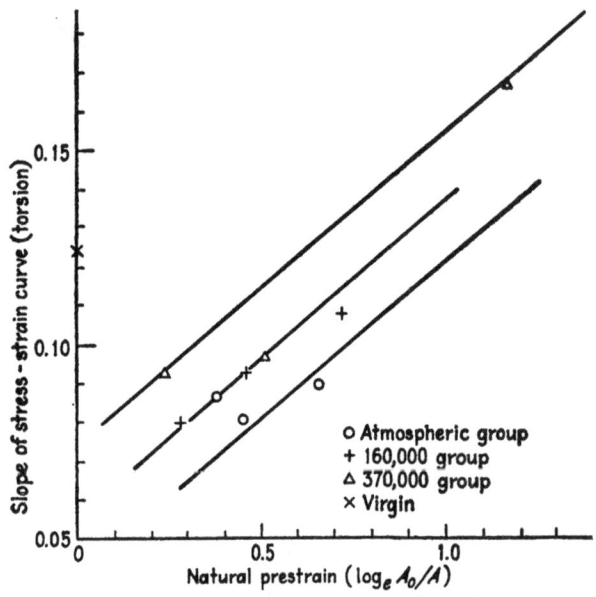

FIG. 166. Slope of the stress-strain curve in torsion against the prestrain for specimens twisted after prestraining in various ways in tension.

the prestrain was imparted. The higher the pressure of the prestrain, the greater the angle of twist at fracture; that is, a given elongation at a higher pressure does less damage to the material. The point for the virgin specimen is now not impossibly out of line with the others.

The data given thus far have been comparative, in arbitrary units, degrees of twist of the wire with which torsion was applied. From the dimensions of the isthmus (length 0.0757 cm and diameter 0.1022 cm) it may be calculated in the first place that one degree of twist means a shearing strain at the outside surface of 0.0106. The arbitrary units in terms of which torque was measured are such that one unit means a torque of 22.3 g-cm. This, combined with the dimensions of the isthmus, means that for one arbitrary unit of torque there is an average shearing stress of 1,140 psi across the section of the specimen. In the full plastic range the shearing stress is approximately constant across the section. With these constants the absolute shearing stress, assumed constant across the section, may be calculated corresponding to any torque. This shearing stress at the torque intercept has been so calculated and is tabulated in column 8 in Table XXVII. In simple tension

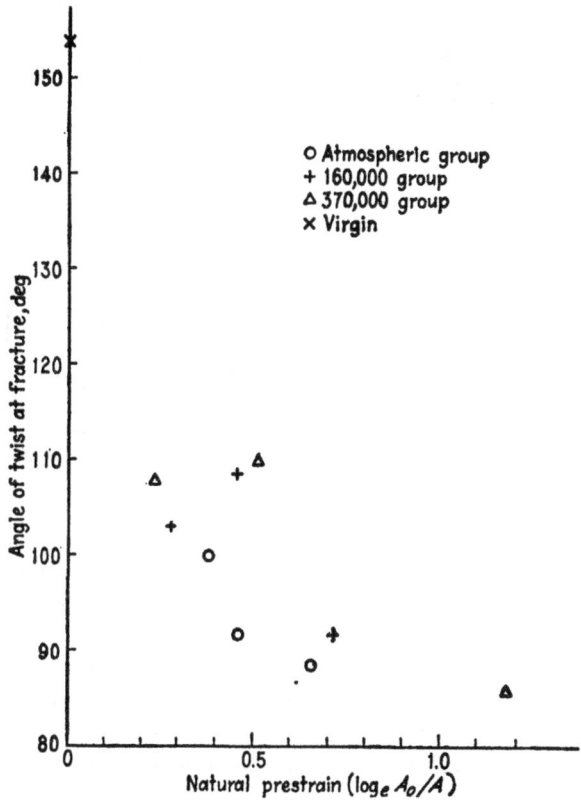

FIG. 167. Angle of twist at fracture as a function of the prestrain for specimens twisted after prestraining in various ways in tension.

the maximum shearing stress is one-half the tension. In accordance with this, a criterion of plastic flow which is often employed is that twice the shearing stress in simple torsion shall be equal to the tensile stress at plastic flow in tension. In Fig. 168 is plotted twice the shearing-stress intercept of Table XXVII against the natural prestrain. In the same diagram the tensile stresses of the prestrains are also plotted, and also the strain-hardening line obtained from the previous measurements on the 9-0 series. The most striking feature of the diagram is the approximate agreement of the two sets of stresses. Twice the extrapolated shearing stress intercept, that is, twice the shearing stress at the "yield point" in torsion, is approximately equal to the maximum tensile stress in the prestrain. This approximate equality is an arresting fact, but its precise significance seems obscure. The shearing-stress intercept gives only a nominal "yield point"; plastic flow and creep are detectable in early stages of the torsion curve at stresses markedly below the intercept.

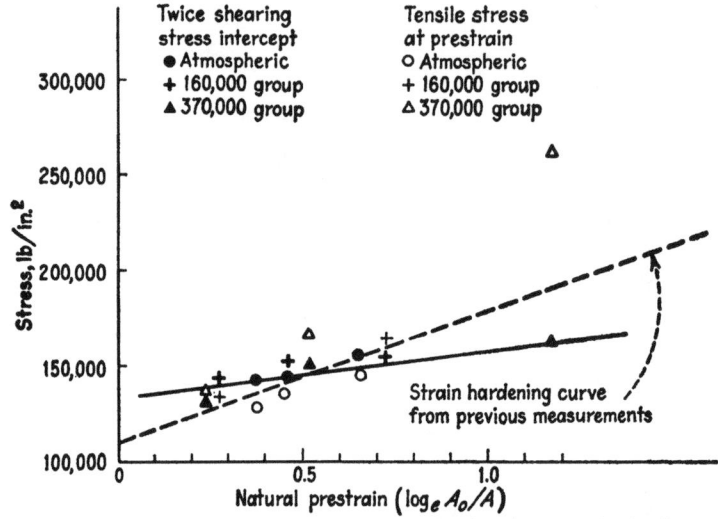

FIG 168. Shows the approximate equality of twice the shearing stress in simple torsion and the tensile stress at plastic flow in tension for various specimens twisted after prestraining in tension.

If the "octahedral" stress criterion of flow were valid, we would have as the connection between shear stress and flow stress in tension $S = \sqrt{1/3}\, F = 0.58F$, instead of $S = 0.5F$ as above. It would seem that the factor 0.5 is definitely better than 0.58, and therefore the simple criterion here is better.

In finer detail, the slope of the best line through the points for twice shearing stress is definitely less than the slope of the line through the tensile stresses of the prestrain. This means that to a second approximation twice the shearing-stress intercept is less than the tensile stress of prestrain for large prestrains. In other words, for large prestrains less of the hardening produced by the prestraining in tension is retained for the subsequent plastic flow under torsion than for smaller prestrains. This seems natural and furthermore is similar to the results already obtained for the second straining in tension after prestrain in tension.

Finally, these results may be used to compare the strain-hardening produced by tension with that produced by torsion or shear. It has already been emphasized in a report by Hollomon and Zener that the two kinds of strain-hardening do not run parallel, the hardening by torsion being less rapid than that by stretch. One question that has to be answered is what is the proper coordinate to take for the strain in torsion. Hollomon and Zener suggest that in simple twisting tests the angle of twist may be expected to be analogous to the natural strain $\log_e (A_0/A)$ in tension, because in both cases the area under the stress-strain curve represents work. Using this coordinate, Hollomon and Zener found that the stress-strain curves in torsion are not linear, but concave downward. They found a linear relation between torque and the natural logarithm of the angle of twist beyond a certain angle and obtained their intercepts by extrapolating the logarithmic curve back to a definite point. The present data, on the other hand, satisfy a linear relation, over a large part of the range, between the torque and the angle itself, not of its logarithm. This difference is perhaps a result of the prestrain. The matter should be investigated further, but in any event it suggests itself that the angle of twist may be taken as analogous to the natural strain in tension even more appropriately for the data of this paper than for those of Hollomon and Zener.

Accepting these variables, we have therefore a first resemblance between strain-hardening in tension and in torsion in that both strain-hardening curves are linear. The next question is as to the slopes of the strain-hardening curves in tension and torsion. The answer to this demands that the next stage in the approximation be made, and correction applied to the torsion experiments for the variation of shearing strain from the outer surface to the axis of the torsion specimen. This correction will involve the strain-hardening curve for shear. We have hitherto neglected such effects and have calculated only the average shearing stress across the section, assuming it constant at all radial distances. As a next approximation, we assume that the shearing stress is a linear function of the shearing strain. Does this mean that it is also a linear function of the distance from the axis? To answer this we have to consider the solution of the general problem in plastic torsion with strain-hardening. We use the general result that, with or without strain-hardening, the circular cylinder twists in such a way that radii of the circular cross sections remain straight lines; that is, there is no drag or advance of the outer part of the cross section with respect to the interior. This result may be proved by considerations of symmetry, by rotating the cylinder through 180° about any axis perpendicular to the longitudinal axis. The same argument shows that plane cross sections remain plane. This applies to an infinite cylinder with no end effects.

The circumferential displacement satisfying these conditions is

$$u_\theta = \text{const } z \cdot r \qquad (23\text{-}1)$$

which is thus valid in the plastic as well as in the elastic range.

Consider now the effect of varying shearing stress across the section arising from strain-hardening. Assume a linear relation between shearing stress and shearing strain; that is, put

$$\widehat{\theta z} = \widehat{\theta z}_0 + \alpha e_{\theta z} \qquad (23\text{-}2)$$

If we put $_a e_{\theta z}$ equal to the strain at the outside surface, $r = a$, then we have

$$e_{\theta z} = \frac{\partial u_\theta}{\partial z} = \text{const } r \qquad (23\text{-}3)$$

and

$$\text{const} = \frac{1}{a} {}_a e_{\theta z} \qquad (23\text{-}4)$$

The total torque is now found at once:

$$\text{Torque} = 2\pi \int_0^a r^2 \widehat{\theta z}\, dr = 2\pi \int_0^a r^2 (\widehat{\theta z}_0 + \alpha\, {}_a e_{\theta z}) \frac{r}{a} dr$$

$$= 2\pi \left(\frac{a^3}{3} \widehat{\theta z}_0 + \alpha\, {}_a e_{\theta z} \frac{a^3}{4} \right) \qquad (23\text{-}5)$$

Now set

$$\text{Torque} = T_0 \quad \text{when } {}_a e_{\theta z} = 0$$
$$\text{Torque} = T_1 \quad \text{when } {}_a e_{\theta z} = 1$$

and the torque equation gives

$$\frac{\alpha}{\widehat{\theta z}_0} = \frac{4}{3} \frac{T_1 - T_0}{T_0} \qquad (23\text{-}6)$$

T_0 is the "torque intercept." T_1 is the torque at a total angle of twist of $0.1/0.00106 = 94.50$ and may be taken from the curves.

The final result is that the coefficient of the linear term in the strain-hardening relation between shearing stress and shearing strain may be obtained by multiplying by 4/3 the relative excess in torque at a shearing strain of unity at the outside surface of the torsion specimen over that at the torque intercept. This coefficient is given in the ninth column of Table XXVII. It is not constant, as we have seen, but increases with the prestrain. At its largest, however, it is still much smaller than the coefficient of the linear term in the strain-hardening in tension. For an increase of natural strain from 0.0 to 1.0 the true tensile stress increases from 110,000 to 179,000 psi, or a relative increase of 0.625, against a maximum increase of the shearing stress of 0.274. Thus, in general, strain-hardening in shear proceeds much more slowly than in tension.

Summary. We may summarize these results for torsion after prestraining in tension under pressure as follows:

1. No appreciable change in the elastic constant for torsion is produced by prestrain in tension.

2. The angle of twist is a linear function of the torque over a large part of the range for most of the specimens.

3. The torque intercept is raised by the prestrain in tension.

4. To a first approximation the raising of the torque intercept and also the raising of the torque at an angle of twist of 70° depends only on the magnitude of the prestrain and is independent of its pressure.

5. The raising of the torque intercept is not linear in the prestrain, but is proportionally less at large prestrains.

6. To a second approximation, the slope of the torque-twist curve is greater the greater the prestrain, and also the higher the pressure at which the prestrain was imparted. This is unlike strain-hardening in tension.

7. The angle of twist necessary to fracture is less the greater the prestrain, and greater the greater the pressure of the prestrain.

8. Twice the shearing stress at the torque intercept is very nearly equal to the maximum tensile stress during the prestrain for the smaller prestrains, but at larger prestrains falls below.

9. The rate of strain-hardening in shear, corrected for the variation in stress across the section due to strain-hardening, is, even at its maximum, less by a considerable factor than the rate of strain-hardening in tension.

Part IV
GENERAL SURVEY

CHAPTER 24
GATHERING UP THE THREADS

Mathematical Background. Let us assume for the present that we shall be able to get along with a description in terms of stresses and strains only, although we know there are cases in which this cannot be adequate, and set ourselves the conventional task of finding what sort of description of the phenomena in mathematical terms is consistent with what we now know about the physics of the situation. We consider first the "idealized" plastic body. Such an idealized body is in the first place in the conventional analysis supposed to be in one of two states, either in the elastic state or in the plastic state. It is in the elastic state if a certain function of the stresses is less than some critical value, and in the plastic state if this function reaches the critical value.

Right here, at the very beginning, there is indicated the necessity for a radical change in our ordinary way of looking at the materials around us. For here we have a change in the fundamental properties of a body brought about by the action of forces. Ordinarily we think of the "properties" as fixed, and in fact what a property is is a description of the way in which the body responds to forces or to other attack. Usually also we have an approximately linear response—twice the force, twice the response. In the domain of linearity we have additive laws and an easy intuitive feeling for the effect of any ordinary change in the kind of attack we make on the body. But if the *properties* of the body themselves change under our attack then we are indeed in strange territory.

Assuming now that there is a plastic state producible by force and a plasticity function, it will be adequate for our purposes to consider only one of the forms which have been proposed for this function, namely, that of von Mises. His condition is that the body is in the plastic state if the following function of the principal stresses, namely,

$$(X_x - Y_y)^2 + (Y_y - Z_z)^2 + (Z_z - X_x)^2 \qquad (24\text{-}1)$$

reaches a critical value. If the externally applied forces are increased until this function reaches its critical value, the body thereupon enters the plastic state and may begin to flow under the action of the stresses. It is this flow which is the primary physical phenomenon of plasticity; in fact, the ordinary implication in the word "plastic" is that of yielding. The rate at which the ideal plastic body flows is subject to the limitation

$$\frac{\dot{e}_x}{X_x - \frac{1}{2}(Y_y + Z_z)} = \frac{\dot{e}_y}{Y_y - \frac{1}{2}(Z_z + X_x)} = \frac{\dot{e}_z}{Z_z - \frac{1}{2}(X_x + Y_y)} \quad (24\text{-}2)$$

These flow equations express in the first place the isotropy of the body, each component of flow being connected with the corresponding stress by relations symmetrical for all directions. In the second place they express the conservation of volume, the sum of the three principal strain rates being identically zero.

The flow equations as written do not determine absolutely any single rate of flow, but merely the ratio of velocities of flow in different directions. The absolute velocity is determined from without and mathematically is in the form of a boundary condition which may be imposed independently of the equations holding at every point inside the body. But this externally imposed velocity is itself subject to very definite restrictions. The assumption of the existence of the idealized plastic body really amounts to the assumption that certain bodies will tolerate the forcible imposition of only certain sorts of behavior, and that we may not do to them certain types of thing that our ordinary physical intuition suggests that we should be able to do. Consider, for example, a body in the form of a cube, and let us apply to its three pairs of opposite faces three different uniform normal force pairs; that is, we apply to this body an arbitrary stress system X_x, Y_y, Z_z. Ordinary experience with ordinary bodies prepares us to expect that we could give any value that we please to all three of these stress components. We know intuitively that we can do this, because we merely have to push as hard as we please on each face and the body can do nothing about it. What we are now saying is that this arbitrary imposition of any force on the faces is possible only within limits. As we increase the forces, we shall presently increase the von Mises plasticity function to its critical value and change the state of the body from the elastic to the plastic state. What we can now do to the body has entirely changed. In the first place, we cannot increase any single one of the stress components. If we attempt to by pushing harder on any one face, the body simply goes back on us by yielding indefinitely in the direction in which we try to push harder. The force has ceased to be an independently controllable parameter, and is now replaced by another, namely, the velocity with which the body deforms. The body in the plastic state is completely moldable; we can displace any chosen one of its faces at any speed that we may care to impart, and to impart that speed requires no addition or change in any of the forces, merely that we move our hand faster or more slowly. Under these conditions the common value of the three ratios in the flow equations (24-2) becomes a function of the particular geometry of the particular plastic system under consideration and is eventually determined by a velocity impressed somewhere at the boundary and is not at all a function of the physical properties of the material.

Only one of the flow velocities can be arbitrarily impressed from outside if the forces on the three pairs of faces are also so maintained that the body is in the plastic state. The other two velocities are then fixed according to the flow equations by the three stresses and the one flow velocity. There is an exceptional case if one of the flow velocities is zero. Thus there is a solution of the flow equations when $\dot{e}_x$ and Y_y are both zero. This is the case of two-dimensional compression considered in Chap. 13. We have $X_x = \frac{1}{2}Z_z$. $\dot{e}_z$ may be impressed at pleasure in spite of the fact that $\dot{e}_x$ has already been fixed at zero. The equations are now satisfied by $\dot{e}_y = -\dot{e}_z$.

The sort of response that the body makes to our attack satisfies in certain respects our physiological feelings of what is proper. The equations say that, if we push with increasing force on one pair of faces and half as hard on another pair and not at all on the third pair, we shall find that as we increase the force we come to a point where we can push no harder, but instead the body yields beneath our hand. But we still are not completely stopped from imposing our will on the yielding faces, for although we cannot push any harder, we can make our hand move as fast or as slowly as we please, still exerting the same force and still pushing half as hard on the other pair of faces. In the free mobility of the face of the body with no increase of force we have something analogous to the perfect liquid (not a viscous liquid which is the analogue of the plastic body which usually occurs to one).

The resemblance to a perfect liquid extends further. For not only does the face under our hand yield at once to the slightest change of speed, but the face at right angles also responds perfectly, for always $\dot{e}_y = -\dot{e}_z$. In other words, we have here a perfect mechanism by which the motion gets transmitted around the corner at right angles, with nothing analogous to the friction which we would surely encounter if we tried to construct a model based on the ordinary properties of solid bodies. We are reminded of the hypothetical "state of slip" which we encountered on page 290 in connection with perfect crystals. Perfect crystal grains of microscopic size arranged with random orientations might provide such an action.

The "idealized" plastic body is at variance with some of our other intuitive ideas of how bodies behave as we have observed them. For if we decide that we want to push harder on one of the sides of a cube of some actual material we know that we can, and that the body has jolly well got to take it. If it does not like it, all it can do is to yield, and the rate at which it yields is determined by its physical properties. I think there can be no doubt that the bodies of almost all our experience correspond to this sort of intuitive demand and the idealized plastic body is at best an abstraction. The fact that it has been found profitable to idealize bodies in this way must imply certain rather common types of behavior when the idealized conditions are departed from. But in the meantime I think more can be said for the idealized plastic body than one might be at first inclined to grant, and if our experience had been broader we would have found the idea more congenial. The fundamental physical phenomenon to which the assumption of an idealized plastic body runs counter is strain-hardening, which means that the critical value of the plasticity function is in general not constant but is an increasing function of the strains. Now we have already encountered situations in which the body ceases to strain-harden, namely, in the experiments on very high shearing strains combined with approximately hydrostatic pressure described in Chap. 16. For such bodies the intuitive prescription of what we can impress on the body fails. All we can do is to apply a shearing force up to the value at which the body becomes plastic, and we cannot push harder than this on the body. Once we are pushing with this critical force all the further control we have of the situation is to move the handle of the shearing lever with any speed that we please, but always the same force is required to do it. It would therefore appear that in the limit, with certain types of deformation at that limit, the assumption of an ideally plastic body approximates to experience. Of course, the behavior at these enormous deformations is far from ideal for other types of deformation. At the other end of our experience the ideally plastic body has proved to be a useful abstraction in the region of small deformations when the elastic limit of the virgin material has been just exceeded and strain-hardening phenomena have hardly begun to enter. Here the range of deformation and stress within which the body behaves in the ideal way is small, and the concept of the ideally plastic body is of only limited applicability.

Accepting now the flow equations, under what conditions may we integrate them and write the more usual formulation for an ideally plastic body in the form

$$\frac{e_x}{X_x - \frac{1}{2}(Y_y + Z_z)} = \frac{e_y}{Y_y - \frac{1}{2}(Z_z + X_x)} = \frac{e_z}{Z_z - \frac{1}{2}(X_x + Y_y)} \quad (24\text{-}3)$$

subject to the restriction $(X_x - Y_y)^2 + (Y_y - Z_z)^2 + (Z_z - X_x)^2 = $ const? It is Eqs. (24-3) that are more commonly understood as the equations for an ideally plastic body. To refer to the equations in this way is really a misnomer. We should say instead "the equations of a body that has been plastic," because by writing the equations in this form we tacitly imply that the body has settled down to a final state and plastic flow has ceased. The reason that we write the equations in this form at all is of course a practical one, because we are usually concerned with the final form into which the body is molded by the action of impressed forces and not so much with the process by which it got into its final form.

Inspection of the flow equations and the integrated form now shows at once that in general the integrated form holds only under special conditions. If at all stages of flow the three stress components remain constant, then the ratio of the three flow components remains the same during all stages of flow and the integrated form follows at once. But if during flow the stress components change, still satisfying the von Mises plasticity condition, then in general the ratio of the flow components alters during flow and we may not integrate. In many of the stress systems of practice the stress system is so simple that the conditions are met and the integrated form is justifiable. This is the case, for example, in all situations where there is only one impressed component of stress, the other two always being zero, as in simple compression or tension or torsion. But in general the integrated form may not be naïvely used, but always inspection is necessary to ensure that the conditions have been such as to permit the integration. In making examination to ensure that the conditions have been suitable we are in effect making a demand on the *history* of the system. Here is an example in the simplest highly idealized case of the entry of a factor which cannot be evaded in the more complex situations of practice; namely, the equations as they stand do not tell the whole story, but concealed in the background there are implications about the history of the system.

The discussion above about the the possibility of using the equations in the integrated form is essentially concerned with the question of whether an "incremental" or a "total" plastic strain theory is preferable. A "total" theory is defined as one which makes the plastic deformations a unique function of the instantaneous values of the stresses. There has been much discussion of this question, particularly by Prager's school at Brown University. The general conclusion seems to be that an "incremental" theory is to be preferred. Mention may also be made of a recent paper by J. L. M. Morrison and W. M. Shepherd[1] in which the same conclusion is drawn on the basis of extensive experimental data.

[1] *Applied Mechanics*, **463**, 1–9, 1950 (W.E.P. No. 55).

It is to be noticed that much of the work in this book permits any "total" theory to be peremptorily discarded as applicable to *large* stresses or strains; for such a theory would make the final strain independent of the order in which the stresses are applied, whereas we have seen that the order is vital. If a large tension and then a large hydrostatic pressure is applied, the body is first ruptured before the pressure is applied, whereas if the order is reversed the body is unbroken.

Leaving now the "idealized" plastic body, what is a closer description of the behavior of actual bodies? The obvious next approximation is to consider the effect of strain-hardening. It is usual to suppose that the flow equations continue to hold (these express isotropy and constancy of volume) and to put the burden on the plasticity function. We still have a plasticity function, but it has become a function of strains as well as of stresses. The body enters the plastic condition when a certain function of both stresses and strains reaches a critical value. Once in the plastic state the manner of response of the body changes qualitatively and discontinuously as compared with the response of the idealized plastic body. Now there is no indefinite yielding of the body when a force is applied slightly in excess of that required to produce the plastic condition; neither is there any possibility of imposing any arbitrarily chosen velocity. What now happens is that if we try to increase the force beyond the limit set by the plasticity condition we find that we can, if we are willing to put up with a certain amount of yielding in the body. The rate of this yielding is determined by the excess force and is not otherwise in our control; that is, the rate of yielding is now determined in some way by the body itself and may not be impressed at will from outside. Or, if we make our hand move with a certain velocity willy-nilly, then we shall find that we have to exert a force over which we have no control if we are intent on maintaining the velocity. The velocity of flow in response to the excess force has become a physical property of the body. Furthermore, the flow does not continue indefinitely in any actual case, but at least for moderately excess forces it presently slows down and stops. This slowing down and stopping is what is conventionally associated with strain-hardening.

For our immediate purposes we simplify the mathematical representation of the situation by what, from other points of view, is a gross oversimplification. We shall consider that we are able to impress simultaneously any desired strain and stress on the body. We further simplify for our immediate purposes by assuming only one generalized strain component and one generalized component of stress. The body, then, is to be represented by a point in the stress-strain plane. Over a certain region of this plane the body is in the elastic state, and over the rest of it is in the plastic state. The two regions are separated by a curve, as indicated in Fig. 169. For any state of the body represented by a point in the elastic region the body is in a steady state under the given stress. For any point in the plastic region the body is flowing, and the speed of flow is fixed at that point; that is, the speed is a function of stress and of strain. The boundary curve between the elastic and plastic regions is a curve of zero flow velocity. Otherwise, it is the "strain-hardening" curve, so called because the curve almost always rises to higher stresses as strains increase.

It is obvious that if we try to get along with only the integrated form of the equations of plasticity we are turning our backs on by far the larger part of the physical phenomena, for we are confining ourselves to points

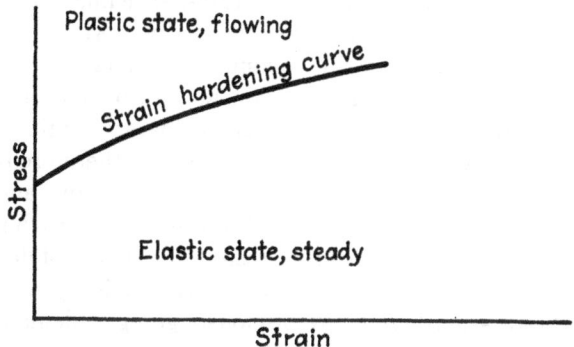

FIG. 169. Indicates in a grossly oversimplified way the most essential characteristics of the elastic and plastic states.

on the strain-hardening curve where the velocity is zero. Not only this, but we are turning our backs on phenomena without which the strain-hardening curve itself could not be determined, for how could we get from one point to another of the strain-hardening curve without a changing strain, and a changing strain means a velocity?

A complete description of the plastic behavior of bodies obviously demands the determination of the velocity of flow as a function of stress and strain for every point within the plastic region. We have an intuitive feeling of what to expect. For short excursions away from the strain-hardening curve into the plastic region we expect in general small flow velocities, the velocity increasing as the distance of our excursion away from the curve increases. The rate of increase with increasing distance from the strain-hardening curve may be very rapid or perhaps even catastrophic in the initial stages of plastic flow near the virgin condition of the material. If this were not the case it would be difficult to understand the success, for some substances, of von Kármán's mathematical analysis for the velocity of propagation of a plastic shock wave. His solution is in some respects analogous to the solution for the velocity of propagation of an elastic disturbance. The velocity is determined by a plastic modulus which is given within experimental error by the *statically* measured strain-hardening curve.

The characteristic velocities of flow within the plastic region are what I have called "primary" creep. It seems to me that the phenomena of primary creep call for much more attention than has apparently been given to them. Certain of these phenomena have been described in simple cases in the chapter on two-dimensional compression. It was found there, for example, that the velocity of creep increases very rapidly with increasing departure from the strain-hardening curve, certainly faster than linearly and probably exponentially. At the same time it appeared in that chapter that this second approximation, which is what we are considering here, does not remain an approximation over a very wide range, but already at comparatively moderate strains time effects begin to enter. To a next approximation we shall therefore have to consider that velocity of flow is a function of stress, strain, and time, and finally a function of stress, strain, and all past history. Such phenomena, for instance, as the delay of the plastic response, the rate of flow accelerating and then falling off, or still more the capricious jerky response in the later stages of strain-hardening, are obviously going to demand the inclusion of past history among the determining parameters.

The "ideal" plastic body is a highly special case of the more general body represented in Fig. 169. The idealized body is one with a horizontal strain-hardening curve, and one for which the flow velocity becomes infinite for any excursion, no matter how short, away from the strain-hardening curve into the plastic region. The diagram suggests that there are various conceivable sorts of "semi-idealized" body. For instance, we could have a rising strain-hardening curve, but still have infinite flow velocity for every point within the plastic region. Or we could have our horizontal strain-hardening line, but a velocity at points within the plastic region which was a function of stress only, or even a function of both stress and strain. I do not know whether any of these possibilities correspond closely enough to any actual materials to make it profitable to examine then further.

The same flow equations, (24-2), which we have applied to the idealized body, are ordinarily carried over to the second stage of approximation as typified by Fig. 169. These equations continue to express the isotropy of the material and the constancy of volume. We may now write them

$$\frac{\dot{e}_x}{X_x - \tfrac{1}{2}(Y_y + Z_z)} = \frac{\dot{e}_y}{Y_y - \tfrac{1}{2}(Z_z + X_x)} = \frac{\dot{e}_z}{Z_z - \tfrac{1}{2}(X_x + Y_y)}$$
$$= f(\text{stress, strain}) \quad (24\text{-}4)$$

The fact that the flow equations retain the same form as for the idealized body means that some of the unusual consequences carry over. In particular is the fact that we have perfect transference of flow velocity in one direction to the direction at right angles, thus simulating the behavior in this respect of a perfect liquid, with nothing corresponding to friction. We may illustrate this by the behavior of a body in two-dimensional compression, for which we have already written the equations in detail. As long as one dimension is maintained fixed, which we choose as the x direction so that $\dot{e}_x = 0$, we have exactly $\dot{e}_y = -\dot{e}_z$, and any change in the one is immediately transferred to the other. The equations also have something to say about deviations from the condition of strict two-dimensionality, deviations which are within the range of the experimental setup described in Chap. 13. The condition $Y_y = 0$ is one which is always exactly maintained, since the y face is free. The flow equations then become

$$\frac{\dot{e}_x}{X_x - \tfrac{1}{2}Z_z} = \frac{\dot{e}_y}{-\tfrac{1}{2}(Z_z + X_x)} = \frac{\dot{e}_z}{Z_z - \tfrac{1}{2}X_x} \quad (24\text{-}5)$$

The additional condition that $\dot{e}_x = 0$ demands that $X_x = \tfrac{1}{2}Z_z$. Let us now consider the effect of allowing X_x to deviate slightly in either direction from this relationship. We notice in the first place that the von Mises function under these conditions, or $X_x^2 + Z_z^2 + (Z_z - X_x)^2$, is a maximum for this value of X_x, for $(\partial/\partial X_x)[X_x^2 + Z_z^2 + (Z_z - X_x)^2] = 0$ when $X_x = \tfrac{1}{2}Z_z$; that is, for small changes in X_x about the value $\tfrac{1}{2}Z_z$, the body does not depart from the plastic state. However, under such changes $\dot{e}_x$ does not remain zero but experiences changes of the first order in terms of the changes of X_x. We may see this by writing $X_x = (\tfrac{1}{2} + \Delta)Z_z$, whereupon

$$\frac{\dot{e}_x}{\Delta \cdot Z_z} = \frac{\dot{e}_z}{\tfrac{3}{4}Z_z - \tfrac{1}{2}\Delta \cdot Z_z}$$

or approximately

$$\dot{e}_x = \tfrac{4}{3}\Delta \cdot \dot{e}_z$$

If, for example, X_x changes by 10 per cent of itself about $\tfrac{1}{2}Z_z$, $\dot{e}_x$ should change through plus or minus $\tfrac{1}{15}\dot{e}_z$. In this linear dependence on the changes of X_x we have again a frictionless response.

The data of Chap. 13 enable an answer to be given as to whether this is the actual behavior. Figure 120 of that chapter puts the answer to this question in another form. Equations (24-5) demand that $\dot{e}_x/\dot{e}_z$ be equal to $(X_z - \frac{1}{2}Z_z)/(Z_z - \frac{1}{2}X_z)$. Figure 120 plots one against the other. If they are equal, all the points should lie on a 45° line. The points are seen on the average to lie below the line with a slope of perhaps one-half as much as expected. This means that $\dot{e}_x$ is too small, or the action does not get transmitted perfectly around the corner from the z to the x direction; that is, there is something of the nature of plastic friction. Another experimental indication in the same direction was that when varying X_x away from equality with $\frac{1}{2}Z_z$ there was always a threshold which had to be exceeded before there was any motion in the x direction at all; after the threshold was exceeded, the velocity of x flow increased at an accelerated rate.

The same remarks may be made with regard to the integration of the flow equations in the strain-hardening case that were made for the idealized body. For many of the cases of practice, as when the stress system has only one component, or when the different stress components increase together, as, for example, when a closed cylinder is collapsed by external pressure, the integration is permissible. The stresses and the strains which are the result of such integration obviously fall on the strain-hardening curve. This is the way the curve will be determined in practice, rather than by fully investigating the velocity of primary creep and picking out the limiting curve on which this velocity goes to zero. In any concrete case, however, it is not safe to use the integrated form without inspection of the past history of the specimen to be sure that it is permissible.

Plastic Flow. What now are the important features of plastic behavior as disclosed by the experiments in this book? In the first place, plastic deformation without fracture may be indefinitely increased with the cooperation of hydrostatic pressure or sometimes other forms of stress (simple compression combined with torsion). The stresses which make large plastic deformations possible are in general of such a character as to push together any incipient fractures.

In the range of large plastic deformation the phenomena of strain-hardening occur. The shape of the strain-hardening curve depends on the material and the type of strain. For steel in tension it rises linearly with pressure until the curve can be followed no farther because of limitations set by grain size. For steel in simple compression the rise accelerates with increasing stress, but the slope of the curve for simple compression is always materially less then for simple tension. For copper in simple compression the curve bends over to a horizontal asymptote. For torsion of steel the curve rises less rapidly than linearly. For practically all substances there is a limiting value for strain-hardening in simple shear, and for most substances ideal behavior is approached in such circumstances in that the force to shear becomes independent of the speed. The generally simpler behavior in simple shear is natural to associate with the geometrically simpler nature of the distortion, whole planes of atoms remaining intact, without the interpenetration of slip on different planes which occurs in simple tension, for example.

The fact that the ultimate course of the strain-hardening curve depends on the type of strain indicates that the usually considered stress functions, such as "octahedral" stresses or the "significant" stresses, cannot have an unlimited validity. We have considered in detail what the range of validity of some of these functions may be.

The time effects accompanying plastic flow are multitudinous. At low strains, near the virgin metal, flow velocity may increase exponentially with the distance of displacement from the strain-hardening curve. At greater strains the velocity does not increase so catastrophically, but is spread out more smoothly over a wider interval. At still greater strains the speed of response loses its smoothness and capricious factors begin to enter. As strain increases it will become increasingly necessary to specify all the parameters of the body in order to determine its behavior, that is, stress, strain, and in addition full history. Even a full history in terms of stress and strain will not ultimately be adequate to cope with the capricious effects. To deal with these it will be necessary to renounce the attempt at description in terms of the macroscopic parameters of stress and strain and to get into the microscopic domain. How far down toward the molecular it will be necessary to go would be difficult to say. There is plenty of complication to be exploited no further down than the domain of grain structure, as will be evident if the significance of Figs. 77 and 78 is properly assessed.

A fully satisfactory theory must provide room for nonisotropic strain-hardening. In every case where we have been able to test this point in this book we have found that the hardening is not isotropic. The effects here are most complicated. At low strain-hardenings we may have the Bauschinger effect, that is, a strain-hardening for the direct strain but a softening for the negative. As strain increases the Bauschinger effect may disappear, it may be in a complicated fashion, and ultimately at high strains a strain-hardening for one type of strain becomes accompanied, in many cases at least, by strain-hardening for other types of strain. This appears to have exceptions, however.

In some cases it is possible to understand the different degrees of hardening for different subsequent types of strain by supposing that there is a reversible component in the process of strain-hardening, something that would seem natural enough from purely geometrical considerations.

A reversible component has been demonstrated in Chap. 12 for the volume change accompanying plastic deformation in simple compression. This volume change begins to be manifest only after the plastic deformation has become fairly large, and may be in the direction opposite to that which would be naturally produced by the stresses in the elastic range. Thus in the last stages of simple compression a reversible component of volume *increase* may enter. A plausible explanation of it is that it is due to the formation of new "dislocations" which are demanded by the great degree of distortion of the crystal lattice. This volume expansion may be a prelude to fracture. It is usual to neglect volume changes in the mathematical solution of problems in plastic flow. It would seem that this factor might in some cases introduce important modifications in the solution, particularly in those situations where the body is partly in the elastic and partly in the plastic state. The boundary between these two states might well prove to be sensitive to the entrance of a new factor in the volume change.

Certain simple qualitative rules are applicable in some cases with regard to the relative strain-hardening in different directions after pre-straining. The rule here is that the strain-hardening is greatest in the direction for which the previous straining in that direction was of the same type. In particular, if a body is strained in simple compression after previous straining in simple compression at right angles, then the strain-hardening in the cross section, where the strain is an extension, is greatest in that direction for which the previous strain was also an extension, that is, for the direction in the cross section which was transverse to the direction of first compression. But even this has exceptions (stainless steel).

The effect of a very small amount of anisotropy introduced by a previous straining may be surprisingly large, as we saw in the difficulty of avoiding the "shearing cross" in two-dimensional compression. The shearing cross arose because of a mathematical singularity at the edge of the compressed cube, and involved infinite stresses, or, in other words, plastic flow confined in a narrow region. It seems to me that the possibilities of mathematical singularities have not been sufficiently exploited. Not only are they present when we have free faces with shearing stress on the surface at right angles, but we have them in the body of the massive material wherever there are discontinuities in the elastic constants, and this must be a pretty common occurrence. Plastic flow may well take its initiation at such mathematical singularities. These would function very much like the "dislocations" which have been so successful in explaining various phenomena of flow and fracture, and in fact this may be to a certain extent the essence of a dislocation.

The considerations of the last paragraph have driven us from the large-scale to the small-scale domain, where the large-scale equations of flow and of stress equilibrium fail because they lose their meaning. It seems to me that we are inevitably driven into this domain when we deal with large deformations. For the structure of actual materials is molecular and atomic, not continuous. What is more, these atoms and molecules function pretty much like little rigid hunks with sharp boundaries. Theoretically of course, an atom has no sharp boundary; under indefinitely great forces the force fields of atoms interpenetrate and there is no sharp separation of atoms. A lattice under indefinitely high hydrostatic pressure is uniformly compressed, and all the atomic nuclei retain their relative positions with distortion of their external force fields. But for the types of stress that result in plastic flow or fracture this is not the picture at all, but the atoms remain at nearly constant distances apart. It is known that fracture takes place for relative displacements of the atoms giving rise to forces of the order of only 1 per cent of the expected maximum. For such relatively small changes in the atomic forces the boundaries of the atoms may be treated as approximately sharp. This means, as we have already seen, that if a body experiences deformations of more than a few per cent the lines of atoms become crumpled. The detailed motion of the atoms as they jostle against each other in finding their new positions of equilibrium can in no way be represented by the smooth macroscopic functions of the macroscopic equations. These crumpling effects may under proper circumstances become visible to the eye, as in the longitudinal wrinkling of the neck of a heavily strained tension specimen, or the slip lines on the inside of a cylinder severely stretched by internal pressure. This may well account for the important role which the surface may play in some of the phenomena of fracture; it would not seem necessary that there be flaws initially on the surface, but these may be created by the distortion.

Fracture. The reader may find it profitable as a preliminary to this discussion of fracture to recall the discussion in the last section of Chap. 5 with regard to the fracture of brittle materials.

Perhaps the single most important insight with regard to fracture to be derived from the experiments in this book is that there is no sharp dividing line between the phenomena of flow and of fracture. Examples can be set up where one would be hard put to say whether the body was flowing or fracturing with continual self-healing. In fact, what is flow in principle in a body with discrete atomic structure except uniformly distributed and quasi-continuous fracture with constant self-healing? Many of the experiments of this book have been made under conditions such that self-healing is particularly easy or even inevitable, as in many of the experiments under hydrostatic pressure, or in shearing under a mean hydrostatic pressure. It is possible to set up all sorts of intermediate cases between the complete and continuous self-healing of many of the shearing experiments of Chap. 16 and the catastrophic separation of the body into distinct parts which is usually connoted by "fracture." In many cases a fracture gets started, propagates itself for a distance, and then is quenched. How far must it be propagated before quenching in order that we would be willing to call it a "fracture"? The experiments on twisting under longitudinal compression presented many examples of incomplete fractures which had been propagated for visible distances but then had run into some sort of snag which made self-healing possible and the fracture stopped. The degree to which the fracture spreads before being stopped by self-healing involves the boundary conditions as well as the conditions internal to the body. Thus in some of the shearing experiments we might have a runaway shearing fracture if the rotating lever was actuated by a weight so as to maintain the force constant, or a chattering yield with constant self-healing if the lever was pushed by hand so that there was an opportunity for a decrease of force and so self-healing after each new microscopic fracture.

It will be agreed, I think, that in all cases in which we want to use the word "fracture" we are concerned with something initiated by some sort of instability which is then propagated to a greater or less extent. The propagation is subject to the demand of mechanics that the potential energy of the system must decrease during the process. This means the total energy of the whole system and embraces the internal elastic or other energy of the stresses within the body which is the seat of potential fracture, as well as the energy of the external forces acting across the exterior boundaries of the system. Although eventually the source of the energy is external, it is not always possible to control the fracture, once it has started, by any manipulation of the external forces, because the local decreases of energy of the internal stresses may dominate the local situation, since the effect of changing external force has to be propagated with a finite velocity to the scene of emergency. We saw examples of this in the difficulty of controlling the fracture of tubes under pressure, even when stops were provided to limit the displacement and so the energy input of the external forces.

The condition that energy must be released during the process of fracture is a necessary condition that applies to any possible fracture, and it has not been sufficiently appreciated. This is the factor that prevents fracture in some cases where other considerations, such as direction and magnitude of stress, suggest that fracture should occur. For instance, it is the reason that a heavy tube under internal pressure does not fracture at the inside surface if the pressure has once been raised so high as to bring the metal at the internal surface into the plastic state. Under these conditions release of stress at the internal surface by fracture would result in an increase of volume of the inner elements, which would do work *against* the pressure and thus operate in the wrong direction. This agrees with the observation that the inside surface of such a cylinder is filled with microscopic slip lines where fracture started and could not propagate because of the energy relations.

It seems to me probable that the requirement that energy be released during fracture is unconsciously back of the intuitive feeling that many people have that it is impossible for fracture to occur against the stress, for, say they, what is there to *make* the fracture go unless it is in the direction of the stress? It is obvious enough that, if the fracture is in the direction of the stress, then *that* component of stress does work during the fracture, and insofar the principle of energy release is satisfied. The principle obviously applies to a tension specimen subject to simple tension. But there may be other components of stress, and these also may do work during the fracture, with the result that the complete situation may become too complicated to be handled in any such simple intuitive way. In fact, in some of the more complicated cases simple intuition may be persuaded to give a different answer. For instance, consider a tensile specimen of some brittle material of such strength that it breaks in tension when a weight of 5,000 lb is hung on a bar of 1 sq in. cross section. We now immerse the specimen and the weight in a medium under hydrostatic pressure; that is, we "pressurize" the laboratory in which we perform the experiment. To be specific, let us pressurize it to the extent of 100,000 psi. Furthermore we suppose that the nature of the specimen is such that it does not lose its brittleness under pressure. I think most people would not feel that their intuitions were outraged if they should find experimentally that the bar now breaks under a weight of 10,000 lb. The difference between the 5,000 lb which produces fracture at atmospheric pressure and the 10,000 which fractures at 100,000 psi would be described in terms of a "pressure coefficient" of strength. Yet in the pressurized laboratory the specimen is breaking at a net *compressive* stress of 90,000 psi, which means that here fracture is *against* the stress. Reflection shows how the principle of energy release applies, because the fluid under hydrostatic pressure, rushing into the void created by the fracture, does more work than is done on the compressive stress. The conclusion is verified of course by observing that the weight has dropped after the fracture. Further reflection on this situation shows that a description in terms of energy only is not in general adequate to determine what the actual behavior will be, but the details must be known. This means, in the case of the tension specimen, the details of the surface action, and the way the surrounding medium under pressure can work its way into any incipient surface crevice. If the surface is completely impervious to the pressure medium, then we cannot have fracture against the stress in the situation above, but have the situation represented by Fig. 58 on page 109 where separation of the two parts does not occur as long as it is against the stress, and does occur so precisely when the net stress allows it that it may be used as a calibrating device. The extreme importance of the surface conditions becomes manifest in such experimental situations as we found for glass where the net stress at the tensile fracture varied enormously as we changed the medium in contact with the surface of the glass from water to oil to rubber to lead to copper. The same sort of thing is exhibited by cast iron, which breaks ductilely in tension under pressure when the outer surface is protected by a sheath of copper, but which breaks brittlely when directly immersed in the pressure liquid. The same sort of thing is doubtless responsible for the brittleness of rods of limestone when subjected to bending forces while directly exposed to the pressure liquid, but which will extend with ductility when sheathed.

The situation depicted in Fig. 58 suggests that tensile fracture will find it difficult to start in the *interior* of a tension specimen at any point where there is a compressive stress. No absolute pronouncement would seem to be possible here, because the total stress system is complex and the other components do work also during fracture. But in general we would expect that the initiation of a tensile fracture on the axis of the specimen, as it is initiated at atmospheric pressure, would become increasingly difficult with increasing hydrostatic pressure. This is doubtless an important factor in the modification of the character of the fracture with increasing pressure, the tensile part of the break eventually disappearing and the break becoming completely shearing in character. A shearing break could easily be initiated at the external surface.

A paradoxical feature of some of the fractures under more unusual pressure conditions is fracture on *release* of stress rather than on its application. The occurrence of such fractures raises the question as to whether a macroscopic description of the body in terms of the displacements of the particles from their *initial* positions can hope to be sufficient, granted that some sort of adequate macroscopic description can be found. In the elastic range, when the stresses are removed, the particles of the body resume their initial positions. In such situations the strain should certainly be calculated from the initial configuration. But, when the deformation involves large permanent deformations, the question forces itself as to what is the proper zero from which to calculate the present deformation of the body and in terms of which to calculate its behavior when the presently acting stresses are removed. The same question presents itself in more extreme form if we consider any piece of cast metal. We take the zero configuration as the position of the atoms in the casting as received, not the initial positions of the atoms as they were before the metal was put into the melting pot. When a metal is plastically deformed something similar to melting, although less drastic, takes place, and it would seem that the initial positions of the particles, as they were before the stresses were applied that produced the plastic deformation, are not particularly pertinent. The problem is to find a more pertinent zero.

It is easy to see qualitatively how some of the fractures on release of stress may occur and how they involve the displaced positions of the atoms. Consider, for example, the fracture of a glass tension specimen, which has been pulled under pressure, into a multitude of disks perpendicular to the tension axis when stresses are released. Under tension the axis was elongated and the transverse dimensions were shortened, so that extra atoms got forced into axial positions. On release of stress these displaced atoms were unable to return to their initial positions and acted like so many wedges of foreign matter more and more out of place as the external compressive force was progressively removed, until eventually they split the material exactly like internal wedges. This state of affairs is easy to visualize. If it can be at all adequately described in macroscopic terms it would seem that the proper zero in terms of which to explain the phenomena must be a zero which recognizes the displaced atoms which are the cause of the fracture on release of stress.

INDEX

A

Abbot, L. H., 5, 47
Adams, L. H., 202
Aluminum tension tests under pressure, 106
Anisotropy in plastic yield, 350
Armor plate, 85
  ballistic behavior of, 131
  Brinell hardness under pressure, 131–134
  penetration, 1
  upsetting by compressive forces, 243–246

B

Ballistic behavior of armor plate, 131
Balsley, J. T., 111, 117
Bauschinger effect, 315, 317, 349
Beeuwkes, R., Jr., 152
Bell, J. B., 123
Beryllium in tension under pressure, 113–114
Bibliography of papers of P. W. Bridgman, 6–7
Birch, Francis, 203, 204
Brandtzaeg, A., 111
Brass, tension tests under pressure, 107
  volume change in simple compression, 211–212
Brinell hardness, 4, 46, 131–133
Brittle fracture, general discussion, 124–131
Brittle materials under pressure, collapse of cavities in, 164–173
  in simple compression, 119–123
  tension tests, 107–118
Bronze tension tests under pressure, 107
Brown, R. L., 111
Brown University, 343

C

Carbides, sintered, in simple compression under pressure, 121–123
Carboloy under pressure, in simple compression, 122
  in tension, 110, 113, 129
Carpenter Steel Co., 62
Cast iron, in tension under pressure, 117
  volume changes of, in simple compression, 210
Catlinite (pipestone) tension under pressure, 115–116
Cavities in brittle materials, collapse of, under pressure, 164–173
Chase, Charles C., 5, 114
Chipman, Prof. John, 114
Collapse, of cavities in brittle materials under pressure, 164–173
  under external pressure, of hollow cylinders, 142–163
  of minerals, 169–172
  of steel cylinders, early results of, 160–163
Compression, simple, combined with torsion (*see* Torsion)
  fracture in, 190
  under pressure, 118–130
    carbides in, 121–123
    diamond in, 124
    ductile materials in, 118–119
    pyrex glass in, 119
    sapphire in, 124
    twinning of, 120
    Teton steel in, 123
    tourmaline in, 124
  after prestraining, in simple compression, 319–322
  in tension, under pressure, 313–318
  in two-dimensional compression, 323–325
  with refiguring, 181–192

357

358   INDEX

Compression, simple, strain hardening in, 189–190
  time effects in, 187, 190
  volume changes in, brass, 211–212
    drill rod, 209
    duralumin, 212
    marble, 201–202
    mild steel, 205–208
    Norway iron, 208–209
    retrograde, 213
    soapstone, 200–201
    spinel, 124
    stainless steel (Type 303), 210
Contour of neck in two-dimensional tension, 89, 96
Copper, tension tests under pressure, 106
  volume changes in simple compression, 210–211
Cored specimen, 17
Correction factor, 24, 36
Criteria of rupture, 81
Cup-cone fracture, 74
  ratio of tensile break to total area, 75
Cylinder, elastic distortion of, under external pressure, 165

D

Davidenkov, N. N., 20
Designation of steels, 60–62
deWald, L. H., 122
Diabase, volume changes in simple compression, 202–204
Diamond in simple compression under pressure, 124
"Dirty" steel, 2, 85
Disking cleavage, 120, 126
Donaldson, Paul, 219
Dow, Richard B., 203
Drill rod (high-carbon steel), volume changes in simple compression, 209
Ductile materials in simple compression under pressure, 118–119
Ductility curves, 66, 72
Duralumin, volume change in simple compression, 212

E

"Elasticity," 11
Ellipticity of section of compression specimens, 312

Energy release in fracture, 124, 352
Equilibrium equations, 11
Extension diagrams, 48
Extrusion under hydrostatic pressure, 177–179

F

Feild, Dr. A. L., 62
Fibrous fracture in tension after prestraining in tension under pressure, 303
Flaking off of interior of collapsing cavities, 170
Flow as determined by complete stress system, 97–102
Flow stress, 22, 23, 48, 67, 74
Fracture, brittle, 124–131
  criteria of, hydrostatic tension, 81
    sum of principal stresses, 82
  effect of molecular structure on, 126
  energy release in, 124, 352
  fibrous, in tension, 303
  instability as prelude to, 91, 352
  on release of stress, 354
  shearing, 76, 104
  in simple compression, 190
  star, 304
  stresses at, 80–85
  theory of, Griffith's, 128
  in torsion combined with simple compression, 264
Frondel, Prof. C., 204
Furry, Prof. Wendell, 150

G

General Electric Co., 113
Generalized strain-hardening curve, 239
Glass, borax, behavior of, in shearing under pressure, 287
  collapse of, under external pressure, 166–169
  effect of water on tensile behavior of, 125
  pyrex, under pressure, in simple compressure, 119
    in tension, 111–113
Grain shape, 21
Grain size, 23

INDEX

Greenspan, M., 194
Grid, 41
Griffith's theory of fracture, 128
Griggs, David, 111, 123
Gurney, C., 111, 128

H

Halcomb Steel Co., 61
Hall, E. H., 208
Hohenhemser, K., 274
Holloman, Dr. John H., 3, 205, 294, 304, 335
Hydrostatic pressure, energy release under, 312
  extrusion under, 177–179
Hydrostatic tension, 25
  as criterion of fracture, 81
Hyperboloid of revolution, elastic solution, 16
Hysteresis in collapsing cylinders, 161

I

Ideal plasticity, 340
Instability, as prelude to fracture, 91, 352
  in two-dimensional compression, 225
Insulated leads, 42
Interlocking of molecules as affecting fracture in glass, 127
Isotropy after torsion combined with simple compression, 277

K

Koehler, James S., 102

L

Laning, J. H., Jr., 152
Lanza, Carmelo, 6, 46
Lead, shape of neck of tension specimens, 31
Legalistic difficulties, 2
Length change of collapsed cylinders, 153
Linde Air Products Co., 115, 120
Long-range program, 2
Love's "Elasticity," 11
Lubrication of simple compression specimens, 183
Lynch, J. J., 174

M

MacGregor, C. W., 142
Manganin coil for measuring pressure, 42
Marble volume changes in simple compression, 201–202
Martensite, tempered, 69
Maximum load, 70
Maximum shearing stress, 102
Maximum stress-difference criterion for plastic flow, 143
Microscopic vs. macroscopic behavior, 531
"Mild steel," volume changes in simple compression, 205–208
Minerals, collapse of, under external pressure, 169–172
"Mixed" compression, 243–246
Molecular structure, effect of, on fracture, 126
Morrison, J. L. M., 343

N

Nadai, A., 223
National Defense Research Committee (NDRC), 1, 46, 85, 243
Natural strain in terms of reduction of area, 49
Naval Research Laboratory, 1
Neuber, H., 16
Newman, S. B., 194
Nonuniformity of stress at neck of tension specimen, 4, 9–37
Norway iron, volume changes in simple compression, 208–209

O

"Octahedral" stress and strain, 99, 153, 334
  in torsion combined with simple compression, 272

P

Pearlite, tempered, 69
Penetration of armor in punching, 134
Phosphor bronze in tension under pressure, 114

INDEX

Pinching-off effect, 108, 125
Pipestone (catlinite) in tension under pressure, 115–116
Plastic flow, maximum stress-difference criterion for, 143
in tension under pressure, 62–70
Polymorphism in shearing under pressure, 286
Poncelet, E., 127
Powders, compacting of, under pressure, 172
Prager, W., 274, 343
Pressure coefficient of tensile strength, 109
"Pressurized" laboratory, 38
Prestraining, effect of, on flow and fracture, 293–338
Primary creep, 346
Princeton Ballistics Laboratory, 243
"Pull pieces," use of, in tension tests, 110
Punching under pressure, 134–141

Q

Quartz, crystal, and glass, compression of, simple, under pressure, 123
infinitesimal flow of, 130
volume changes in simple compression, 204–205
Quartz glass shearing under pressure, 288
Quinney, H., 4, 181, 196

R

Radial component of tension, 18
"Radial pressure," 128
Radius of curvature of contour of neck, 11, 25
Ratio of tensile fracture to neck area, 305
"Reduced" principal stress, 102
Refiguring neck of tension specimen, 79
Retrograde volume change in simple compression, 213
Reversible component, of plastic flow, 310
of volume change, 350
Richart, F. E., 111
Ripling, E. J., 174
Rock salt in tension under pressure, 116
Rockwell C hardness of simple-compression specimens, 184, 190

Rowe, P. W., 111, 128
Rubber behavior in shearing under pressure, 287
Rupture in two-dimensional tension, 103

S

Sachs, G., 174
Sapphire, natural, in simple compression under pressure, 124
synthetic, in simple compression under pressure, 120
in tension under pressure, 115
Seitz, Frederick, 2, 47, 102
Self-healing, 283, 351
"Semi-idealized" body, 346
Shape of complete contour, 30
Shearing cross in two-dimensional compression, 222, 223, 226
Shearing fracture, 76, 104
Shearing strength, under approximately hydrostatic pressure, 279, 292
limiting, for indefinite flow, 289
Shearing stress, maximum, 102
"true," 135
Shepherd, W. M., 343
Siebel, E., 21
"Significant" stress and strain, 221
Simple compression (see Compression, simple)
Singularity at edge in two-dimensional compression, 223, 226
Soapstone, volume changes in simple compression, 200–201
Solenhofen limestone in tension under pressure, 117
Sosman, R. B., 123
Speed effects, in shearing under pressure, 285
in torsion combined with compression, 258–260
Spendelow, Howard R., Jr., 178
Sphere, elastic distortion of, under external pressure, 164
Spinel in simple compression under pressure, 124
Spridonova, N. I., 20
Stainless steel (Type 303), volume changes in simple compression, 210
Stang, A. H., 194

INDEX

"Star" fracture, 304
"State of slip," 290
Strain, natural, in terms of reduction of area, 49
  uniformity of, 17, 18
Strain hardening, 13
  in collapse of cylinders, 154
  correlation of, with ductility, 73
  in shear, compared with tension, 337
  in simple compression, 189–190
Strain-hardening curve, idealized, 345
  in shear, 141
Strain-hardening curves, 49, 63
Strain softening, 143
Sum of principal stresses as criterion of fracture, 82
Superintendent of Public Documents, 1
Surface conditions, 354
  as affecting tensile properties, 110

T

Tate pressure gauge, 160
Taylor, G. I., 4, 181, 196, 206
Tensile strength, under pressure, effect of transmitting medium on, 125
  pressure coefficient of, 109
Tension, hydrostatic, 25
  as criterion of fracture, 81
  under pressure, fibrous fracture in tension after plastic flow in, 62–70
    prestraining in, 303
    of steel, 38–86
    tension at atmospheric pressure after, 294–306
    tests, aluminum, 106
      beryllium, 113–114
      brass, 107
      brittle materials, 107–118
      bronze, 107
      carboloy, 110, 113, 129
      cast iron, 117
      catlinite, 115–116
      copper, 106
      use of "pull pieces" in, 110
  radial component of, 18
  simple, after prestraining, in simple compression, 307–310
    in two-dimensional compression, 311–312

Teton steel in simple compression under pressure, 123
Thin-walled tubing, 32
Time effects, in collapse of thick cylinders, 144, 161
  in simple compression, 187, 190
Tolman, Dr. Richard C., 46
Torsion, 5
  combined with simple compression, 247–278
    isotropy after, 277
    "octahedral" stress and strain in, 272
    after prestraining in tension under compression, 326–338
"Total" theory of plasticity, 344
Tourmaline in simple compression under pressure, 124
Transient effects in shearing under pressure, 284
Transmitting medium, effect of, on tensile strength under pressure, 125
True stress, 21
Twinning of sapphire in simple compression, 120
Two-dimensional compression, 215–246
  instability in, 225
  shearing cross in, 222, 223, 226
  simple compression after prestraining in, 323–325
  singularity at edge in, 223, 226
  velocity of flow in, 233
Two-dimensional tension, contour of neck in, 89, 96
  correction of stress distribution at neck, 31
  rupture in, 103

U

Uniformity of strain, 17, 18
Union Carbide and Carbon Corp., 178
Uniqueness of solution, 16

V

Velocity of flow in two-dimensional compression, 233
Voigt, W., 117

Volume change, associated with simple compression, 193–214
   reversible component of, 350
Volume conservation, 13
von Mises plasticity function, 12, 17, 20
   23, 31, 69, 143, 151, 339

W

Warping of sections of thick cylinders, 145
Water, effect of, on tensile behavior of glass, 125

Watertown Arsenal, 2, 3, 46, 243
Westergaard, H. M., 223
Williamson, E. D., 202
Wire drawing under pressure, 174–177

Y

Yoke for tension specimens, 39

Z

Zener, Dr. Clarence, 3, 294, 335
Zornig, Col. H. H., 2

Bei Fragen zur Produktsicherheit wenden Sie sich bitte an:
If you have any questions regarding product safety,
please contact:

Walter de Gruyter GmbH
Genthiner Straße 13
10785 Berlin
productsafety@degruyterbrill.com